ANATOMICA'S

Body
Atlas

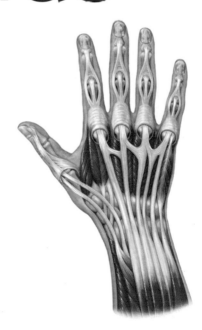

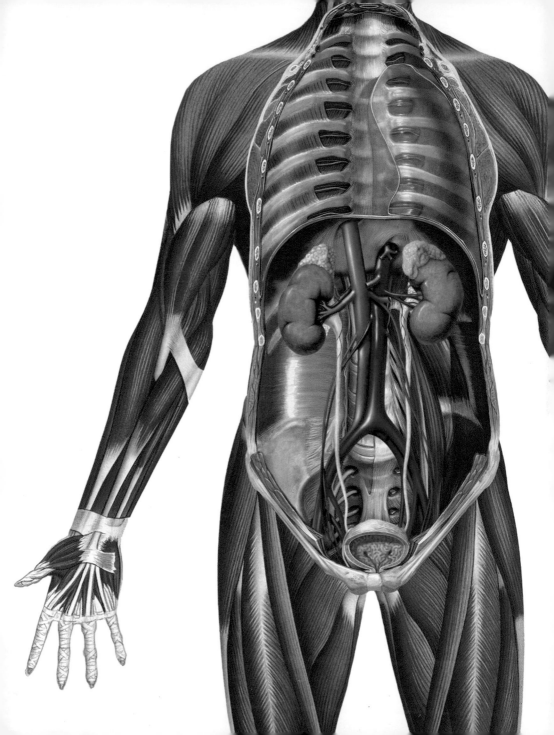

ANATOMICA'S

Body
Atlas

THUNDER BAY
P · R · E · S · S

San Diego, California

Managing director	Chryl Campbell
Publishing manager	Sarah Anderson
Art directors	Stan Lamond Kylie Mulquin
Managing editor	Janet Parker
Chief consultants	Kurt H. Albertine, Ph.D. The Honorable Emeritus Professor Peter Baume, A.O., M.D., B.S., Hon.D.Litt., F.R.A.C.P., F.R.A.C.G.P., F.A.F.P.H.M. Dr R. William Currie, B.S.A., M.Sc., Ph.D. Laurence Garey, M.A., D.Phil., B.M., B.Ch. Gareth Jones, B.Sc. (Hons), M.B., B.S., D.Sc., C.Biol., Fl.Biol. David Tracey, B.Sc., Ph.D.
Contributors	Robin Arnold, M.Sc.; Ken Ashwell, B.Med.Sc., M.B., B.S., Ph.D.; Deborah Bryce, B.Sc., M.Sc.Qual.; M.Chiro., Gr.Cert.H.Ed.; Carol Fallows, B.A.; Martin Fallows; John Frith, M.B., B.S., B.Sc. (Med.), Grad.Dip.Ed., M.C.H., R.F.D.; John Gallo, M.B., B.S. (Hons), F.R.A.C.P., F.R.C.P.A.; Brian Gaynor, M.B., B.S., F.R.A.C.P., F.R.A.C.G.P., M.R.C.G.P., D.C.H.; Rakesh Kumar, M.B., B.S., Ph.D.; Peter Lavelle, M.B., B.S.; Lesley Lopes, B.A. Communications (Journalism); Karen McGhee, B.Sc.; Michael Roberts, M.B., B.S., LL.B. (Hons); Emeritus Professor Frederick Rost, B.Sc. (Med.), M.B., B.S., Ph.D., D.C.P., Dip.R.M.S.; Elizabeth Tancred, B.Sc., Ph.D.; Dzung Vu, M.D., M.B., B.S., Dip.Anat., Grad.Cert.H.Ed.; Phil Waite, B.Sc. (Hons), M.B.Ch.B., Cert.H.Ed., Ph.D.
Chief illustration consultant	Dzung Vu, M.D., M.B., B.S., Dip.Anat., Grad.Cert.H.Ed.
Illustration consultants	John Frith, M.B., B.S., B.Sc. (Med.), Grad.Dip.Ed., M.C.H., R.F.D. David Jackson, M.B., B.S., B.Sc. (Med.)
Illustrators	David Carroll, Peter Child, Deborah Clarke, Geoff Cook, Marcus Cremonese, Beth Croce, Wendy de Paauw, Levant Efe, Hans De Haas, Ian Faulkner, Mike Golding, Jeff Lang, Alex Lavroff, Ulrich Lehmann, Ruth Lindsay, Richard McKenna, Annabel Milne, Tony Pyrzakowski, Oliver Rennert, Caroline Rodrigues, Otto Schmidinger, Bob Seal, Vicky Short, Graeme Tavendale, Jonathan Tidball, Paul Tresnan, Valentin Varetsa, Glen Vause, Spike Wademan, Trevor Weekes, Paul Williams, David Wood
Symptoms table text	Jenni Harman Melanie George, M.B.B.S., Dip. Paed. Annette Kifley, M.B.B.S. Robyn McCooey, B.App.Sci. Sue Markham, B.App.Sci.

Text editors	Denise Imwold, Janet Parker
Illustration editor	Heather McNamara
Cover design	Stan Lamond
Page design	Dee Rogers
Index	Diane Harriman
International rights	Kate Hill
Publishing assistant	Christine Leonards
Production	Ian Coles

Thunder Bay Press
An imprint of the Advantage Publishers Group
10350 Barnes Canyon Road
San Diego, CA 92121
THUNDER BAY
P·R·E·S·S www.thunderbaybooks.com

Text © Global Book Publishing Pty Ltd 2002
Illustrations from the Global Illustration Archives
© Global Book Publishing Pty Ltd 2002
Reprinted 2003, 2006, 2007

Global Book Publishing Pty Ltd
Level 8, 15 Orion Road, Lane Cove, N.S.W., 2066, Australia
Phone: +61 2 9425 5800 Fax: +61 2 9425 5804
E-mail: rightsmanager@globalpub.com.au

ISBN-13: 978-1-59223-743-2
ISBN-10: 1-59223-743-6

The Library of Congress has cataloged the Lauren Glen edition as follows:
Anatomica's body atlas / text editors, Denise Imwold, Janet Parker
  p. cm.
  ISBN: 1-57145-923-5
  1. Human anatomy--Atlases. 2. Human physiology. I. Imwold, Denise
II. Parker Janet

  QM25 .A486 2003
  611--dc21
                                        2002035361
Printed in Hong Kong by Sing Cheong Printing Co. Ltd, Hong Kong
Film separation Pica Digital Pte Ltd, Singapore

 2 3 4 5 10 09 08 07

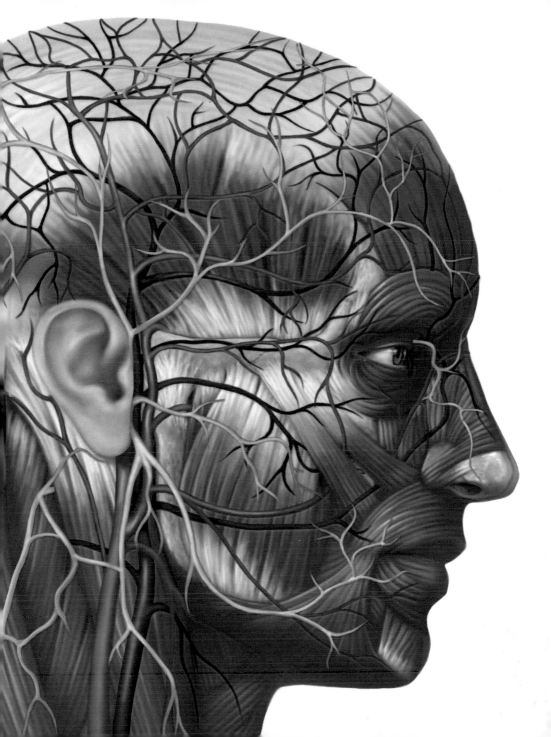

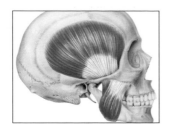

C o n t

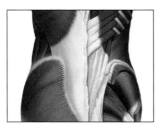

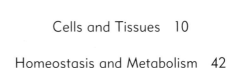

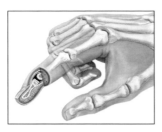

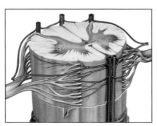

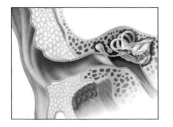

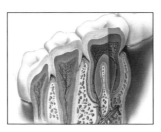

e n t s

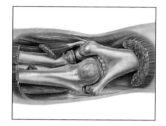

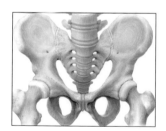

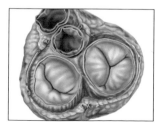

Introduction

Since ancient times, the human body and its complexities have been a source of fascination and intrigue. While many of the early theories have been proven incorrect, they have provided the catalyst for further investigation and discovery—a task that continues in these modern times. Today, research, investigation, and discovery continue, using the most up-to-date techniques and equipment available to uncover the remaining mysteries of this amazing structure—the human body.

The human body is a unique masterpiece, comprising a multitude of components that come together to form a complete working unit.

This volume reveals the wonders of the human body, taking the reader on a journey through the body, system by system. The journey begins with the fundamental units of cells and tissues, and continues through the body, detailing the individual role of each system, and the interaction of the systems, which work together to form the complete person. From microstructure to major organs, each component of the body is explained with informative text, accompanied by superbly detailed illustrations.

How each of the body systems works, and how it complements the other systems, is essential to an understanding of the body. Each system is covered, beginning with the largest organ of the body, the skin, along with its accessory structures, which comprise the integumentary system. From the composition of bone tissue to the major bones that make up the body's framework, the skeletal system encompasses all of the bone structures of the body. The muscular system produces movement—from facial expressions to walking—and comprises muscles, tendons, and associated structures.

The vital elements of the brain and spinal cord make up the central nervous system, while the peripheral nervous system includes all the remaining

nerves in the body. Composed of billions of nerve cells, the nervous system is designed to communicate messages between the brain and the skin and organs. The autonomic nervous system controls the automatic internal functions of the body. The intricate network of blood vessels and the amazing pump of the heart form the circulatory system, carrying nutrients, hormones, and waste products through the body.

How our bodies fight invading organisms, raise our defenses, and ward off attack, is determined by the lymphatic/immune system. Breathing—the restoration of oxygen to the body, and the disposal of unwanted carbon dioxide—is the task of the organs of the respiratory system—the lungs and respiratory airways. The organs of the digestive system break down food into essential nutrients required for the maintenance, repair, and regeneration of cells. The reproductive system—formed during the embryonic stages of life—matures during adolescence, in readiness for its function during adult life. The components of the urinary system operate to maintain fluid levels, filtering blood in order to return necessary elements to the body, and eliminating wastes. The function of the endocrine system is important to the smooth operation of the body, with the hormones produced contributing to the maintenance of a constant state in the body.

The growth process, and the many stages it passes through from conception to old age, is chronicled in the chapter on the human life cycle.

Finally, to shed a different perspective on the relationship of neighboring organs, various levels of the body are shown in cross section in the last chapter.

At the end of each chapter, several of the more common diseases and disorders associated with each particular body system or function are discussed.

All in all, this useful reference book provides information on the myriad structures and components of the body, presented in an accessible format, and further enhanced with stunningly illustrated visual references. This book will provide a handy reference to anyone interested in how the human body works, from the merely curious to the medical student, and is a must for the family library.

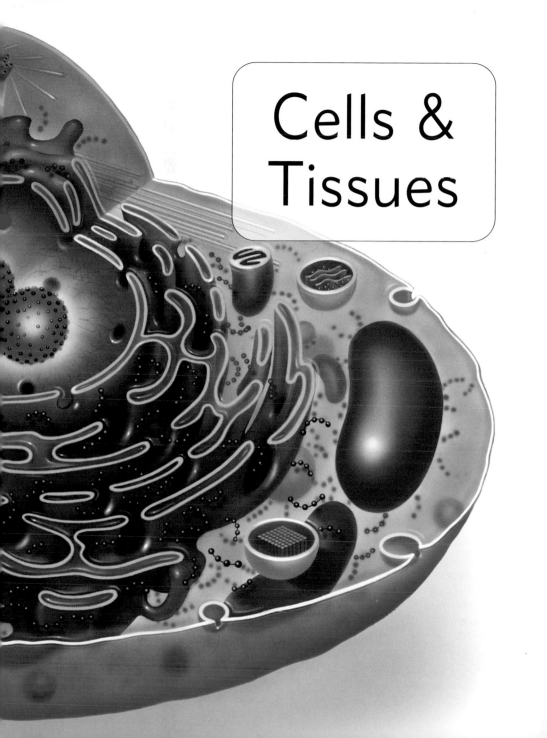

Cells & Tissues

CELLS AND TISSUES

Cells

Every living organism is made up of cells. With the capacity to perform all the essential life functions, the cell is the basic functional unit of all tissues. The organs of the body are composed of various tissue types, and these tissues are made up of cells.

Typical cells have an outer membrane, and held within this membrane, in a fluid known as cytoplasm, are many important structural units called organelles.

While each of the organelles plays a key role in cell activity and function, perhaps the most important component, DNA, is held within the cell nucleus. DNA holds the genetic information of every individual.

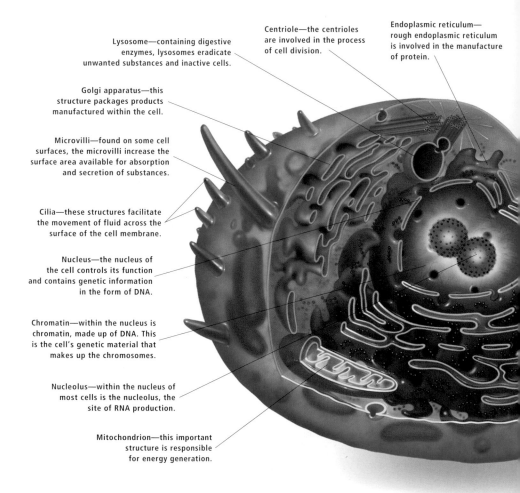

Lysosome—containing digestive enzymes, lysosomes eradicate unwanted substances and inactive cells.

Golgi apparatus—this structure packages products manufactured within the cell.

Microvilli—found on some cell surfaces, the microvilli increase the surface area available for absorption and secretion of substances.

Cilia—these structures facilitate the movement of fluid across the surface of the cell membrane.

Nucleus—the nucleus of the cell controls its function and contains genetic information in the form of DNA.

Chromatin—within the nucleus is chromatin, made up of DNA. This is the cell's genetic material that makes up the chromosomes.

Nucleolus—within the nucleus of most cells is the nucleolus, the site of RNA production.

Mitochondrion—this important structure is responsible for energy generation.

Centriole—the centrioles are involved in the process of cell division.

Endoplasmic reticulum—rough endoplasmic reticulum is involved in the manufacture of protein.

Cell division

Cell division, or mitosis, is a process whereby a mother cell divides into two daughter cells. In some regions of the body, such as the skin and the gut, cell division occurs rapidly.

A different type of cell division, known as meiosis, takes place in the male and female sex cells (sperm and ovum). At fertilization, the male sperm and female ovum merge to form a zygote. As it moves along the fallopian tubes, this zygote divides rapidly, several times over, to become a cluster of cells. This cluster is known as a blastocyst, which implants into the uterine wall, and develops into the embryo.

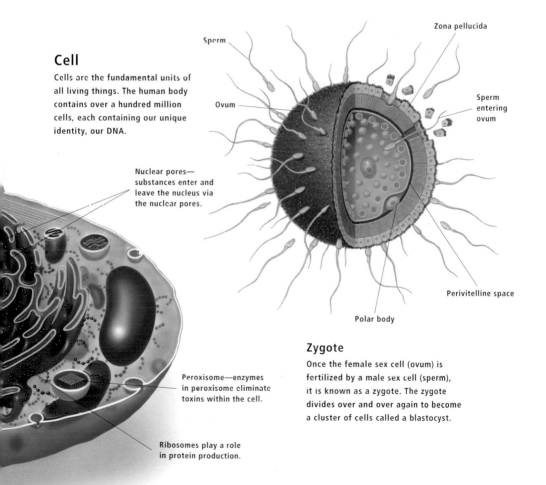

Cell

Cells are the fundamental units of all living things. The human body contains over a hundred million cells, each containing our unique identity, our DNA.

Sperm

Ovum

Zona pellucida

Sperm entering ovum

Nuclear pores— substances enter and leave the nucleus via the nuclear pores.

Perivitelline space

Polar body

Peroxisome—enzymes in peroxisome eliminate toxins within the cell.

Ribosomes play a role in protein production.

Zygote

Once the female sex cell (ovum) is fertilized by a male sex cell (sperm), it is known as a zygote. The zygote divides over and over again to become a cluster of cells called a blastocyst.

Specialized cells

There are many specialized cell types that perform various functions in the body. These include red blood cells, white blood cells, macrophages, neurons, muscle cells, and skin cells.

Developing in the bone marrow, red blood cells transport oxygen and carbon dioxide through the body via the bloodstream. As the cells develop, the nucleus and cytoplasm are replaced by hemoglobin that gives blood its characteristic color.

Lymphocyte

Macrophage

Red blood cell

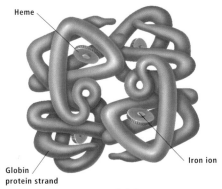

Heme

Globin
protein strand

Iron ion

Hemoglobin

There are several types of white blood cells. White blood cells that have granules in their cytoplasm are known as granulocytes; those without granules are known as agranulocytes—this group includes lymphocytes and monocytes. The two types of lymphocytes, B lymphocytes and T lymphocytes, are involved in immune response, with B lymphocytes playing a role in humoral response and T lymphocytes involved in cell-mediated immune response.

Macrophages usually travel through the bloodstream, but can also enter and move through tissue. These cells defend the body against foreign material, either by engulfing or destroying the offending substance.

Neurons, or nerve cells, communicate using electrical impulses to transmit information to and from all parts of the body. Information passes along the nerve fibers at rapid speeds, with the neurons meeting at junctions called synapses, where the information is relayed.

Muscle cells operate using electrical activity, responding to nervous or hormonal stimulus.

The skin is in a constant state of re-newal, continuously producing new skin cells to replace old dead cells.

Neuron

The unique structure of nerve cells, or neurons, comprises several dendrites and a single axon projecting from the cell body. Nerve impulses are received by the dendrites and carried to the cell body, while the axon carries impulses away from the cell body.

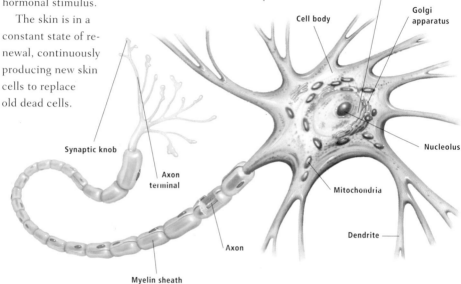

Nuclear membrane

Cell body

Golgi apparatus

Synaptic knob

Axon terminal

Nucleolus

Axon

Mitochondria

Dendrite

Myelin sheath

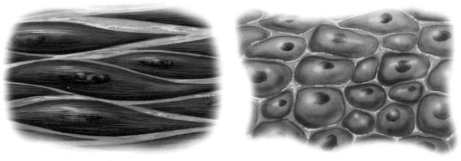

Muscle cells

Skin cells

Cartilage tissue

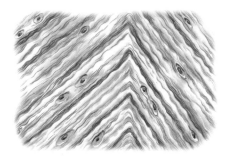

Fibrocartilage

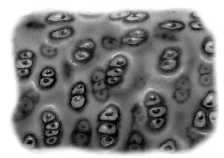

Hyaline cartilage

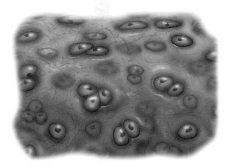

Elastic cartilage

Tissue

Tissues are formed by a layer or grouping of similar cells, which collectively perform a specialized function. The four major tissue types in the human body are: connective, epithelial, muscular, and nervous tissue.

Connective tissue

Comprised of cells and protein fibers, connective tissue provides support for other body tissues, often separating or surrounding other tissues or organs. Connective tissue is comprised of cells and protein fibers—the main proteins are collagen and elastin. Connective tissue often contains components, including: macrophages, plasma cells, and mast cells. These components are held in a substance known as "ground substance."

There are five principal types of connective tissue: loose connective tissue, dense connective tissue, blood, bone, and cartilage. Loose connective tissue includes adipose tissue, a fat storage tissue. Dense

Loose connective tissue

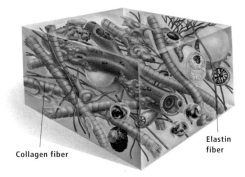

Collagen fiber

Elastin fiber

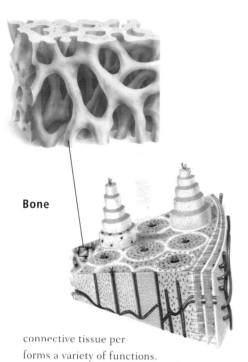

Bone

Adipose tissue

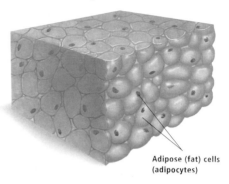

Adipose (fat) cells
(adipocytes)

Tendon tissue

connective tissue per-
forms a variety of functions.
One type, dense regular connective
tissue, forms tendons and cartilages. Elastic
connective tissue is found in the lungs, vocal
cords, and some vessel walls, where its elastic properties are required. Yet another type
of connective tissue is cartilage—a tough,
semi-transparent, flexible tissue comprised
of cartilage cells. There are three different
types of cartilage tissue: hyaline cartilage,
which provides a surface coating to the
bones in synovial joints, forms the tip of the
nose and the ends of the ribs, and supports
the larynx and trachea; fibrocartilage, which
includes the intervertebral disks; and elastic cartilage, which is found in the outer ear
and the epiglottis.

Ligament tissue

Epithelial tissue

Epithelial tissue, or epithelium, is made up of tightly packed cells, arranged in sheets from one to several layers thick. Epithelium performs a variety of functions, including protection, excretion, absorption, sensory reception, secretion, and reproduction. The cells of epithelium are in constant production, enabling rapid replacement of dead or inactive cells. With no direct blood supply of its own, epithelium is nourished by vessels from surrounding tissues.

There are two main types of epithelial tissue—glandular epithelium, found in exocrine and endocrine glands; and covering and lining epithelium, which forms the outer layer of the skin and some internal organs, and lines the digestive, respiratory, reproductive and urinary systems, the body cavities, and blood vessels.

Nervous tissue

The nervous system, comprising the central nervous system (CNS) and the peripheral nervous system (PNS) is made up of neural or nervous tissue. With the CNS encompassing the brain and spinal cord, and the PNS incorporating all the remaining nerves in the body, the neural or nervous tissue of these two systems transmits messages to and from the brain, implementing the body's response to stimuli and controlling muscles, glands, and sense organs.

Muscle tissue

Muscle tissue comprises up to 60 percent of the body's mass. Muscle tissue enables body movement, providing stability to the skeletal frame and internal organs, and is responsible for generating a large percentage of body heat. There are three principal types of muscle tissue: cardiac, skeletal, and smooth. Cardiac muscle tissue is found in the heart and is an involuntary muscle type, that is, it operates without our conscious effort. Skeletal muscle tissue makes up the majority of muscle tissue found in the body. The long fibers of skeletal muscle are arranged parallel to each other and are usually attached to bone. These are the

Nuclei   **Epithelial tissue**

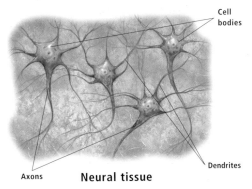

Cell bodies

Axons   **Neural tissue**

Dendrites

Muscle tissue

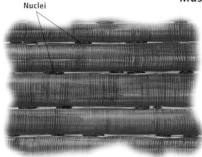

Nuclei

Skeletal muscle

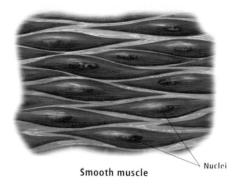

Nuclei

Smooth muscle

Nuclei

Cardiac muscle

muscles that allow body movement and are called voluntary muscles, as they are usually under our conscious control. Smooth muscle is another involuntary muscle type; it is found in the walls of blood vessels, in the digestive and respiratory systems, and inside the eye.

Lymphatic tissue

Lymphoid tissue is found at the entrances to the respiratory system and in the digestive and urogenital tracts. These masses of lymphoid tissue provide protection to these regions and act as a line of defense against infection and bacteria.

Membranes

There are four main types of membrane: cutaneous, mucous, serous, and synovial. Membranes are thin sheets of tissue that provide a pliable surface lining or protective layer to organs and structures in the body.

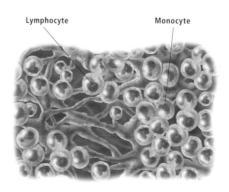

Lymphocyte   Monocyte

Lymphatic tissue

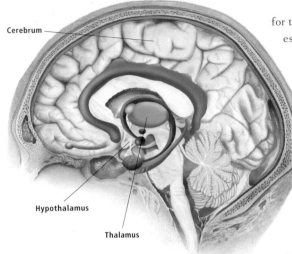

Cerebrum

Hypothalamus

Thalamus

for the production of proteins. Non-essential amino acids are provided either by dietary intake, or can be produced by the body. Essential amino acids can only be provided by dietary intake, and it is important to include foods rich in essential amino acids, such as meat, fish, and dairy produce, in the daily diet.

Serotonin

Serotonin serves a variety of functions in the body. It stimulates muscle contractions in the intestine and triggers blood clotting at the site of tissue damage. It is also a vital neurotransmitter in the limbic system and is related to mood and alertness levels.

Proteins and Fats

Essential to life, proteins play a vital role in the building and repair of cells in every part of the body. Proteins also carry oxygen from the lungs to cells, and act as chemical messengers. The body requires only a small amount of fats and oils for use as an energy source. Foods provide the fatty acids that the body cannot manufacture, such as linoleic acid, used in building cell structure.

Amino acids

Proteins allow the body to build, grow, and repair cells, and amino acids are essential

Serotonin

Synthesized from the amino acid tryptophan, serotonin performs several functions in the human body, acting as both a hormone and as a neurotransmitter in the limbic system. It affects muscular contractions in the digestive system and plays a role in the blood clotting process. The body's levels of serotonin influence the dilation and constriction of blood vessels.

Collagen

Made up of chains of amino acids, collagen is an important structural protein and a major component of connective tissue, providing strength and flexibility to ligaments, tendons, and internal organs.

Lipids

Found throughout the body, lipids include fats and cholesterol. Lipids provide a fuel source for the body, and play an important role in the formation of cell membranes and the myelin sheaths of nerve cells.

Cholesterol

The liver manufactures cholesterol, and some is acquired through dietary intake. Circulated through the blood bound to lipoproteins, cholesterol has a role in building and repairing cells, the production of sex hormones, and the bile acids necessary for digestion.

Cholesterol

Traveling through the bloodstream, cholesterol, in the form of free cholesterol or cholesteryl esters, is bound to lipoprotein.

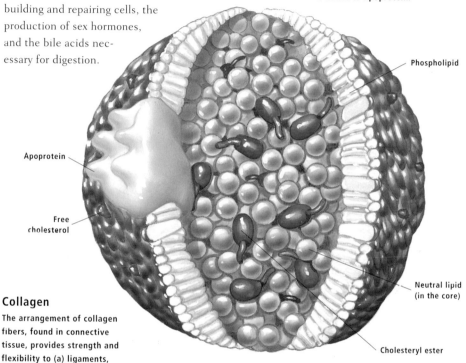

Phospholipid

Apoprotein

Free cholesterol

Neutral lipid (in the core)

Cholesteryl ester

Collagen

The arrangement of collagen fibers, found in connective tissue, provides strength and flexibility to (a) ligaments, (b) tendons and (c) supporting capsules of internal organs.

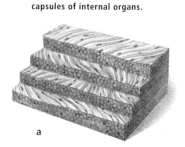

a

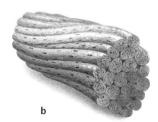

b

c

Glands

The specialized cells of glands are instrumental in the production of fluid secretions, with the secretions dependent on the purpose of the gland. Glands are distinguished by two types: exocrine and endocrine.

Exocrine glands consist of secretory cells and ducts. The ducts transport the secretions to the surface of the body, the digestive system, the reproductive organs, and the lungs. Exocrine glands include the sweat and sebaceous glands found in the skin, the mammary glands, and the salivary glands.

Endocrine glands are responsible for the production of hormones, which are then circulated through the body via the bloodstream. The principal endocrine gland is the pituitary gland, which controls hormone production in other endocrine glands, including the thyroid, parathyroid, and adrenal glands, and the ovaries and testes.

The pancreas is unique in that it is both an exocrine and endocrine gland. Primarily an exocrine gland, it contains clusters of acinar cells that produce enzymes for use in the digestive system. These enzymes are transported to the digestive system via the pancreatic ducts. The endocrine function of the pancreas is to produce the hormones glucagon, insulin, and somatostatin. These hormones are produced by alpha, beta, and delta cells which are clustered together to form the islets of Langerhans. These hormones are released into the bloodstream for distribution through the body.

Endocrine glands

The endocrine glands of the body include the pituitary, pineal, thyroid, parathyroid, and adrenal glands. These ductless glands secrete their hormones into the bloodstream.

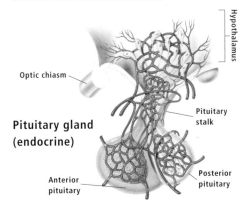

Hypothalamus

Optic chiasm

Pituitary gland (endocrine)

Anterior pituitary

Pituitary stalk

Posterior pituitary

Glands

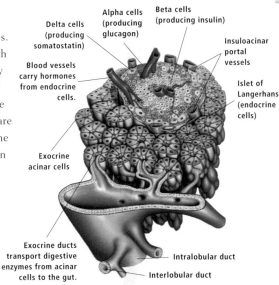

Delta cells (producing somatostatin)

Alpha cells (producing glucagon)

Beta cells (producing insulin)

Insuloacinar portal vessels

Islet of Langerhans (endocrine cells)

Blood vessels carry hormones from endocrine cells.

Exocrine acinar cells

Exocrine ducts transport digestive enzymes from acinar cells to the gut.

Intralobular duct

Interlobular duct

Pancreas (endocrine and exocrine)

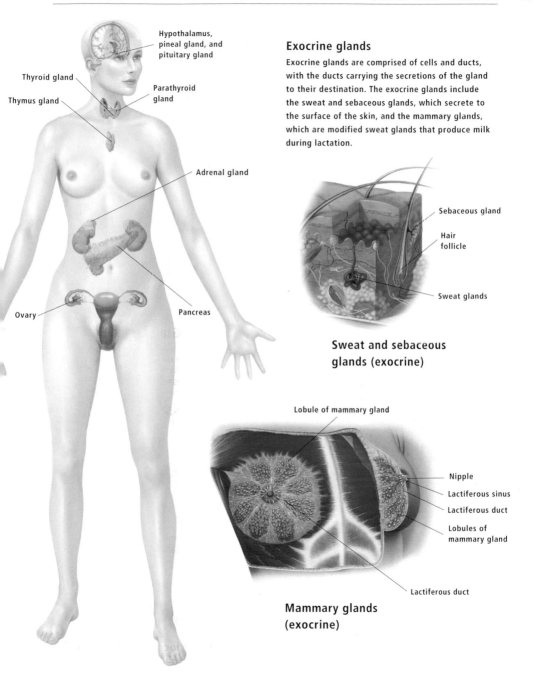

Hypothalamus, pineal gland, and pituitary gland

Thyroid gland

Parathyroid gland

Thymus gland

Adrenal gland

Ovary

Pancreas

Exocrine glands

Exocrine glands are comprised of cells and ducts, with the ducts carrying the secretions of the gland to their destination. The exocrine glands include the sweat and sebaceous glands, which secrete to the surface of the skin, and the mammary glands, which are modified sweat glands that produce milk during lactation.

Sebaceous gland

Hair follicle

Sweat glands

Sweat and sebaceous glands (exocrine)

Lobule of mammary gland

Nipple

Lactiferous sinus

Lactiferous duct

Lobules of mammary gland

Lactiferous duct

Mammary glands (exocrine)

Scar

Wounds are filled with granulation tissue, which may form a scar.

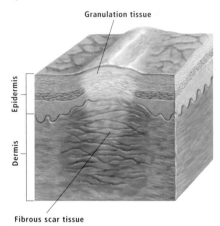

Granulation tissue

Epidermis

Dermis

Fibrous scar tissue

Healing

The body has various procedures in place to promote healing and repair of tissue. How the body approaches this restorative process is dependent on the injury.

Inflammation

Inflammation is characterized by redness, heat, swelling, pain, and tenderness. These are the classic symptoms of inflammation, indicating the body's response to tissue damage such as infection and burns. The body initiates its response by dilating the blood vessels to allow an increased volume of blood to the area. These dilated vessels allow leaching of plasma into the extracellular fluid, causing the area to swell.

Leukocytes are also released and discharge their chemicals to eliminate invading organisms—a flow that continues until all foreign organisms and dead tissue have been eradicated. Once this has been achieved, the fluid is drained by the lymphatic system, the blood vessels return to original size, and the swelling subsides.

Any necessary cellular repair is achieved by cell division of the remaining tissue, although cell division is not always possible, such as in the case of brain or nerve cells.

Wound healing

The process of wound healing depends on whether the wound is narrow or wide. If it is narrow, a blood clot forms in the wound. This clot then contracts, drawing the two surfaces of the wound closer together. Granulation tissue is supplied from the edges of the wound by fibroblasts, and is gradually replaced by connective tissue. This healing process is known as "primary intention."

A slightly different process, known as "secondary intention" takes place if the wound is open and wide. In this case, the wound area is filled with granulation tissue beginning from the base and sides of the wound, gradually filling in the wound area.

Bone healing

Special bone forming cells, known as osteoblasts, initiate healing of bone following damage or fracture. The osteoblasts produce callus, a tough binding material, which knits the bones together, and once this preliminary repair has been achieved, the callus is then replaced by true bone.

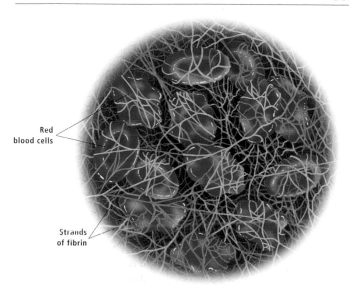

Blood clot

A blood clot is formed when red blood cells, platelets, and strands of fibrin enmesh and coagulate.

Red blood cells

Strands of fibrin

Blood clotting

(a) Injury which involves damage to a blood vessel causes platelets and red blood cells to leak out into damaged tissue.

(b) Platelets release chemicals to attract more platelets, which then attach to the damaged tissue. Strands of fibrin are also formed.

(c) A blood clot is formed by the conglomeration of blood cells, platelets, and fibrin fibers.

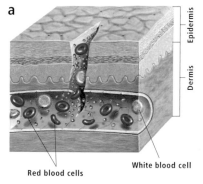

a

Epidermis

Dermis

Red blood cells

White blood cell

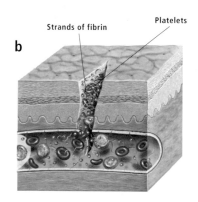

Strands of fibrin

Platelets

b

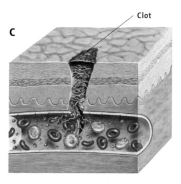

Clot

c

DNA

Within the nucleus of every cell are the chromosomes, which contain deoxyribo-nucleic acid, or DNA. Discovered in 1953 by Francis Crick and James Watson, the DNA molecule is a double helix, resembling a spiral ladder. The sides of the "ladder" are comprised of sugar and phosphate units. Attached to the sugar units are the "rungs," which are comprised of various combinations of nucleotide bases.

There are four bases: adenine, cytosine, guanine, and thymine (A, C, G, and T), with two bases making up each "rung." Compatibility of the bases limits the possible combinations to A-T, T-A, C-G or G-C, with the bases forming different patterns along the length of the "ladder." Three bases together form a codon, which encodes a single amino acid of a protein.

By encoding a messenger ribonucleic acid (mRNA) with information for the ribosomes in the cell, DNA also determines the proteins made by the cell.

The sequence of amino acids produced results in a protein which correlates directly to a specific sequence of bases in the DNA.

Before a cell divides, the DNA duplicates, so that when the cell divides the two new cells have identical DNA molecules.

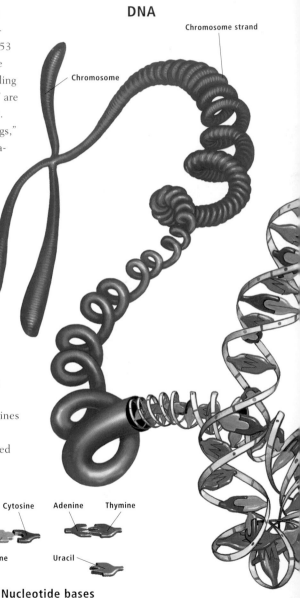

DNA

Chromosome strand

Chromosome

Cytosine   Adenine   Thymine

Guanine   Uracil

Nucleotide bases

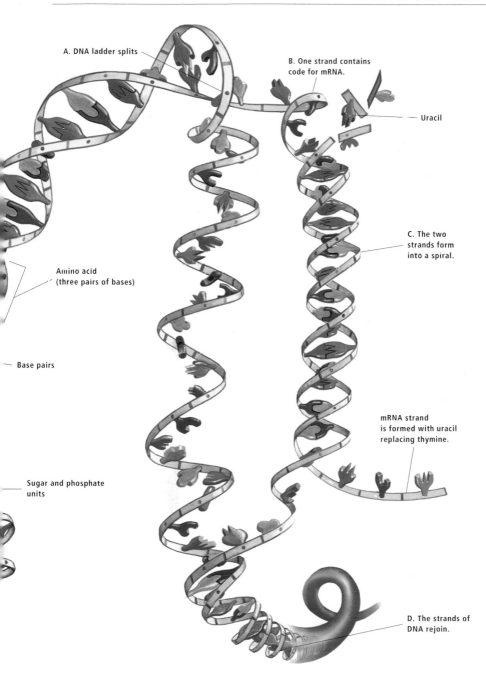

A. DNA ladder splits

B. One strand contains code for mRNA.

Uracil

C. The two strands form into a spiral.

Amino acid (three pairs of bases)

Base pairs

mRNA strand is formed with uracil replacing thymine.

Sugar and phosphate units

D. The strands of DNA rejoin.

Chromosomes

With the exception of the sex cells, the nucleus of every cell in the human body contains 46 chromosomes, arranged as 23 pairs. The thread-like chromosomes consist of DNA, with each chromosome containing hundreds of genes. We each have around 100,000 genes, with the unique combination of these genes giving each of us our individual characteristics.

The sex cells—the female ovum and the male sperm—differ from the majority of the body's cells. Each of the sex cells contains 23 single chromosomes. When fertilization takes place, the sperm and ovum merge to form an embryo

Chromosomes

cell with 23 pairs of chromosomes. Of these 23 pairs of chromosomes, one pair is the sex chromosomes, which will determine the sex of the embryo, with the male sperm providing the determining chromosomes.

Males have X and Y sex chromosomes, while females have only X. The Y chromosome contains fewer genes and is shorter than the X, being roughly one-third its length. If, at fertilization, the pairing male and female sex chromosomes are both X then the result will be a girl, and if the pairing sex chromosomes are a female X and a male Y, then the result will be a boy.

Cell detail

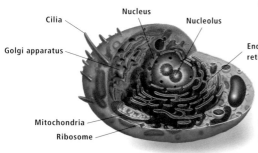

Cilia
Golgi apparatus
Nucleus
Nucleolus
Endoplasmic reticulum
Mitochondria
Ribosome

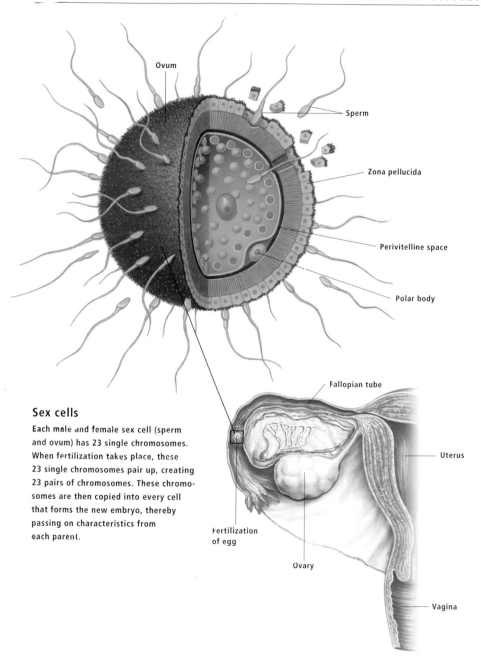

Ovum

Sperm

Zona pellucida

Perivitelline space

Polar body

Fallopian tube

Uterus

Sex cells

Each male and female sex cell (sperm and ovum) has 23 single chromosomes. When fertilization takes place, these 23 single chromosomes pair up, creating 23 pairs of chromosomes. These chromosomes are then copied into every cell that forms the new embryo, thereby passing on characteristics from each parent.

Fertilization of egg

Ovary

Vagina

Genes and Heredity

Passed from parent to child, genes consti-
tute the genetic makeup of each individual.

At conception, 23 chromosomes from
each parent join to make 23 pairs. These
chromosomes, containing DNA, are du-
plicated in each cell in the body.
Each gene is a particular sequence of nucleotide bases
in the DNA, and this se-
quence determines the
production of a par-
ticular protein. The
genes are responsible
for initiating chemical
reactions; building,

Genetic code

Each individual has about 100,000 genes,
with each gene made up of a combination
of nucleotide bases that form the code for
a particular protein.

Nucleotide base

Base pairs

maintenance and repair of cells; and creating the unique individual characteristics
of each person.

Thousands of genes contained in DNA
determine our genetic characteristics.
These characteristics are passed down from
one generation to another, with the genes
carrying our hereditary code. Genes govern
traits such as tissue and organ development, blood type, hair type, and color vision,
and some characteristics are determined by
more than one gene.

Some genes are dominant and some are
recessive. Depending on the parental genes,
characteristics such as eye color are influenced either by a dominant gene or are the
result of two recessive genes.

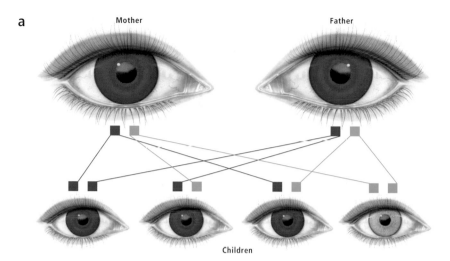

a

Mother

Father

Children

Dominant and recessive genes

Most characteristics result from a mixture of two sets of genetic instructions contained in one or more gene pairs, but some features, such as eye color, are determined by a single gene. This dominant gene overrides the instructions of the other recessive gene. A recessive trait can only emerge when two recessive genes for that trait are inherited. The gene for brown eyes is dominant over the recessive gene for blue eyes. This means that two parents with brown eyes can only have a child with blue eyes if the child inherits a recessive blue gene from each parent (a). If one parent with brown eyes has two dominant brown-eye genes, all children will inherit at least one dominant gene (even if one parent is blue eyed) and will all have brown eyes (b). If both parents have blue eyes, neither will have the dominant gene and all children will have blue eyes.

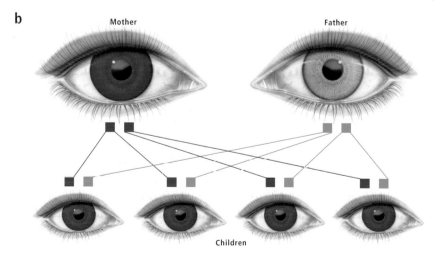

b

Mother

Father

Children

Tissues and Cells in the Ageing Process

As we age, the effectiveness of cells and tissues in the body systems is reduced. Some systems maintain steady levels, such as the endocrine system, which generally suffers no decrease in hormone production, although often the target organs do not function at full capacity or decline in activity. An exception to this steady hormonal level is the reproductive hormones, which decrease in males over time, and decrease markedly in females at menopause.

Other systems slow or decline in efficiency as we age. For instance, while the nerve cells of the nervous system do not

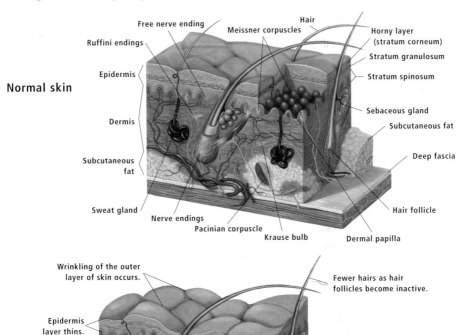

Normal skin

Free nerve ending

Ruffini endings

Epidermis

Dermis

Subcutaneous fat

Sweat gland

Nerve endings

Pacinian corpuscle

Krause bulb

Hair

Meissner corpuscles

Horny layer (stratum corneum)

Stratum granulosum

Stratum spinosum

Sebaceous gland

Subcutaneous fat

Deep fascia

Hair follicle

Dermal papilla

Wrinkling of the outer layer of skin occurs.

Fewer hairs as hair follicles become inactive.

Epidermis layer thins.

Reduced amount of elastic fibers results in wrinkling of the skin.

Sebaceous glands become less active, resulting in dry, fragile skin.

Sweat glands become less active, making body temperature regulation less efficient.

Ageing skin

The regenerative properties of skin decline as we age. Cell production diminishes, causing wrinkled skin. Gland activity slows—the amount of sebum produced at the skin surface is reduced, and the body is less tolerant of heat.

individually lose efficiency, any damaged or dead neural cells are not replaced, resulting in an overall reduced number of nerve cells to cope with specific tasks.

In other systems, cell production or cell division slows, resulting in cell production numbers being unable to keep pace with cell losses. For instance, the skin cells are produced at a much slower rate as we age, resulting in thinner skin and a propensity for wrinkles due to loss of elasticity.

The reduced production of melanocytes in the skin causes a lowered tolerance to the effects of the sun. In conjunction with these changes, the growth of hair follicles begins to fail or slow, and the sweat and sebaceous glands become less active.

Bone density often decreases as we age, sometimes resulting in osteoporosis and other bone-related diseases. This occurs when bone production by osteoblasts is outstripped by the rate of reabsorption of bone by osteoclasts.

As a result, structural strength is lost as the outer protective layer of bone thins and the core of spongy bone loses substance, becoming thin and brittle and filled with airy pockets. These deficiencies make the bone vulnerable to fracture and damage.

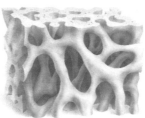

Normal bone

Osteoporosis

In osteoporosis, bone strength is lost as bone mass reduces. The inability of bone to rebuild at the same pace as bone reabsorption results in brittle, weak bones.

Osteoporotic bone

Dementia

Degeneration of cells in the cerebral cortex results in deterioration of memory and reasoning capacity, coupled with changes in personality, such as is seen in dementia.

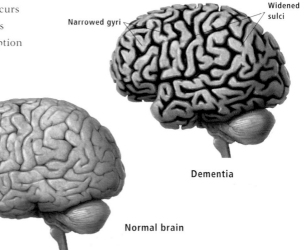

Narrowed gyri

Widened sulci

Dementia

Normal brain

Bacteria

The body is constantly being infiltrated by bacteria. Many of these bacteria are beneficial, being essential to the smooth running of the body's operations, and working in harmony with the body systems. This is particularly so in the digestive system, where bacteria such as certain Lactobacillus bacteria are essential to the effective breakdown and absorption of food.

These naturally occurring bacteria fight off foreign invaders and maintain the healthy balance of intestinal flora required for good health.

However, there are also many bacteria that are harmful to the body, causing disease and infection by invading the body and multiplying.

Microscopic in size, bacteria are single-celled organisms, without a nucleus or major organelles. They contain the DNA required to duplicate and multiply, a process they carry out at a rapid rate, thereby infesting tissues or organs very quickly. In the case of harmful bacteria, the body must initiate a response to the invading bacteria. This process is called an immune response. The body develops antibodies to neutralize the invading bacteria.

Antibiotics are also often prescribed to fight bacterial infections; however, this solution can also result in the destruction of beneficial bacteria along with the offending bacteria. Overuse of antibiotics has, in some cases, resulted in certain bacteria becoming resistant to treatment.

Immunization has proven a successful approach to many bacterial infections. Many previously fatal or life-threatening diseases have now been eradicated through systematic immunization of the population.

Bacteria are classified by their shape, with the major types being spherical (coccus), coil-shaped (spirochete), comma-shaped (vibrio), or rod-shaped (bacillus).

Bacteria

Classified by shape, bacteria usually fall into one of the following classifications: rod-shaped (bacillus), spherical (coccus), coil-shaped (spirochete), or comma-shaped (vibrio).

Coil-shaped (spirochete)

Syphilis

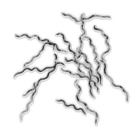

Lyme disease

Rod-shaped (bacillus)

Tetanus

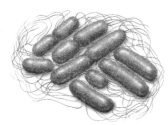

Salmonella

E. coli

Spherical (coccus)

Streptococcus

Meningococcus

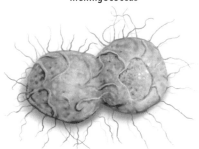

Gonococcus

Comma-shaped (vibrio)

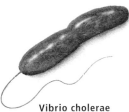

Vibrio cholerae

Viruses

Viruses are tiny organisms that invade the body, attaching themselves to cells within the body in order to reproduce. Viruses have no nucleus or organelles, they contain only the DNA required to duplicate.

Because the virus requires a host cell in order to reproduce, it is difficult to eradicate without disturbing the balance of cells in the body.

While viruses can infect all body tissues, some attack specific systems or tissues. For instance, the polio virus attacks the nervous system, the wart virus infects skin tissue, and the common cold virus attacks the upper respiratory tract.

Once a virus has entered the system— usually via the respiratory tract, and then infiltrating the bloodstream—it can manifest in several ways. Some cause short-term acute disease, others cause recurring or chronic disease, while others do not cause any disease.

The acute viral infections are of two types—local and systemic—as the result of the effect of the invading virus on the host cell.

Some viruses can remain dormant in the body for many years before producing symptoms; the herpes zoster virus, for example, can lie dormant after chickenpox and causes later attacks of shingles at times of reduced immunity.

Immunization has proven successful in eliminating some viral infections, with smallpox and polio being notable examples.

Viruses

Varying in shape and size, viruses invade other cells in order to reproduce. Viruses can infect all body tissues, but individual viruses generally target a specific part of the body; for example, the herpes virus infects the skin.

Influenza virus

Polio virus

Cold virus

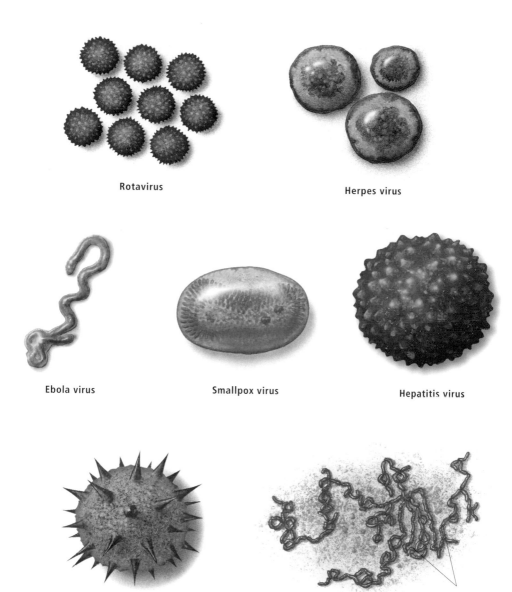

Rotavirus

Herpes virus

Ebola virus

Smallpox virus

Hepatitis virus

Wart virus

Infectious mononucleosis

Epstein-Barr
viral particles

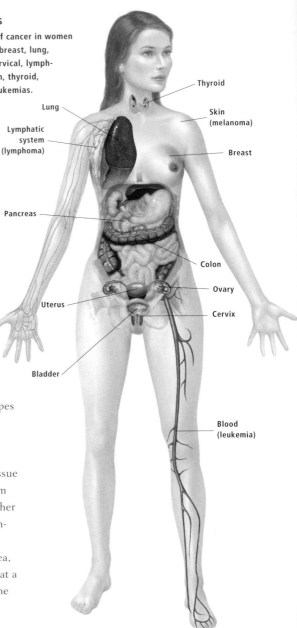

Female cancer sites

The most common types of cancer in women are, in descending order, breast, lung, colorectal, uterine and cervical, lymphomas, melanomas, ovarian, thyroid, pancreas, bladder, and leukemias.

Thyroid

Lung

Skin (melanoma)

Lymphatic system (lymphoma)

Breast

Diseases and Disorders of Cells and Tissues

There are many diseases and disorders affecting cells and tissues, with cancer being one of the most common in developed countries.

Pancreas

Colon

Cancer

Cancer is a disease in which normal cells become abnormal and grow uncontrollably, often spreading from the site of origin to affect other areas. Its cause is unknown, though some types of cancer would appear to run in families, while others can be linked to certain risk factors. There are over 100 different types of cancer, and it is a principal cause of death in developed countries, second only to heart disease.

Ovary

Uterus

Cervix

Bladder

Blood (leukemia)

Cancer spreads by infiltrating the tissue around it or by entering the bloodstream or lymphatic system and spreading further afield. When the disease enters the lymphatic system, it is often carried to the lymph nodes that drain the affected area, forming a tumor. When a tumor forms at a site removed from the initial point of the disease, this is known as metastasis.

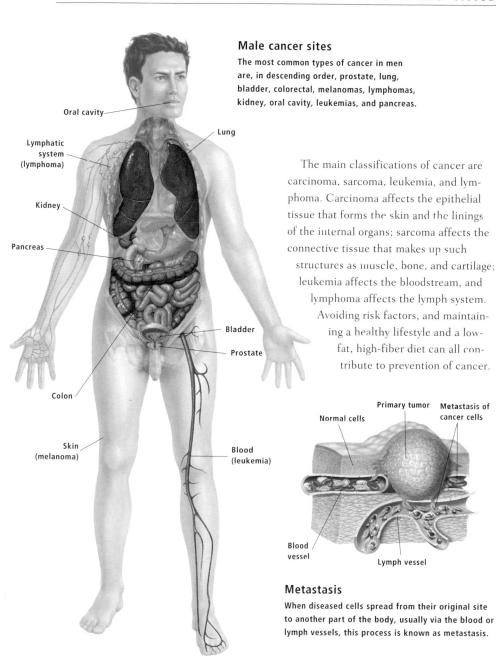

Male cancer sites

The most common types of cancer in men are, in descending order, prostate, lung, bladder, colorectal, melanomas, lymphomas, kidney, oral cavity, leukemias, and pancreas.

Oral cavity

Lymphatic system (lymphoma)

Kidney

Pancreas

Lung

Bladder

Prostate

Colon

Skin (melanoma)

Blood (leukemia)

The main classifications of cancer are carcinoma, sarcoma, leukemia, and lymphoma. Carcinoma affects the epithelial tissue that forms the skin and the linings of the internal organs; sarcoma affects the connective tissue that makes up such structures as muscle, bone, and cartilage; leukemia affects the bloodstream, and lymphoma affects the lymph system. Avoiding risk factors, and maintaining a healthy lifestyle and a low-fat, high-fiber diet can all contribute to prevention of cancer.

Normal cells

Primary tumor

Metastasis of cancer cells

Blood vessel

Lymph vessel

Metastasis

When diseased cells spread from their original site to another part of the body, usually via the blood or lymph vessels, this process is known as metastasis.

Genetic and Inherited Diseases and Disorders

Birthmarks

The cause of distinguishing marks present on some babies at birth is unknown. These birthmarks are usually harmless, and some types of birthmarks will gradually disappear over time. Permanent birthmarks can sometimes be removed using surgical techniques such as plastic surgery or cryosurgery (freezing).

Among the more common types of birthmarks are salmon patches and port wine stains; these are both types of hemangiomas—clusters of blood vessels—which form the colored markings. Cafe au lait spots are permanent coffee-colored patches of skin, while congenital pigmented nevi are light brown to black moles.

Down's syndrome

Trisomy 21 is a congenital defect where a person has three number 21 chromosomes instead of two. This causes Down's syndrome, for which there is no cure, but which with specialized education and training can result in a full and useful life.

Down's syndrome is characterized by several physical features such as a round, flat head; small, low-set ears; and eyes that slant upwards. Intellectual development

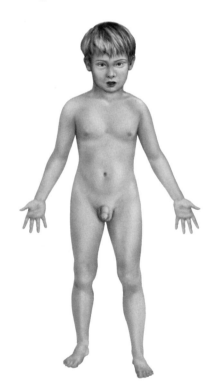

Down's syndrome

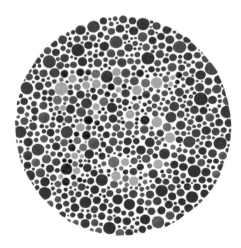

Color blindness

Cystic fibrosis

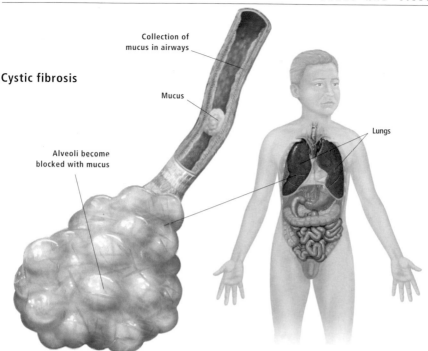

Collection of mucus in airways

Mucus

Alveoli become blocked with mucus

Lungs

is retarded, and other congenital disorders, such as heart, respiratory, or visual problems often occur.

Color blindness

Primarily an inherited disorder, color blindness is caused by a gene linked to the X chromosome, and is more prevalent in males. Color blindness occurs when certain cone cells are missing from the retina of the eye, causing difficulty in distinguishing red from green, or blue from yellow. There is no cure for color blindness, and impaired color vision can impact on career choices where full color vision is a vital component.

Cystic fibrosis

An inherited disorder caused by a defective gene, cystic fibrosis causes mucous secretions to become thick, interfering primarily with the normal function of the airways. The mucous secretions of other organs, such as the pancreas and liver, can also be affected, with the ducts leading to the digestive tract becoming blocked and unable to absorb adequate nutrients. Signs of cystic fibrosis usually manifest between infancy and childhood. There is no cure, but the symptoms can be eased with breathing and dietary treatment—however, life expectancy is still short.

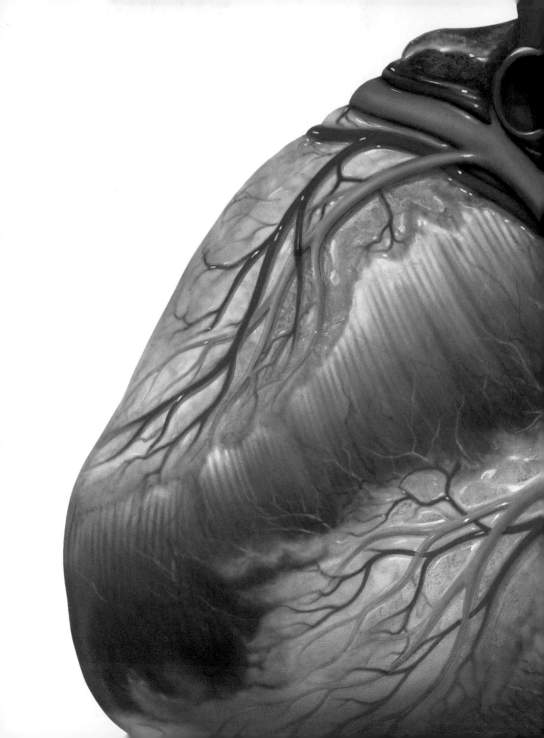

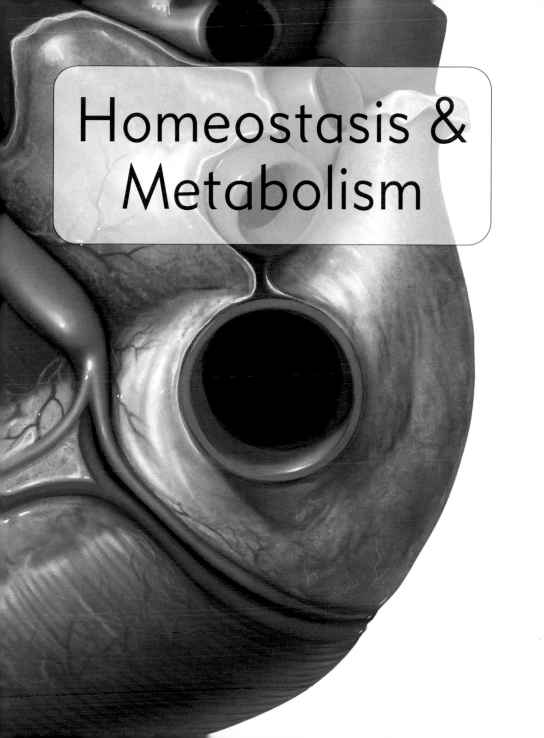

Homeostasis & Metabolism

HOMEOSTASIS AND METABOLISM

"Homeostasis" refers to the tendency to stability in the normal physiological state of the human body, while "metabolism" refers to all the physical and chemical processes that take place in the body.

Hormones

The endocrine organs—pituitary, thyroid, parathyroid, adrenal, pancreas, testes in males, and ovaries in females—produce the hormones that control a wide range of the body's functions, including growth, reproductive activity, and metabolism. The hormonal substances produced fall into three main types, and they are defined by their structure.

Hormones made from the amino acid tyrosine, such as epinephrine (adrenaline) and thyroxine, are made by specific chemical reactions. Peptide and protein hormones, such as growth hormone and insulin are made from an mRNA (messenger RNA) sequence and stored ready for release. Steroid hormones, which include the sex hormones and corticosteroids, are produced from cholesterol.

Most hormones are released into the bloodstream and transported to their target organs, where they trigger the required response. Some, such as thyroxine, have target cells throughout the body. Others trigger response only from specific tissues; for example, thyroid-stimulating hormone, as its name suggests, activates the thyroid gland.

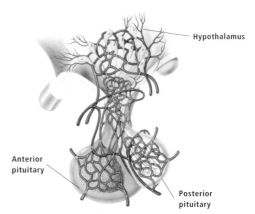

Hypothalamus

Anterior pituitary

Posterior pituitary

Pituitary gland

The pituitary gland regulates the glandular activity of the organs of the endocrine system, and is itself controlled by the hypothalamus.

Hormones

The endocrine glands are responsible for the production of hormones—some are produced for use throughout the body, while others are produced to activate a response from a specific target area. The hormones bond with receptors on the target cells, stimulating a response to body requirements. The response may involve an increase in hormone or enzyme production, or inhibit production in the case of an oversupply.

Male and female sex organs are triggered into activity at the onset of puberty. These hormones stimulate follicle activity in the ovaries in females, and initiate the production of estrogen and progesterone; in males, sperm production and testosterone secretion commences. The sex hormones are also involved in the development of external genitalia and secondary sexual characteristics.

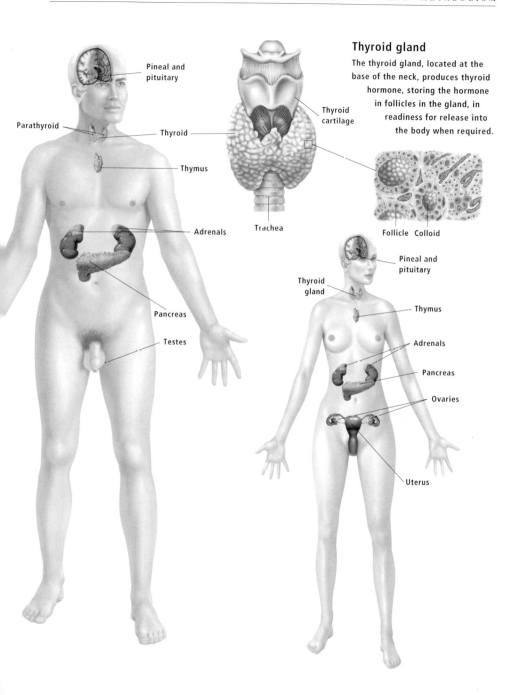

Pineal and pituitary

Parathyroid

Thyroid

Thymus

Adrenals

Pancreas

Testes

Thyroid cartilage

Trachea

Thyroid gland

The thyroid gland, located at the base of the neck, produces thyroid hormone, storing the hormone in follicles in the gland, in readiness for release into the body when required.

Follicle   Colloid

Pineal and pituitary

Thyroid gland

Thymus

Adrenals

Pancreas

Ovaries

Uterus

Metabolism

The smooth, reliable operation of the body is maintained by chemical reactions produced in the cells. The chemical processes that influence all of the body's systems are collectively known as metabolism.

Metabolism involves two phases—anabolism and catabolism. Anabolism is the building-up phase, where complex molecules and substances are created from a foundation of simple molecules. Anabolism is energy-driven, requiring steady cellular activity. Catabolism is the process of creating energy by breaking down complex molecules into simple structures.

This breakdown process produces energy to ensure that the body's cells work efficiently and correctly. The energy produced is stored, ready for release when needed, in a substance known as ATP (adenosine triphosphate).

Our dietary intake provides nutrients that are involved in the catabolic process. These nutrients include carbohydrates, lipids, proteins, minerals, and vitamins. Equally essential for effective chemical reactions to occur are enzymes, which are manufactured by the body.

The enzymes and nutrients necessary for the anabolic and catabolic processes to occur must be in harmony—any imbalance can often be addressed by strict nutritional or dietary regimes.

Hormones regulate our overall metabolism, in particular the hormones produced by the thyroid gland. An over- or under-productive thyroid gland can cause metabolic disorders and poses serious health problems.

Pituitary gland

Thyroid gland

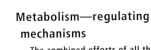

Metabolism—regulating mechanisms

The combined efforts of all the chemical processes in the body are referred to as metabolism. Hormones play an integral role in metabolism, with the hormones produced by the thyroid gland regulating our metabolic rate.

Chemical metabolism

Chemical metabolism occurs in various parts of the body. Enzymes produced by the body are utilized in the digestive system to systematically break down food into viable nutrients. Once broken down, the nutrients are carried via the hepatic portal system to the liver. The liver metabolizes these nutrients and stores glucose to provide energy for body cells and muscles.

Pancreas produces digestive enzymes for food breakdown.

Muscles are powered by glucose.

Stomach produces gastric juices that react with food.

Liver metabolizes nutrients derived from food, and stores reserves of glucose in the form of glycogen.

Intestines absorb nutrients from food as it travels through their channels.

Negative Feedback Mechanism

The body has various methods in place to control hormone levels, but a common mechanism is called negative feedback. Negative feedback is the body's way of keeping hormone levels stable. Whenever there are excess or reduced amounts of a particular hormone, the body triggers a response to restore normal levels—a fine balance is necessary, as any imbalance of a particular hormone can impact on the function and efficiency of other hormones.

An example of negative feedback is seen in the interaction between the parathyroid and thyroid glands, which regulate calcium levels in the blood. Normal calcium levels are maintained by the parathyroid gland in conjunction with the kidneys. When levels fall, parathyroid hormones stimulate the release of calcium from bone to bolster levels. When levels rise, the parathyroid decreases hormone secretions, and the thyroid releases calcitonin, prompting the kidneys to increase excretion of calcium, thereby restoring the normal balance.

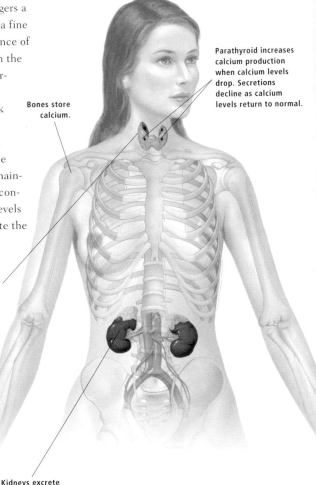

Bones store calcium.

Parathyroid increases calcium production when calcium levels drop. Secretions decline as calcium levels return to normal.

Thyroid stabilizes calcium levels, when oversupply occurs. Hormones released by the thyroid cause the kidneys to excrete more calcium.

Negative feedback mechanism

The body's negative feedback mechanism keeps hormone levels in check. The illustration shows the organs involved in calcium production and regulation.

Kidneys excrete calcium.

Insulin

Insulin is a hormone produced by the pancreas, and plays a role in the body's ability to use sugars, facilitating absorption of glucose into body cells. Within the pancreas are clusters of cells known as the islets of Langerhans. Beta cells in the islets of Langerhans produce and release insulin to aid absorption of glucose, which is acquired from the breakdown of carbohydrates during the digestive process.

Insulin production

Produced in the pancreas, insulin aids absorption of glucose by the body. Glucose is an essential element for energy production. Any excess glucose is stored in the liver and muscle tissue, ready for release when required.

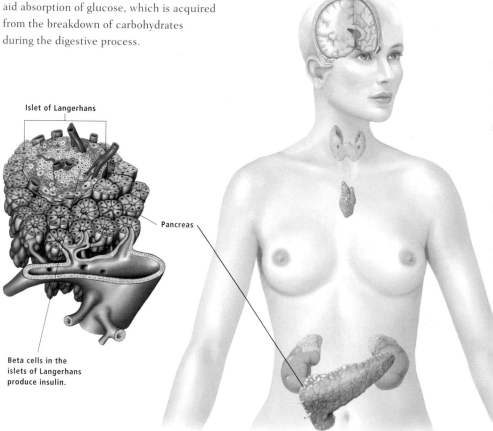

Islet of Langerhans

Pancreas

Beta cells in the islets of Langerhans produce insulin.

Prostaglandins

Prostaglandins are naturally occurring chemicals produced in many tissues in the body. Responsible for contractions of smooth muscle such as that found in the uterus, these chemicals also play a role in inflammation and pain response. Prostaglandins in platelets cause them to clump together, forming a blood clot or thrombus.

Steroids

Produced from cholesterol, steroids are a large component of the body's hormones. Glucocorticoids, mineralocorticoids, and sex steroids are the three main types of naturally-occurring steroids.

The glucocorticoids include cortisol, also known as the "stress hormone," which is released in response to physical and mental stresses on the body, including exercise, infection, fear, and depression.

Mineralocorticoids play a role in maintaining salt and water levels in the body.

The sex steroids are androgens and estrogens. Androgens, which increase muscle mass and strength, include androsterone and testosterone. Androgens produced by the adrenal glands are converted to testosterone in the testes. Estrogens are produced in the female ovaries.

Androgens

Androgens are produced in the adrenal glands and testes in men, with small amounts produced in the adrenals and ovaries of women. Androgens are responsible for muscle mass and strength.

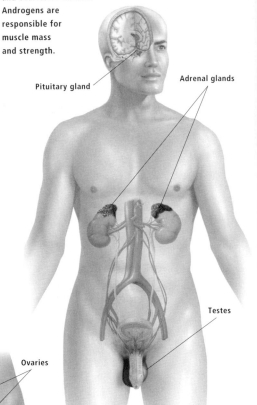

Pituitary gland

Adrenal glands

Testes

Ovaries

Electrolytes

Sodium, potassium, calcium, phosphate, chloride, and bicarbonate are the major electrolytes in the body. Found in the bloodstream and body fluids, electrolytes play an essential role in body functions. Sodium maintains blood pressure and volume. Potassium is involved in muscle and tissue growth and muscle contractions, and along with phosphate, maintains the balance of the acid-base in blood and tissues.

Calcium and phosphate form bone and teeth. Bicarbonate maintains the pH level in blood and body fluids. Sodium, potassium, and calcium are all involved in maintaining heart rhythm and the body's water balance.

Electrolytes

Different organs control the electrolyte balance in the body. The adrenal glands regulate sodium and potassium; the parathyroid glands regulate calcium and phosphate; the kidneys and lungs regulate bicarbonate; and the kidneys regulate chloride.

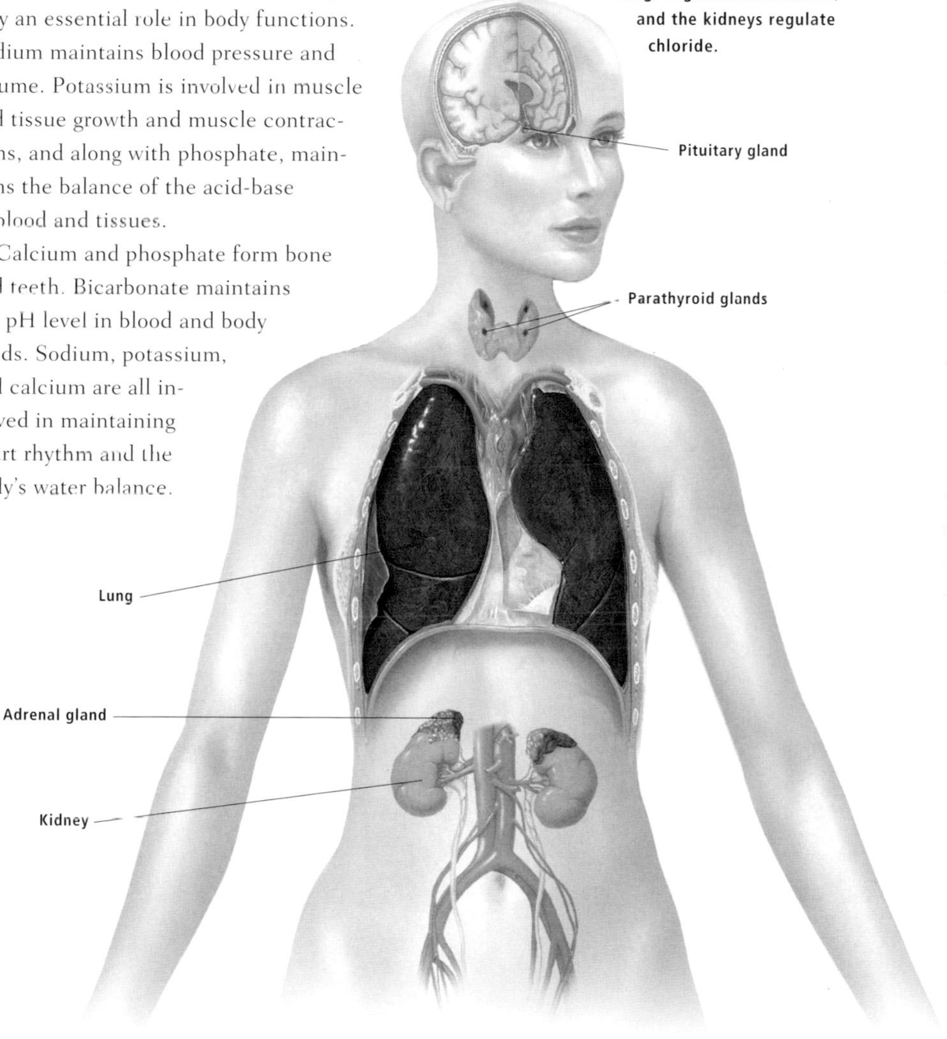

Pituitary gland

Parathyroid glands

Lung

Adrenal gland

Kidney

Temperature

Generally, the normal body temperature is regarded to be 98.6°F (37°C), but it can range between 97.2–100°F (36.2–37.8°C). An accurate temperature reading can only be achieved with a thermometer, as often a person can feel hot to the touch even when their core body temperature is normal.

Temperature Regulation

Body temperature is regulated by the hypothalamus. The normal core temperature of the human body ranges between 97.2–100°F (36.2–37.8°C), and varies with the circadian rhythm of the body, physical exertion, food intake, and emotional levels.

When the body becomes heated, the hypothalamus initiates responses to restore normal body temperature including dilation of the blood vessels and activation of the sweat glands.

In this way, body heat can be lowered through evaporation, radiation, conduction, or convection. When the body is cold, the metabolic rate is raised, increasing heat production,

the blood vessels and pores of the skin contract, and the hairs rise up, trapping body heat close to the skin surface, thus creating an insulating layer.

Temperature regulation

Body temperature is regulated by the hypothalamus, which is alerted to changes in external temperature by nerve endings in the surface of the skin.

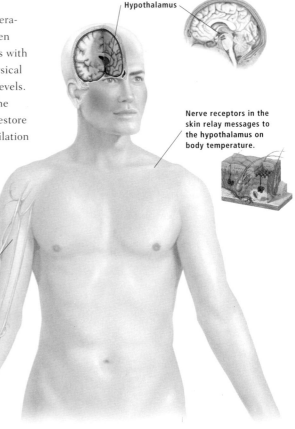

Hypothalamus

Nerve receptors in the skin relay messages to the hypothalamus on body temperature.

Signals sent by nerve cells direct blood flow to organs or the skin, depending on the temperature of the body.

Skin reaction to cold

When the body is cold, the arteries and
pores contract, the hairs rise up
and trap body heat close to
the skin surface, thus
insulating the body.

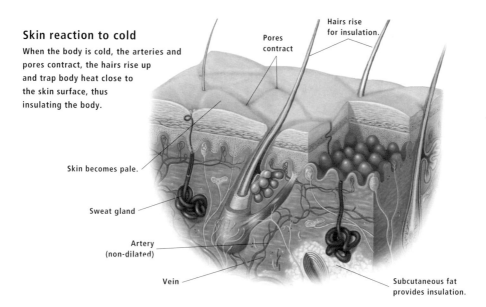

Pores
contract

Hairs rise
for insulation.

Skin becomes pale.

Sweat gland

Artery
(non-dilated)

Vein

Subcutaneous fat
provides insulation.

Skin and temperature

Responding to changes in body temperature,
the hypothalamus activates responses to heat
and cold.

Skin reaction to heat

When the body is hot, the arteries dilate, blood flow
to the skin surface is increased and heat loss is
maximized. The sweat glands are stimulated, releasing
fluid that evaporates and cools the body temperature.

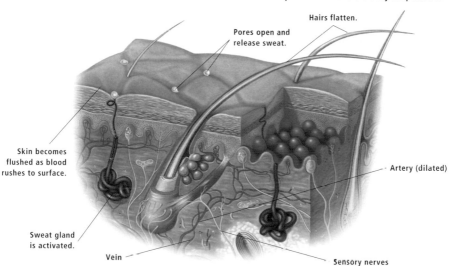

Hairs flatten.

Pores open and
release sweat.

Skin becomes
flushed as blood
rushes to surface.

Artery (dilated)

Sweat gland
is activated.

Vein

Sensory nerves

Metabolic Imbalances and Disorders of Homeostasis

Diabetes

Diabetes mellitus is caused by a reduced production of insulin in the pancreas or the reduced ability of the body to utilize insulin. Glucose, the body's major energy source, is derived from nutrients extracted from digested food. Insulin aids in the transport of glucose from the blood to the liver and muscle cells. Under-production of the required amount of insulin for this task results in diabetes.

The inability to transport glucose to body tissues and cells, caused by insufficient insulin production in the pancreas, means that glucose instead circulates in the bloodstream, disturbing the body's chemical balance.

There are two major types of diabetes. Type I, often called juvenile-type or insulin-dependent diabetes, usually affects those under 25 years of age. Type II, often referred to as non-insulin dependent, maturity onset or adult-type diabetes, usually affects those over 40 years of age.

Diabetes

Diabetes mellitus is caused by a disruption in the body's processing of glucose. Long term, this can cause various disorders to arise.

Eyes

Glaucoma often occurs in diabetes. In glaucoma, pressure inside the eyeball rises, damaging the retina and causing loss of vision.

Heart

Diabetes can cause coronary artery disease, which in turn can cause death of cardiac tissue known as myocardial infarction.

Kidneys

Diabetic nephropathy damages the glomeruli and small blood vessels of the kidneys, due to high levels of blood glucose. This results in the loss of necessary proteins through the urine, swelling of body tissues, and eventually renal failure.

Feet

Diabetes slows the healing of body tissues and also causes degeneration of the peripheral nerves (neuropathy). These two factors act together to cause foot ulcers in diabetics.

Arteries

Diabetes is one of the risk factors for atherosclerosis, a disease in which fatty deposits build up under the lining of the artery and block off its blood flow.

In both types of diabetes, reduced production of insulin causes raised glucose levels in the blood. In Type II diabetes, as well as abnormal insulin secretion, there is a resistance to insulin action in target tissues. Insulin is also involved in the metabolism of other substances in the body, and these functions are also affected. In the case of type I diabetes, daily injections of insulin are required to augment the body's naturally-occurring supply.

Fever

Fever is an abnormally elevated body temperature. Normal body temperature is 98.6°F (37°C), and fever is registered by an increase of at least 0.5°F (around 0.3°C). Often triggered by viral or bacterial infections, a fever and resulting high temperature can also be associated with exercise, dehydration, or can occur following immunization. Associated symptoms of fever can include shivering, sweating, headache, restlessness, weakness, and loss of appetite.

Dehydration

The pituitary gland works in conjunction with the kidneys to maintain correct water levels in the body—lean body mass normally comprises around 70 percent water.

Caused by insufficient fluid intake, gastrointestinal disorders, overheating, or heat stress, dehydration can cause a range of symptoms, including vomiting, overheating, and fever, and can particularly affect children, as the water levels in their bodies are higher than that of adults.

When fluid levels fall and dehydration occurs, the hypothalamus triggers the posterior pituitary gland to release antidiuretic hormone (ADH), which acts on the kidneys, making them reabsorb water into the blood and decrease urine production. Once the body is rehydrated, normal functioning of the kidneys resumes as production of ADH in the pituitary returns to normal.

Hypothalamus
Pituitary gland
Arteries
Posterior pituitary gland
Network of capillaries
Kidneys
Bladder

Dehydration

Water balance is regulated by the pituitary gland and the kidneys. Fluctuations in water levels are addressed by the posterior pituitary gland, which releases hormones to control kidney function.

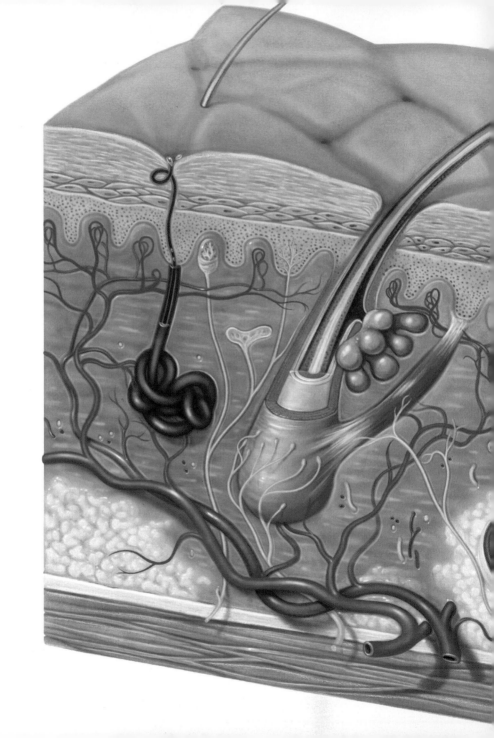

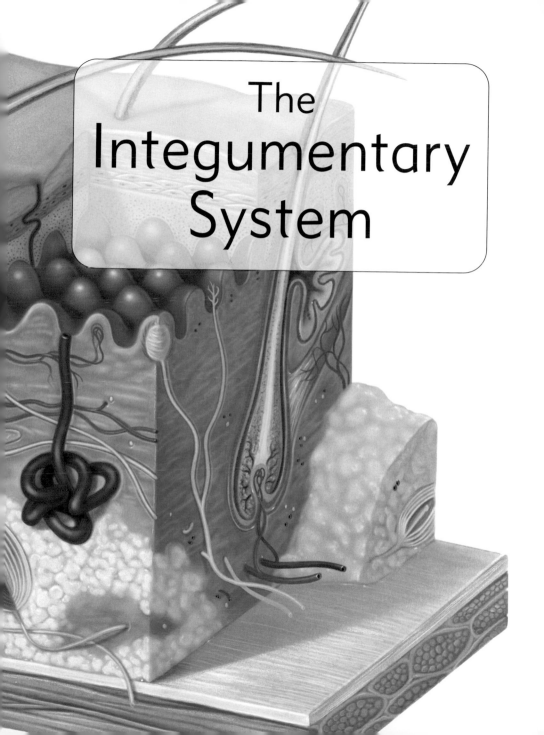

The Integumentary System

Hairs covering the scalp
provide protection to the skull.

The hairs of the eyebrows and
eyelashes protect the eyes.

Thin delicate skin covers the lips.
Underlying blood vessels give color
to this transparent layer.

The nipple and its surrounding
area, the areola, are areas of
pigmented skin.

Veins run just below the skin
at the elbow, providing a
useful site for venepuncture.

The nails serve a pro-
tective function for
the tips of the fin-
gers and toes.

The skin of the eyelids
is extremely thin.

The skin surface extends into
the ear canal and nasal cavity,
where it merges into the
internal membranes.

Sweat and sebaceous glands
lie in the upper layers of the
skin. Sweat glands play a role
in temperature regulation;
sebaceous glands lubricate
the skin.

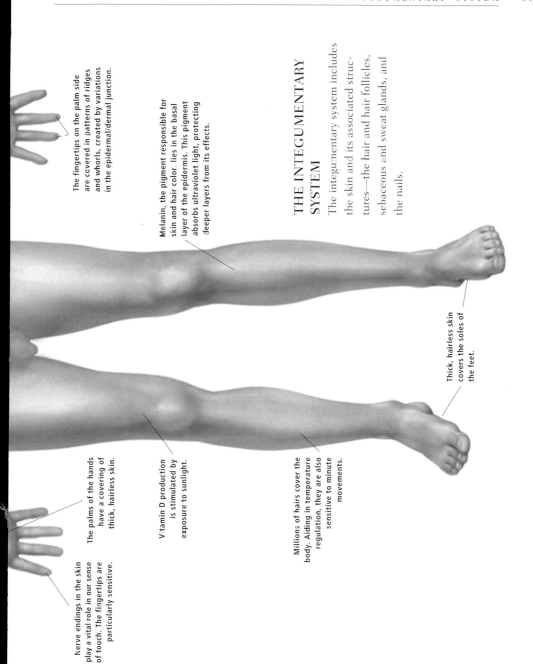

The fingertips on the palm side are covered in patterns of ridges and whorls, created by variations in the epidermal/dermal junction.

Melanin, the pigment responsible for skin and hair color, lies in the basal layer of the epidermis. This pigment absorbs ultraviolet light, protecting deeper layers from its effects.

THE INTEGUMENTARY SYSTEM

The integumentary system includes the skin and its associated structures—the hair and hair follicles, sebaceous and sweat glands, and the nails.

Thick, hairless skin covers the soles of the feet.

Nerve endings in the skin play a vital role in our sense of touch. The fingertips are particularly sensitive.

The palms of the hands have a covering of thick, hairless skin.

Vitamin D production is stimulated by exposure to sunlight.

Millions of hairs cover the body. Aiding in temperature regulation, they are also sensitive to minute movements.

The Skin

Covering our entire body, the skin is the largest organ, forming a protective layer over the internal organs against external elements.

Nerve receptors in the skin are extremely sensitive, allowing the body to sense heat and cold, pain, pressure, and touch. The skin also plays a role in temperature regulation and vitamin D manufacture, and provides protection against invading organisms and ultraviolet light.

The three layers of tissue that make up the skin are the epidermis, the dermis, and subcutaneous tissue.

Epidermis

The epidermis has five layers in its structure, which are (in ascending order): *stratum basale* or *stratum germinativum, stratum spinosum, stratum granulosum, stratum lucidum,* and *stratum corneum.*

The epidermal layers are comprised primarily of cells called keratinocytes. The outermost layer (*stratum corneum*) is comprised of dead cells that are continually shed and replaced by cells from the layers beneath, as the skin constantly renews itself.

This renewal process begins in the basal layer (*stratum basale* or *stratum germinativum*), where the cells divide (mitosis), slowly reaching the top layer over several weeks.

Melanocytes in the basal layer produce melanin, which is responsible for absorbing ultraviolet light, and for skin pigmentation. The next layer, *stratum spinosum*, is comprised of cells with spiny projections.

Above this is *stratum granulosum*, the site of keratin production. Overlying this

Free nerve ending

Ruffini endings

Epidermis

Dermis

Subcutaneous fat

Sweat gland

Nerve endings

Skin

is the clear layer of *stratum lucidum*; this keratin-rich layer is usually found only on the palms of the hand and soles of the feet. This layer is overlaid by the *stratum corneum*.

Dermis

The inner layer of skin, the dermis, is comprised of collagen and elastin fibers, interwoven with blood vessels, nerves, and fat lobules. Finger-like projections in the dermis, called papillae, run up into the epidermis.

Subcutaneous layer

Lying beneath the dermis is the subcutaneous layer. This is comprised of sub-cutaneous fat that insulates the body against extremes of temperature, connective tissue, and a small amount of blood vessels.

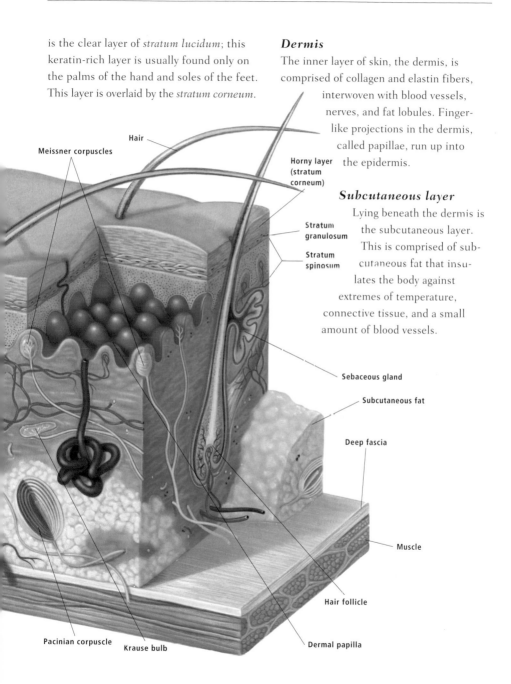

Hair

Meissner corpuscles

Horny layer (stratum corneum)

Stratum granulosum

Stratum spinosum

Sebaceous gland

Subcutaneous fat

Deep fascia

Muscle

Hair follicle

Dermal papilla

Pacinian corpuscle

Krause bulb

Hair

The root of the hair terminates in a bulb, which
is lodged in a follicle. Extending from the root,
through the dermis and epidermis
to project out from the skin,
is the hair shaft.

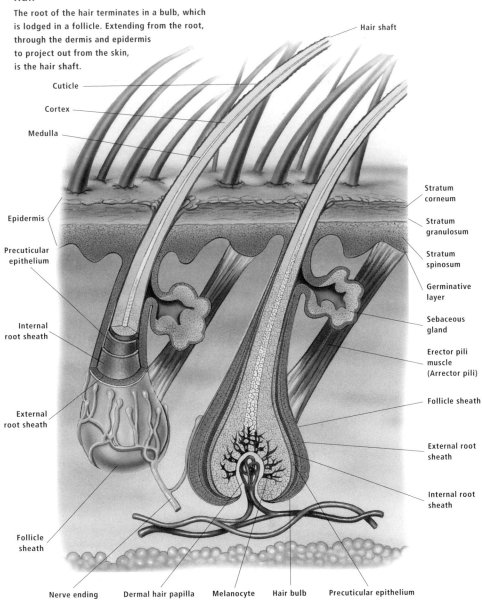

Hair shaft

Cuticle

Cortex

Medulla

Stratum corneum

Stratum granulosum

Stratum spinosum

Germinative layer

Epidermis

Sebaceous gland

Precuticular epithelium

Erector pili muscle (Arrector pili)

Internal root sheath

Follicle sheath

External root sheath

External root sheath

Internal root sheath

Follicle sheath

Nerve ending   Dermal hair papilla   Melanocyte   Hair bulb   Precuticular epithelium

Hair

Hairs cover the skin, with their structures extending down into the dermis. The root of the hair sits in a follicle, with blood supply to the root and follicle provided via a papilla. The hair grows upwards from the follicle base, with the hair shaft extending out from the skin surface. Each follicle has a small muscle attached, the arrector pili, which causes the hair to become erect, providing a layer of insulation in cold conditions.

Melanin is responsible for providing hair color, though hair color is genetically determined, with dark hair generally dominating over light hair. As we age, less melanin is produced, resulting in gray or white hair.

All body hair falls out gradually, to be replaced by new hair. The hair on some parts of the body serves a protective purpose. The eyebrows provide the eyes with some protection from light. The hairs of the ears and nose trap dust and foreign particles, preventing them from entering the body.

Scalp

The scalp contains around 100,000 hair follicles, which serve to protect against heat loss, ultraviolet light, and abrasions. Covering the skull, the scalp is comprised of skin and connective tissue. Tiny muscles in the front of the scalp contribute to facial movements and expressions.

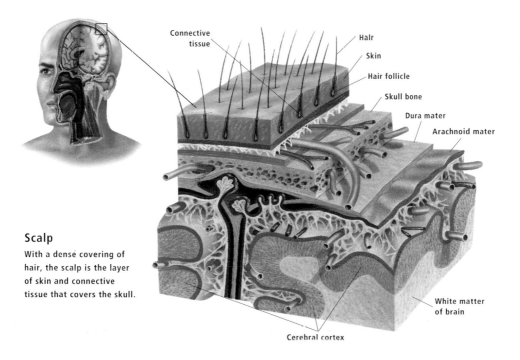

Scalp
With a dense covering of hair, the scalp is the layer of skin and connective tissue that covers the skull.

Connective tissue

Hair

Skin

Hair follicle

Skull bone

Dura mater

Arachnoid mater

White matter of brain

Cerebral cortex

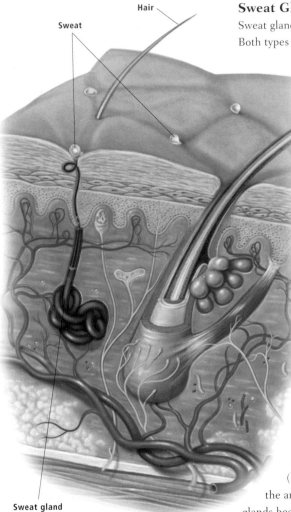

Hair

Sweat

Sweat gland

Sweat gland

Involved in temperature regulation, the sweat
glands release sweat to the surface of the
skin, which cools the skin as it evaporates.

Sweat Glands

Sweat glands are found in the skin layers.
Both types of sweat gland—eccrine and
apocrine—secrete per-
spiration onto the
surface to the skin.

The eccrine glands
are found all over the
body, with the excep-
tion of the lips and
around parts of the
external genitalia.

When the body be-
comes overheated, or
heat-affected by emo-
tional stress, the eccrine
sweat glands are acti-
vated, secreting sweat
through pores to the
surface of the skin.

As the sweat evaporates,
it cools the body, thus playing a role in temperature
regulation.

Secreting into hair follicles, rather than directly onto
the skin surface, apocrine
glands are found in the axillary
(armpit) and pubic regions, and in
the areolae of the nipples. Apocrine
glands become active during puberty, and
their pungent secretions are associated with
sexual attraction.

Body odor generally occurs as a result
of the combination of apocrine gland
secretions and bacteria.

Sebaceous Glands

Usually allied with hair follicles, sebaceous glands are found over most of the body, with the notable exception of hairless areas, such as the soles of the feet and the palms of the hands.

Triggered by hormones (testosterone in men, ovarian and adrenal androgens in women), the sebaceous glands become more active after puberty.

Sebaceous glands secrete sebum, an oily substance designed to protect the skin from damaging elements, and to lubricate and soften the skin.

Sebum often travels along the hair shaft to the skin surface, though some sebaceous glands release directly to the skin surface.

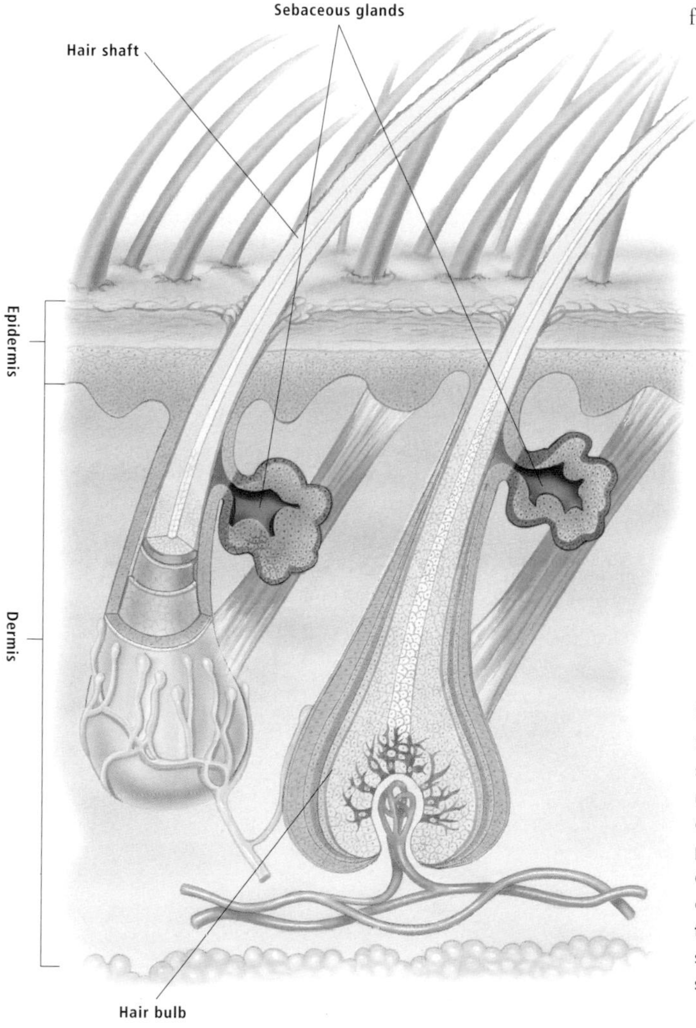

Sebaceous glands

Hair shaft

Epidermis

Dermis

Hair bulb

Sebaceous glands

Sebum produced by the sebaceous gland is released onto the surface of the skin, to form a protective film, preventing damage from water, chemicals, bacteria, and fungi. The oily base of sebum also lubricates and softens skin.

Nails

Nails are composed of keratin, a tough fibrous protein. The visible part of nail is mostly comprised of dead cells, with the only living part of the nail being the root, located at the base of the nail.

Nail growth commences at the nail root. Here, keratin is deposited in cells at the nail root; as these cells die off, they become part of the visible nail body. Although the nail body is comprised of dead cells, some sensation can be detected in the nail region, due to the network of nerves serving the end of the finger.

The half-moon shaped lunula, near the base of the nail, is a thickened area of nail. This thickening prevents the underlying blood vessels from showing through, making the area appear white in contrast to the body of the nail. While the nail itself is colorless, the nail body appears pink due to the abundance of blood vessels coursing beneath it.

Where the skin of the fingers and toes meets the nail, is a fold of skin, created by the skin folding back upon itself. This fold of skin is the eponychium, or cuticle. The underside of the fold is the "true cuticle,"

Nail

The nails are designed to protect the sensitive areas of the tips of the fingers and toes.

Capsule

Distal phalanx

Root of nail

Cuticle

Lunula

Nail

Auricular cartilage

Palmar ligament

Synovial cavity

Nail

The nails are often used as an overall indicator of health and well-being. The condition and color of the nails can often be a guide to underlying illness or disease.

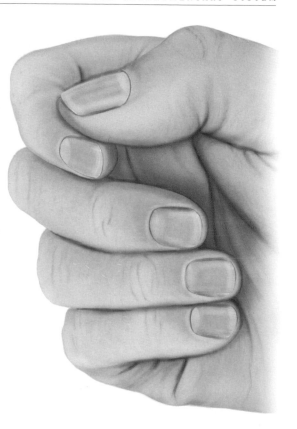

which effectively seals off and protects the nail area where new growth occurs.

Growth is slower in the toenails than in the fingernails—the fingernails grow at a rate of about 2 inches (5 centimeters) per year.

While the purpose of the nails is to provide protection to the tips of fingers and toes, they also provide indications of overall health. Changes in the nail appearance can indicate a variety of problems or disorders.

A yellow tinge to the nails can indicate hepatitis; a brown tinge can suggest kidney disease; extremely white nails can be a sign of iron deficiency; and white spots may indicate zinc deficiency. Change in texture or shape of the nails can also indicate deficiencies or internal disorders.

Touch

Motor area

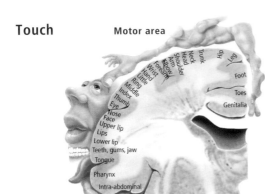

Sensory area

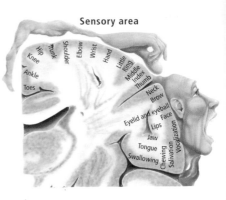

Touch pathways

Nerve receptors in the skin transmit information to the brain by sending nerve impulses along the peripheral nerves. The illustration uses proportionate sizing of body parts to indicate the amount of involvement of each of the primary cortices in that particular part of the body.

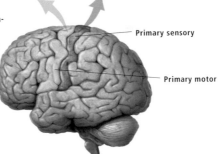

Primary sensory

Primary motor

Peripheral nerves

Spinal cord

Sensitivity

Our sense of touch is heightened in some areas of the body. Much of the information received by our brain is related to the sense of touch, with areas such as the fingertips, lips, and tongue being particularly sensitive, as they contain large concentrations of nerve receptors. Once the brain has received the impulses transmitted from the nerve receptors, the information is processed in the sensory cortex of the brain. The brain then initiates a response to the information, often involving the motor cortex, as in a reflex movement in response to touching a sharp or hot object.

Touch

Levels of sensitivity in the body vary, dependent on the concentration of nerve endings in each area. The nerve endings, known as cutaneous receptors, lie in the surface of the skin and allow us to sense heat and cold, pain, and pressure. The concentration of nerve endings results in some areas of the body being more sensitive than others—the fingertips, lips, and tongue, each with a large concentration of nerve endings,

are particularly sensitive. Our sense of touch is also called the tactile sense.

The nerve receptors in skin relay information in the form of nerve impulses. These impulses travel from the peripheral nerves to the spinal cord and on to the brain, where the information is processed, and an appropriate response initiated.

Dorsal funiculus

Spinal gray matter

Spinothalamic tract

Dorsal horn

Central canal

Ventral horn

Dorsal rootlets

Ventral rootlets

Spinal ganglion

Spinal nerve

Spinal ganglion

Spinal nerve

Spinal cord cross section

When stimulated, the nerve receptors in the skin relay messages along the peripheral nerves to the spinal cord. Sensory cells in the gray matter receive and relay the information to the brain.

Diseases and Disorders of the Integumentary System

Skin cancer

Skin cancers are one of the most common cancers. Various factors can cause skin cancers, including exposure to sun, x-rays, and some industrial chemicals. Certain hereditary disorders also predispose some individuals to skin cancer.

There are three major types of skin cancer, which originate in the epidermis: basal cell carcinoma, squamous cell carcinoma, and melanoma.

Basal cell carcinomas are the most common form of skin cancer, and usually occur following exposure to sun over an extended period of time.

Usually localized, these tumors rarely spread and can usually be removed by non-invasive procedures, such as cautery, or the application of liquid nitrogen.

Squamous cell carcinomas often occur as the result of extended exposure to sun, triggering changes to cells in the epithelium. Unlike basal cell carcinomas, these carcinomas can spread, often invading nearby lymph nodes and body organs.

Melanomas are a dangerous form of skin cancer, and are also often triggered by extended exposure to sunlight; however, other factors can trigger the onset of a melanoma, and they can even arise from ordinary moles. Melanomas usually spread radially initially, before spreading inwards, increasing in depth and affecting other areas.

Detection in the early stages can minimize or eliminate the threat of further spreading, If undetected, this type of skin cancer can spread to the lymphatic system and the blood, so early detection and intervention is vital.

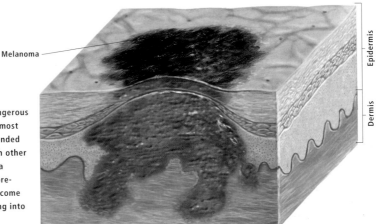

Melanoma

Epidermis

Dermis

Melanoma

Melanomas are a dangerous form of skin cancer, most often related to extended sun exposure, though other factors can produce a melanoma. Often a pre-existing mole can become malignant, developing into a melanoma.

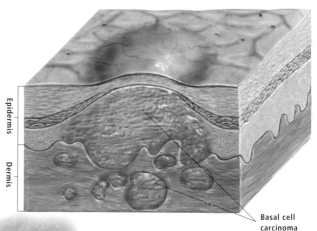

Epidermis

Dermis

Basal cell carcinoma

Basal cell carcinoma

The most common form of skin cancer, basal cell carcinomas first appear as a small nodule, which eventually change in outward appearance to form a small scab.

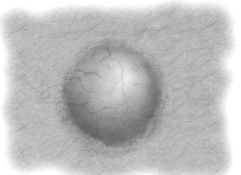

Squamous cell carcinoma

A squamous cell carcinoma forms in the surface layer of the skin. This particular type of skin cancer can spread to neighboring organs and lymph nodes.

Skin cancer

Skin cancer is one of the most common forms of cancer. Basal cell and squamous cell carcinomas commonly develop in those who spend a great deal of time outdoors. Melanomas are a particularly aggressive and potentially dangerous type of skin cancer. While prevention is the best solution, early detection and treatment of all types of skin cancer is crucial.

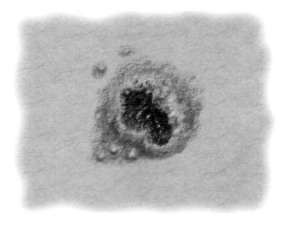

Acne

Most commonly occurring during adolescence and often continuing into adulthood, acne (cystic acne and acne vulgaris) is caused by over-production of sebum by the sebaceous glands. Over-active hormones during puberty are responsible for triggering this excess pro-duction. The excess oil is an ideal medium for bac-teria to breed, with flaked skin and bacteria blocking the glands, causing the red inflamed area of the pimple. Trapped bac-teria causes pus to form in the blocked gland. The areas most susceptible to acne are the forehead, face, nose, chin, chest, and back.

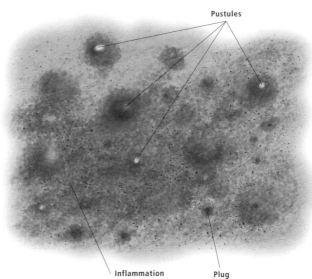

Pustules

Inflammation

Plug

Acne

Overproduction of sebum by the sebaceous gland can cause the outlet to become blocked. The trapped sebum can cause a blackhead to form, which often results in the area becoming inflamed. If pus forms under the skin, a pustule develops.

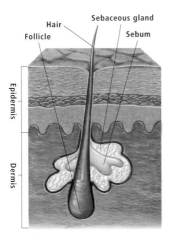

Hair
Follicle
Sebaceous gland
Sebum
Epidermis
Dermis

Clear skin

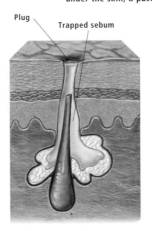

Plug
Trapped sebum

Blackhead

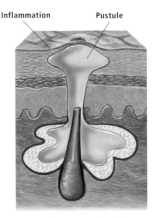

Inflammation
Pustule

Infected follicle

Warts

Warts are caused by the papillomavirus and are mildly contagious. They most commonly occur on the hands, legs, and feet.

Warts

Caused by the papillomavirus, warts most commonly occur on the hands and feet. These relatively harmless, raised growths are usually skin colored, and usually disappear spontaneously within 2–3 years. Warts that occur on the hands, arms, and legs are referred to as common warts; those found on the soles of the feet are called plantar warts; and those confined to the genital area are called genital warts.

Impetigo

Highly infectious, impetigo is a bacterial infection of the skin, usually affecting young children. Caused by the bacteria *Streptococcus pyogines* or *Staphylococcus aureus*, impetigo causes blistering of the skin, which then ruptures to an open sore before forming an itchy crust. Antibiotics, often in conjunction with antibiotic creams, can be administered to treat the condition. To minimize the risk of infecting others, the sufferer should avoid contact with others, use personal linen and towels, and keep the affected area clean and dry.

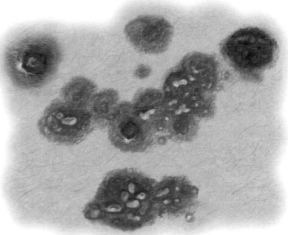

Impetigo

Commonly affecting young children, impetigo causes blisters that erupt into sores, eventually healing over with an itchy scab.

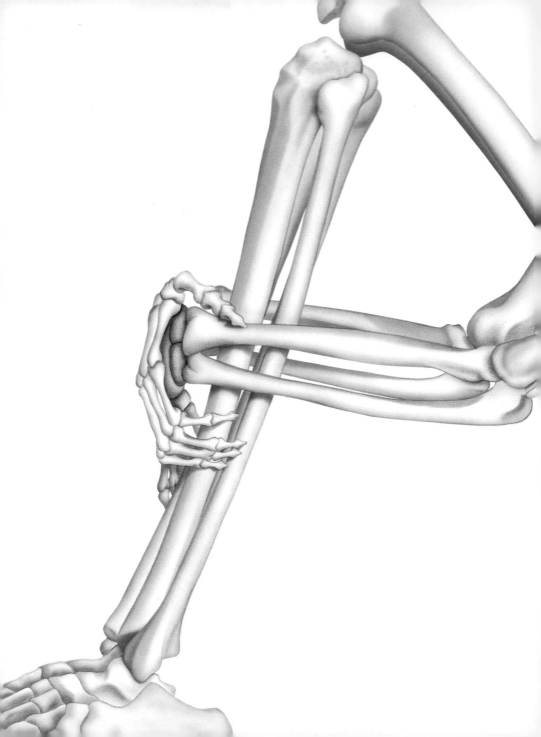

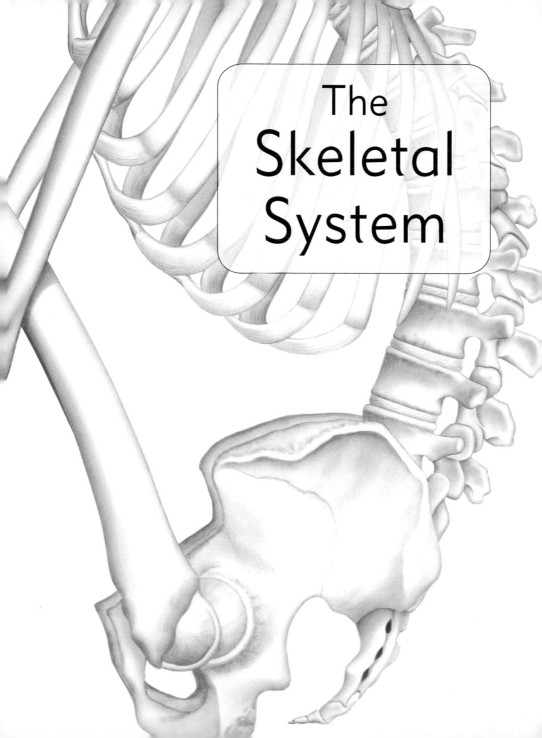

The
Skeletal
System

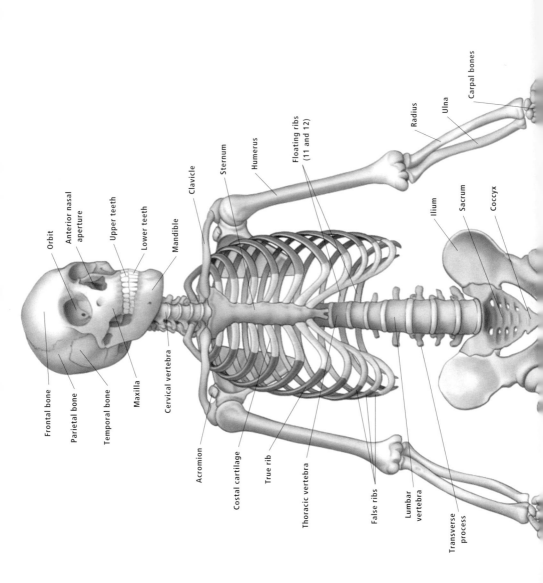

Frontal bone

Parietal bone

Temporal bone

Maxilla

Cervical vertebra

Orbit

Anterior nasal
aperture

Upper teeth

Lower teeth

Mandible

Clavicle

Sternum

Humerus

Floating ribs
(11 and 12)

Radius

Ulna

Carpal bones

Acromion

Costal cartilage

True rib

Thoracic vertebra

False ribs

Lumbar
vertebra

Transverse
process

Ilium

Sacrum

Coccyx

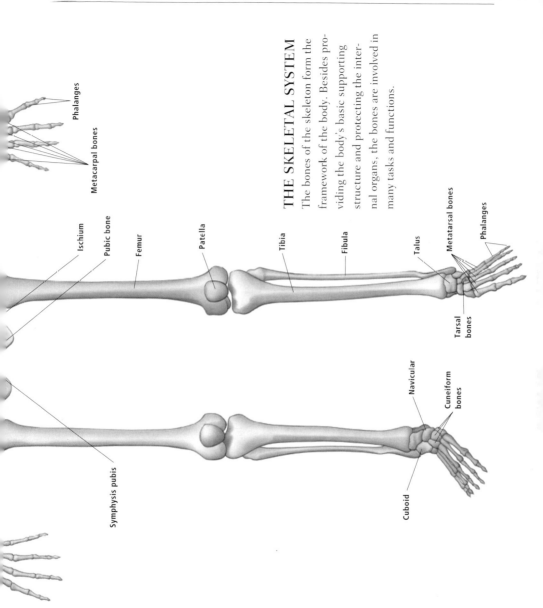

THE SKELETAL SYSTEM

The bones of the skeleton form the framework of the body. Besides providing the body's basic supporting structure and protecting the internal organs, the bones are involved in many tasks and functions.

Phalanges

Metacarpal bones

Ischium

Pubic bone

Femur

Patella

Tibia

Fibula

Talus

Metatarsal bones

Phalanges

Tarsal bones

Symphysis pubis

Navicular

Cuneiform bones

Cuboid

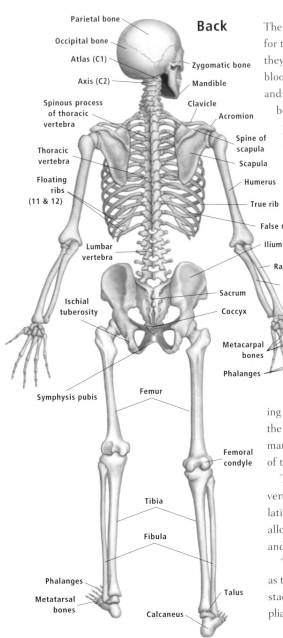

Back

Parietal bone
Occipital bone
Atlas (C1)
Axis (C2)
Spinous process
of thoracic
vertebra
Thoracic
vertebra
Floating
ribs
(11 & 12)
Lumbar
vertebra
Ischial
tuberosity
Symphysis pubis

Zygomatic bone
Mandible
Clavicle
Acromion
Spine of
scapula
Scapula
Humerus
True rib
False rib
Ilium
Radius
Ulna
Sacrum
Coccyx
Carpal
bones
Metacarpal
bones
Phalanges
Femur
Femoral
condyle
Tibia
Fibula
Phalanges
Metatarsal
bones
Talus
Calcaneus

The bones provide of a point of attachment for the muscles, enabling body movements; they are the site of production of red blood cells for the circulatory system and; controlled by the endocrine system, bones store the body's calcium and phospate requirements.

At birth, the body contains over 300 bones at various stages of ossification. During infancy and childhood many of the bones fuse together, and by adulthood the skeletal frame comprises 206 bones. The skeleton is usually described in two parts: the axial skeleton and the appendicular skeleton.

The Axial Skeleton

The axial skeleton comprises the skull, the vertebral column (spine or backbone), and the thoracic cage (chest).

Protecting the brain and forming the face, the skull is the skeleton of the head, and comprises the cranium, the mandible, and the hyoid bone at the base of the tongue.

The base of the skull joins to the first vertebra of the spine—the atlas. The articulation process between these two structures allows head movements, including nodding and sideways movement.

The vertebral column, commonly known as the spine or backbone, is a tower of bones, stacked one upon the other, with spongy, pliant disks of cartilage between each vertebra.

The column itself is divided into sections, with 7 vertebrae in the neck (cervical vertebrae), 12 in the chest region (thoracic vertebrae), and 5 in the lower back (lumbar vertebrae). The sacrum and coccyx lie beneath the lumbar vertebrae.

The thoracic cage (chest) is composed of the 12 thoracic vertebrae, the ribs, and the sternum. The 12 ribs extend from each of the vertebrae, with the top 7 ribs (true ribs) encircling the vital organs and joining the sternum at the front.

The next 3 ribs (false ribs) do not extend around fully—instead they are connected to each other and then to the lowest true rib by costal cartilage. The last two ribs (floating ribs) do not extend to the front of the body.

The Appendicular Skeleton

The appendicular skeleton comprises the limb bones of the arms and legs and the shoulder and pelvic girdles.

The shoulder girdle connects the arm bones to the axial skeleton. The arm is composed of 3 bones: the humerus, radius, and ulna—below this is the wrist that has 8 bones, the palm that has 5 bones, and finally the 14 bones that make up the fingers.

The pelvic girdle connects the leg bones to the axial skeleton. The 3 bones of the leg—the tibia, fibula, and femur—connect to the ankle, which contains 7 bones. The foot has 5 bones and the toes contain 14 bones.

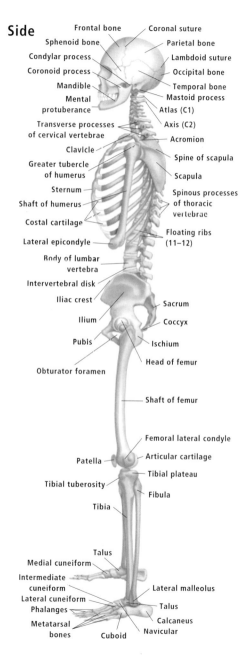

Side

Frontal bone
Coronal suture
Sphenoid bone
Parietal bone
Condylar process
Lambdoid suture
Coronoid process
Occipital bone
Mandible
Temporal bone
Mental protuberance
Mastoid process
Atlas (C1)
Transverse processes of cervical vertebrae
Axis (C2)
Acromion
Clavicle
Spine of scapula
Greater tubercle of humerus
Scapula
Sternum
Spinous processes of thoracic vertebrae
Shaft of humerus
Costal cartilage
Floating ribs (11–12)
Lateral epicondyle
Body of lumbar vertebra
Intervertebral disk
Iliac crest
Sacrum
Ilium
Coccyx
Pubis
Ischium
Head of femur
Obturator foramen
Shaft of femur
Femoral lateral condyle
Patella
Articular cartilage
Tibial plateau
Tibial tuberosity
Fibula
Tibia
Talus
Medial cuneiform
Intermediate cuneiform
Lateral malleolus
Lateral cuneiform
Talus
Phalanges
Calcaneus
Metatarsal bones
Navicular
Cuboid

Bones

Bone tissue is a type of connective tissue, with its major components comprising mineral salts and collagen fibers. There are four types of bone cell found in bone tissue: osteoprogenitor cells, osteoblasts, osteocytes, and osteoclasts. Bone hardness comes from the mineral salts, primarily calcium phosphate, while bone strength comes from collagen.

Bone is made up of layers of different bone tissue. At the center of bone is the bone marrow. It is surrounded by spongy (cancellous) bone, then compact (cortical) bone, with a membranous outer layer called the periosteum. This arrangement of layers gives a structure with a light, airy center and strong, hard, outer shell—essential qualities, as bone must be strong enough to bear the weight of the body, yet light enough to allow movement by the muscles.

Bone marrow is the site of red blood cell production. There are two types of bone marrow: red bone marrow and yellow bone marrow. Initially all marrow is red, but as we age, some of this marrow becomes yellow. Yellow marrow, comprised mainly of fat cells, does not produce red blood cells, but can be converted back to red marrow in emergency situations, when extra supply of red blood cells is required. The bone marrow runs through spongy (cancellous) bone—a honeycomb-like formation comprising bands of bone called trabeculae. The random arrangement of the trabeculae provides strength in all directions.

Compact (cortical) bone consists of densely packed, cylindrical units called osteons, which form around blood vessels. The osteons are arranged parallel to one another, which provide bone strength in one direction only.

Periosteum, the outer layer of most bone surfaces (with the exception of synovial joints) is a thin, fibrous membrane with two layers. Fibers in the inner layer join the periosteum to the bone within, while the outer layer contains the blood vessels and nerves that nourish the bone cells.

There are several types of bone, classified by shape, including long bones, short bones, flat bones, irregular bones, sesamoid bones, and sutural bones.

Found primarily in the limbs, and somewhat misleading in name, long bones range from the classic long bone of the femur to the long bones (though short by comparison) of the phalanges (finger bones). The term long bone describes the design, rather than the length, of the bone. The design of long bone incorporates a long cylindrical shaft, called the diaphysis, with the ends of the bone known as the epiphyses. The point where the diaphysis and an epiphysis meet is known as the metaphysis. Running through the core of the shaft is the marrow cavity.

Short bones include the carpal bones of the wrist; the scapula is an example of a flat bone; irregular bones include the vertebrae; the patella is an example of a sesamoid bone; and sutural bones are found in the skull.

Bone

The different layers of bone comprise the inner core of bone marrow, surrounded by a layer of spongy bone, then cortical or compact bone, and an outer casing of periosteum.

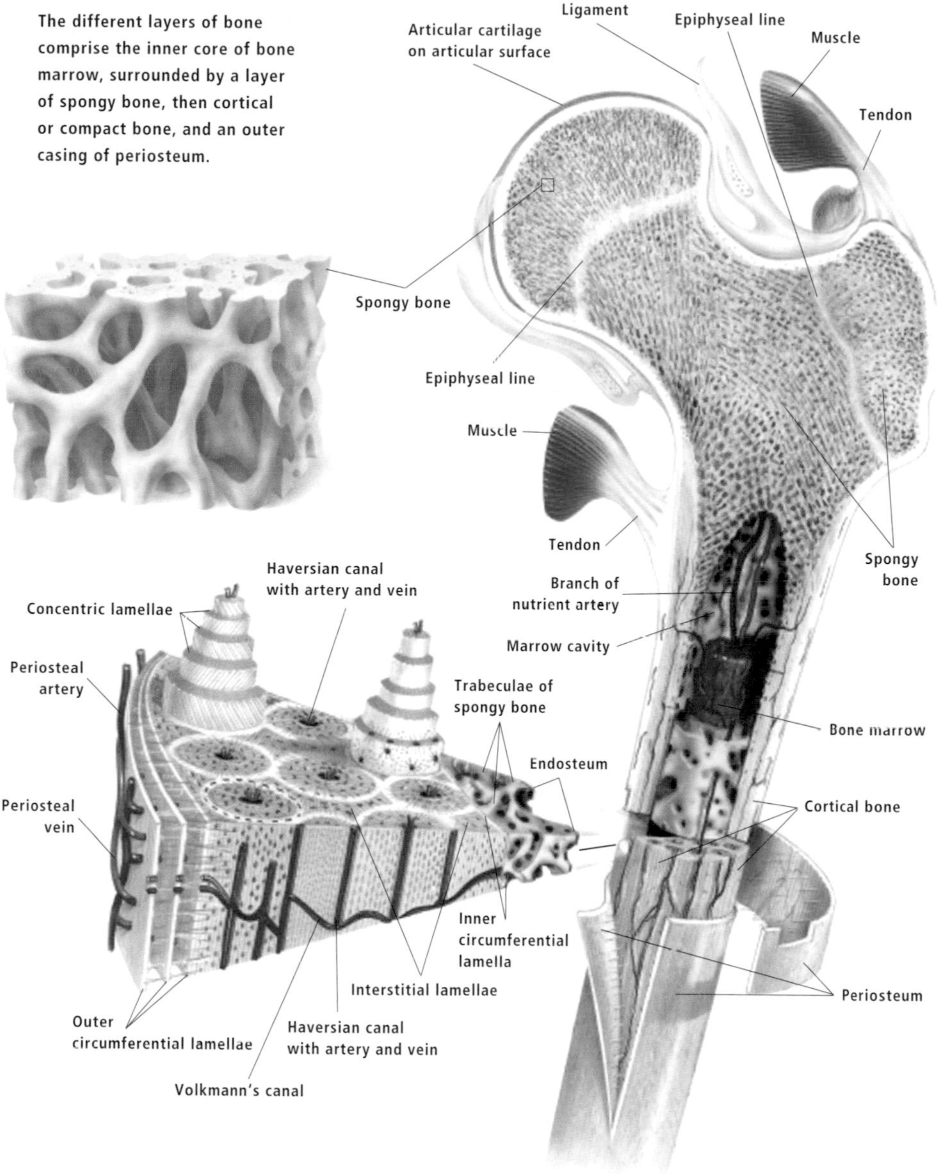

Ligament

Articular cartilage on articular surface

Epiphyseal line

Muscle

Tendon

Spongy bone

Epiphyseal line

Muscle

Tendon

Branch of nutrient artery

Marrow cavity

Spongy bone

Bone marrow

Concentric lamellae

Haversian canal with artery and vein

Periosteal artery

Trabeculae of spongy bone

Endosteum

Cortical bone

Periosteal vein

Inner circumferential lamella

Outer circumferential lamellae

Haversian canal with artery and vein

Interstitial lamellae

Volkmann's canal

Periosteum

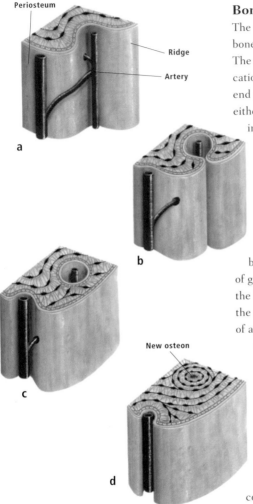

Periosteum

Ridge

Artery

a

b

c

New osteon

d

Bone formation

Bone is continually being formed,
enclosing around each blood vessel
(a) and (b). As the vessel becomes completely
surrounded, and more bone is laid down around
it, an osteon is formed (c). The process is continually
repeated, resulting in bone growing in width (d).

Bone Formation and Growth

The cartilage template of subsequent long
bone is laid down during fetal development.
The cartilage gradually ossifies, with ossification eventually extending almost to the
end of the bone. Growth then begins at
either end of the bone, with cartilage forming the junction between the existing
bone and new bone growth.

This cartilage layer is known as
the growth plate (epiphyseal plate),
and is where new growth occurs. The
cartilage gradually moves away from
the new center of growth, ossifying as
it does. When fully ossified, growth in
bone length is complete. Completion
of growth in long bone length varies with
the individual, but generally ceases around
the age of 18 years in females, and 21 years
of age in males.

Growth in bone width occurs as new
bone is laid down. Osteons, the basic
structural component of compact bone,
are added to the outer surface. As the
outer surface of bone is increased, the
internal marrow cavity increases.

Osteons are formed as bone cells
are produced and laid down around
blood vessels. Firstly following the
contours of the blood vessel, bone gradually encircles the blood vessel, eventually
filling the area around the blood vessel
with bone to form an osteon.

Bone tissue is continually recycled,
gradually breaking down and being
reabsorbed to produce new tissue.

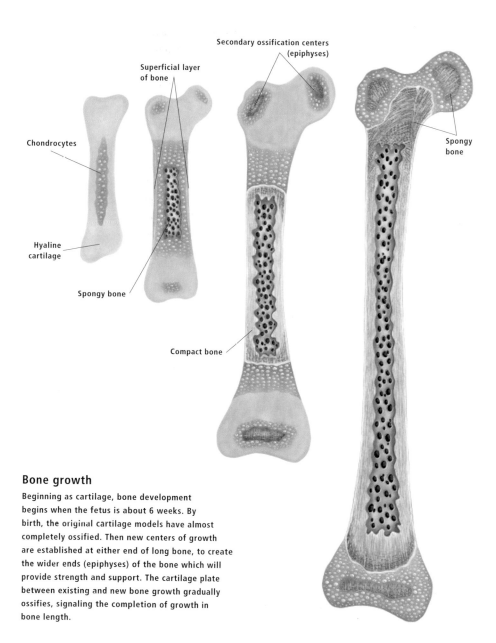

Secondary ossification centers
(epiphyses)

Superficial layer
of bone

Chondrocytes

Spongy
bone

Hyaline
cartilage

Spongy bone

Compact bone

Bone growth

Beginning as cartilage, bone development
begins when the fetus is about 6 weeks. By
birth, the original cartilage models have almost
completely ossified. Then new centers of growth
are established at either end of long bone, to create
the wider ends (epiphyses) of the bone which will
provide strength and support. The cartilage plate
between existing and new bone growth gradually
ossifies, signaling the completion of growth in
bone length.

Joints

There are various joint types in the body —a joint is where two bones meet and articulate. In some places, such as where the bones of the skull meet, they are rigidly fixed and immoveable. In other places where very limited movement is required, the bones are connected by a layer of cartilage and held firm by strong fibrous ligaments, such as in the front of the pelvis. The mobile joints in the body require the synergy of muscles, ligaments, and tendons to enable their full range of movement.

The articulation surfaces of these mobile joints, called synovial joints, have a smooth layer of cartilage to allow free movement. The joint capsule surrounding the articulation is lined with synovial membrane, which secretes synovial fluid to lubricate the joint.

The synovial joints include several different types, each providing a different range of movement. A ball-and-socket joint, such

Joints

Gliding joint

Gliding joints allow bones to slide across each other. These joints are found in the carpal bones of the wrist and the tarsal bones of the ankle.

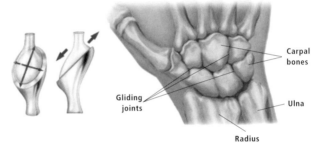

Carpal bones

Gliding joints

Ulna

Radius

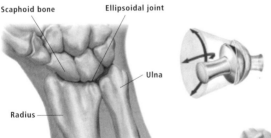

Scaphoid bone

Ellipsoidal joint

Ulna

Radius

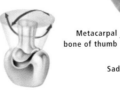

Ellipsoidal joint

This mobile joint type allows a wide range of movement. Ellipsoidal joints are found at the wrist, where the carpal bones meet the forearm bones.

Saddle joint

This flexible joint type moves in several directions. The joint between the base of the thumb and the trapezium of the wrist is an example of a saddle joint.

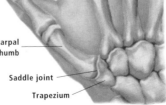

Metacarpal bone of thumb

Saddle joint

Trapezium

as in the hip and the shoulder, is the most mobile type of joint, allowing free movement in all directions. A hinge joint, such as in the elbow and knee, allows movement in one plane only, such as in bending and straightening of the knee. A saddle joint, found for instance in the thumb, is a highly mobile joint, allowing sliding movement in two directions. A gliding joint, such as is found in the bones of the wrist and ankle, provides limited movement of the bones, allowing them to slide across each other. An ellipsoidal joint allows a good range of movement in two directions, such as the movement at the junction of the wrist and forearm. A pivot joint allows rotational movement, such as is found in the first two vertebrae.

A mobile joint may have more than one type of joint movement, such as is found at the elbow, which comprises both a hinge joint and a pivot joint.

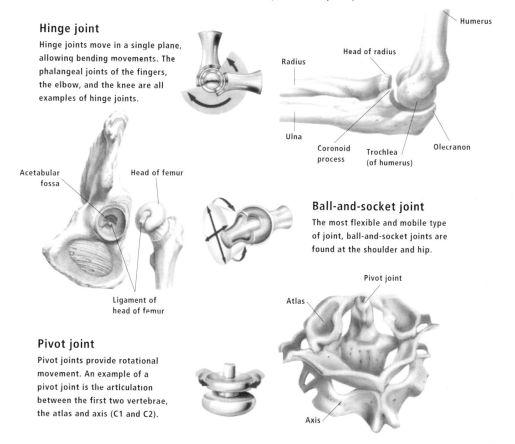

Hinge joint

Hinge joints move in a single plane, allowing bending movements. The phalangeal joints of the fingers, the elbow, and the knee are all examples of hinge joints.

Humerus

Radius

Head of radius

Ulna

Coronoid process

Trochlea (of humerus)

Olecranon

Acetabular fossa

Head of femur

Ligament of head of femur

Ball-and-socket joint

The most flexible and mobile type of joint, ball-and-socket joints are found at the shoulder and hip.

Pivot joint

Atlas

Pivot joint

Pivot joints provide rotational movement. An example of a pivot joint is the articulation between the first two vertebrae, the atlas and axis (C1 and C2).

Axis

Synovial Membrane and Synovial Fluid

Synovial membrane is a smooth membrane, found in synovial joints, bursae, and tendon sheaths. It provides a lining to the joint capsule and any exposed areas of bone that do not have a cartilage surface. The membrane produces synovial fluid, which serves to lubricate the joint surfaces, and provide nourishment to cartilage areas of the joint.

The cells that produce the synovial fluid, synoviocytes, also reabsorb the fluid in a constant recycling process. The smooth articular surfaces, combined with the lubricating qualities of the synovial fluid, provide almost friction-free movement of the joints.

Cartilage

Cartilage serves several purposes: it provides a covering for joint surfaces, provides shock-absorbing qualities, and is a structural component of the skeleton, where it joins two bone surfaces, such as is found in the ribs.

Cartilage is a resilient connective tissue containing cartilage cells and collagen in a matrix of glycoprotein. There are three different types of cartilage: elastic cartilage, fibrocartilage, and hyaline cartilage, each with different physical properties. Apart

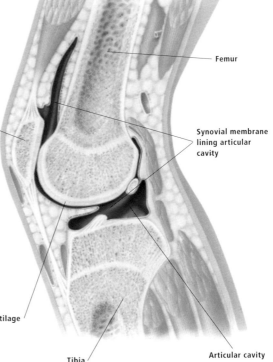

Synovial membrane and fluid

Lining noncartilage areas of joints, synovial membrane provides a smooth surface for ease of movement. Synovial fluid produced by cells in the membrane acts as a lubricant, and provides nutrients to the cartilage surface.

Femur

Patella

Synovial membrane lining articular cavity

Articular cartilage

Tibia

Articular cavity

from its function in joints, cartilage is found in the outer ear, the voice box, and the tip of the nose.

Ligaments

Providing support and strength to joints and supporting internal organs, ligaments are sturdy fibrous tissues, with some elastic properties.

Ligaments supporting joints keep the joint stable and prevent excessive movement, thus minimizing the risk of injury.

Ligaments support internal organs including the uterus, liver, and bladder.

Ligaments

The tough fibers of ligaments stabilize and strengthen joints, preventing excessive movement, thereby reducing the likelihood of injury. In this illustration of the knee, the patella has been reflected to expose the inner ligaments and bones of the knee joint.

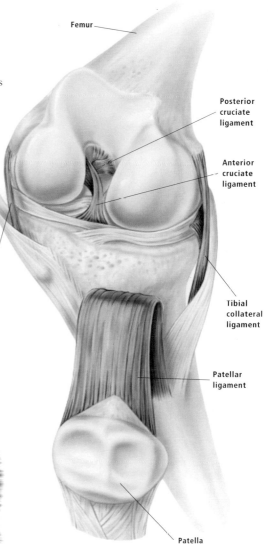

Femur

Posterior cruciate ligament

Anterior cruciate ligament

Fibular collateral ligament

Tibial collateral ligament

Patellar ligament

Patella

Skull

A component of the axial skeleton, the skull forms the skeleton of the head, protecting the brain and delicate sense organs, and forming the face, including the upper and lower jaws. As a general rule, the female skull and associated paranasal sinuses are smaller and lighter than that of the male.

The skull is often described in two parts: the cranium, surrounding the brain, and the face.

With the exception of the jaw bones, the bones of the skull are joined to each other by joints called sutures. Where the bones meet, their irregular edges lock together and are further secured by connective tissue. The stability of these joints, coupled with the shape of the cranium, the strong arching formation of the facial bones, the resilience of the outer bones,

Skull

The unique combination of bones in the skull makes it the most complex bone structure in the body.

Skull—side view

External acoustic meatus
Coronal suture
Frontal bone
Parietal bone
Temporal line
Squamous part of temporal bone
Greater wing of sphenoid bone
Lambdoid suture
Supraorbital notch
Ethmoid bone
Lacrimal bone
Nasal bone
Nasolacrimal duct
Zygomatic bone
Zygomatic process of temporal bone
Pterygoid process of sphenoid bone
External occipital protuberance
Maxilla
Coronoid process of mandible
Occipital bone
Styloid process
Mental protuberance
Mastoid process
Tympanic plate of temporal bone
Condylar process of mandible
Mandibular notch
Mental foramen
Angle of mandible
Ramus of mandible
Wisdom tooth
Body of mandible

and the cushioning effect of spongy bone within, all contribute to the protective qualities provided by the skull.

Fitting together to form a protective shell, the skull is comprised of paired bones on each side, and unpaired bones running through its midline.

The seven unpaired bones are: the frontal, ethmoid, occipital, sphenoid, mandible, vomer, and hyoid bones. The ten paired bones are: the temporal bones, including the tiny middle ear bones (ossicles) of the malleus, incus, and stapes (hammer, anvil, and stirrup), parietal, lacrimal, nasal, zygomatic, palatine, and maxillary bones.

Each of the lobes of the brain is named to correspond with the overlying bone: occipital, parietal, temporal, and frontal. These bones, with the exception of the parietal bone, also form the floor of the cranium: this floor is divided into three

Skull—cross section

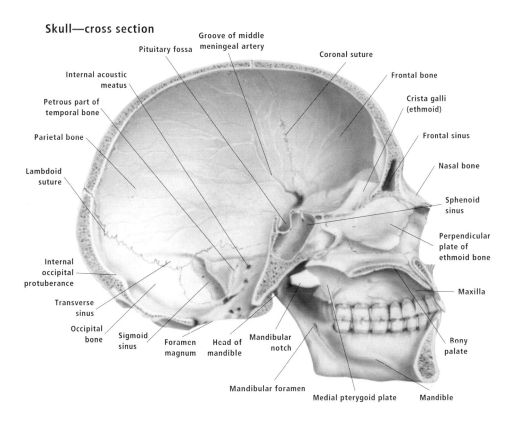

Groove of middle meningeal artery
Pituitary fossa
Coronal suture
Internal acoustic meatus
Frontal bone
Petrous part of temporal bone
Crista galli (ethmoid)
Parietal bone
Frontal sinus
Lambdoid suture
Nasal bone
Internal occipital protuberance
Sphenoid sinus
Perpendicular plate of ethmoid bone
Transverse sinus
Maxilla
Occipital bone
Sigmoid sinus
Foramen magnum
Head of mandible
Mandibular notch
Bony palate
Mandibular foramen
Medial pterygoid plate
Mandible

regions, called the anterior, middle, and posterior cranial fossae. Within the anterior fossa are the frontal lobes of the brain; the temporal lobes lie in the middle fossa; and the cerebellum and brain stem lie in the posterior fossa. The spinal cord, cranial nerves, and blood vessels enter and leave the brain through holes in the floor of the cranium known as foramen.

Beneath the skull are the meninges—three layers of membrane—comprising the dura mater, arachnoid mater, and pia mater. It is through these two lower layers, the arachnoid and pia maters, that the cerebro-spinal fluid flows, providing a cushioning layer around the brain. The blood vessels supplying the skull lie between the dura mater and the skull.

Skull—back view

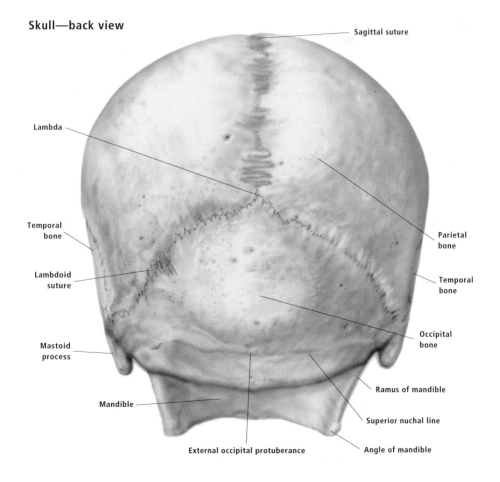

Sagittal suture

Lambda

Temporal bone

Lambdoid suture

Mastoid process

Mandible

Parietal bone

Temporal bone

Occipital bone

Ramus of mandible

Superior nuchal line

External occipital protuberance

Angle of mandible

Bones of the face

The face is formed by the frontal bone, zygoma, maxilla, and the mandible.

The frontal bone forms the forehead, and the zygoma forms the cheek bone. The maxillary and palatine bones form the upper jaw and bony roof (hard palate) of the mouth, while the mandible forms the lower jaw. The temporomandibular joints connect the mandible and the temporal bones on each side, allowing the mandible to move up and down, in and out, and side to side. These movements are involved in speech and chewing. Both the maxilla and mandible have sockets (dental alveoli) to accommodate the roots of the teeth.

The orbital cavity in the front of the skull is designed to house the eyes and their

Skull—front view

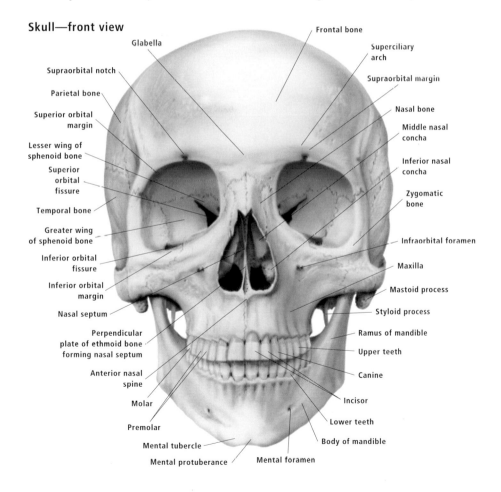

Glabella

Frontal bone

Superciliary arch

Supraorbital notch

Supraorbital margin

Parietal bone

Nasal bone

Superior orbital margin

Middle nasal concha

Lesser wing of sphenoid bone

Inferior nasal concha

Superior orbital fissure

Zygomatic bone

Temporal bone

Greater wing of sphenoid bone

Infraorbital foramen

Inferior orbital fissure

Maxilla

Inferior orbital margin

Mastoid process

Nasal septum

Styloid process

Perpendicular plate of ethmoid bone forming nasal septum

Ramus of mandible

Upper teeth

Anterior nasal spine

Canine

Molar

Incisor

Premolar

Lower teeth

Mental tubercle

Body of mandible

Mental protuberance

Mental foramen

associated musculature, providing protection to the delicate organ of sight. The cavity is created by the union of several of the skull bones including the sphenoid, frontal, zygomatic, lacrimal, palatine, and ethmoid bones.

Cartilage and bone make up the nasal cavity. The nasal bone and maxilla comprise the bony areas of the nose; these give way to cartilage which makes up the remainder of the nose. The nasal septum is made up of both bone and cartilage. The bones in the walls of the nasal cavity create the nasal conchae. The nasal cavity connects to the paranasal sinuses located in the adjacent bones.

Skull—top

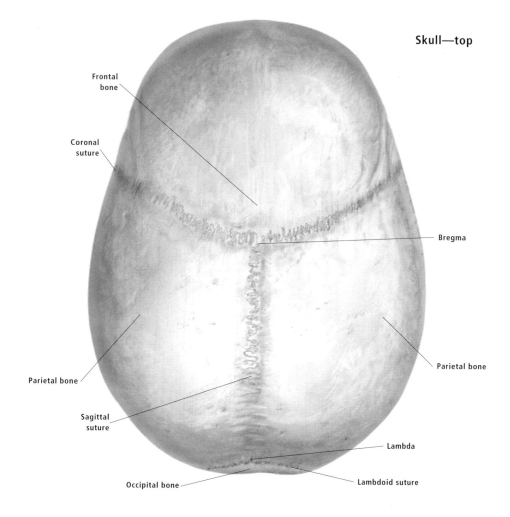

Frontal bone

Coronal suture

Bregma

Parietal bone

Parietal bone

Sagittal suture

Lambda

Occipital bone

Lambdoid suture

Base of the skull

The temporal, occipital, and frontal bones join to form the base of the brain. Joints between the occipital bone and the first vertebra of the spine—known as the atlas—support the skull. It is this joint with the atlas that allows nodding movements.

While there are a number of holes, or foramina (singular, foramen), in the base of the skull to accommodate the passage of blood vessels and nerves, the largest of these, the foramen magnum, accommodates the passage of the medulla oblongata (part of the brainstem) and the meninges.

Skull—base

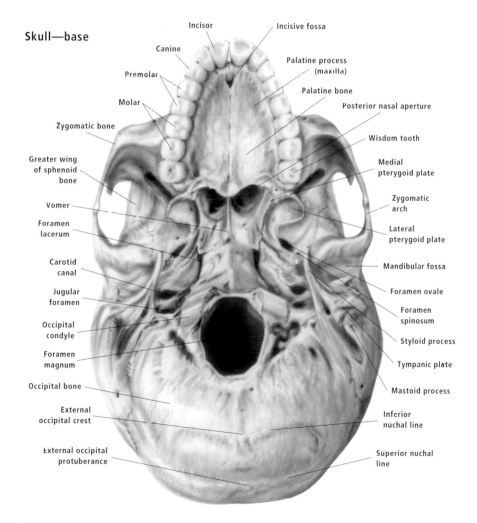

Incisor

Incisive fossa

Canine

Palatine process (maxilla)

Premolar

Palatine bone

Molar

Posterior nasal aperture

Zygomatic bone

Wisdom tooth

Greater wing of sphenoid bone

Medial pterygoid plate

Vomer

Zygomatic arch

Foramen lacerum

Lateral pterygoid plate

Carotid canal

Mandibular fossa

Jugular foramen

Foramen ovale

Occipital condyle

Foramen spinosum

Foramen magnum

Styloid process

Occipital bone

Tympanic plate

External occipital crest

Mastoid process

External occipital protuberance

Inferior nuchal line

Superior nuchal line

Orbit

The orbit is the bony housing of the eye and its associated musculature, blood vessels, nerves, and lacrimal gland. The housing itself is formed by the frontal, ethmoid, lacrimal, zygomatic, nasal, palatine, sphenoid, and maxillary bones. The thickened outer edges of each orbit are strong and resilient, protecting the intricate structure of the eye from damage and injury. The inner walls, however, are thin and more delicate, making them more susceptible to injury.

Bones of the ear

The temporal bone encloses the auditory ossicles of the middle ear and houses the structures of the inner ear (labyrinth). The three auditory ossicles include the malleus (hammer), incus (anvil), and stapes (stirrup). These tiny bones—the stapes is the smallest bone in the body—play an important role in hearing. The malleus is attached to the eardrum (tympanic membrane), while the incus connects the malleus and the stapes. When sound waves hit the eardrum the three ossicles vibrate, amplifying the vibrations. The base of the stapes covers the vestibular window of the cochlea; when vibrations reach the stapes they are then transmitted to the vestibular window, and on to the major organ of hearing, the cochlea.

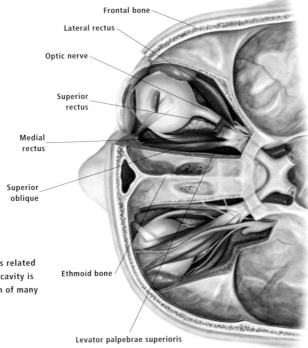

Frontal bone

Lateral rectus

Optic nerve

Superior rectus

Medial rectus

Superior oblique

Ethmoid bone

Levator palpebrae superioris

Orbital cavity

Housing the eye and its related structures, the orbital cavity is created by the junction of many of the skull bones.

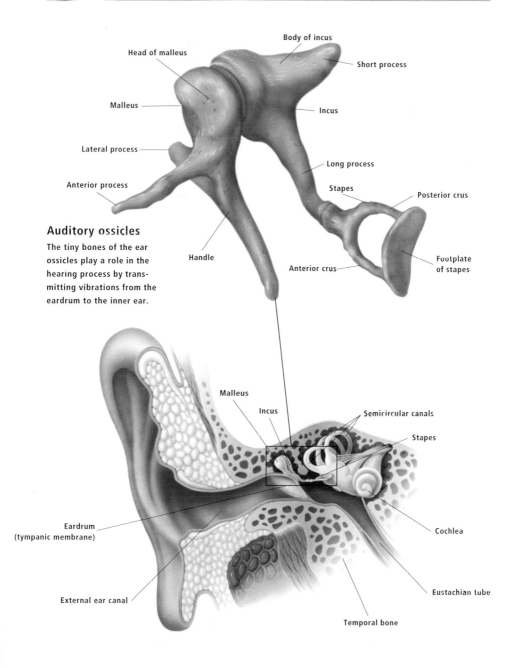

Body of incus

Head of malleus

Short process

Malleus

Incus

Lateral process

Long process

Anterior process

Stapes

Posterior crus

Auditory ossicles

The tiny bones of the ear ossicles play a role in the hearing process by transmitting vibrations from the eardrum to the inner ear.

Handle

Footplate of stapes

Anterior crus

Malleus

Incus

Semicircular canals

Stapes

Eardrum
(tympanic membrane)

Cochlea

External ear canal

Eustachian tube

Temporal bone

Fetal Skull Development

As with other bone structures in the body, the templates for the skull bones begin as cartilage in the embryonic stages of development. Bone gradually spreads out from ossification centers in the cartilage. By week 4 the face and jaws begin to develop. These structures begin life as thickened areas of cartilage that join at the midline to form the beginnings of the facial features of the jaws, mouth, and nose.

At birth, the individual bones of the skull have ossified, but are not joined together, being held together by fontanelles. Joining the frontal and parietal bones is the anterior fontanelle, and joining the occipital and parietal bones is the smaller posterior fontanelle.

The fontanelles gradually close over with ossified bones, with the posterior fontanelle closing over at around 3 to 4 months of age, and the anterior fontanelle usually closing over by around 12 months of age. There are also several smaller fontanelles between the skull bones.

At birth the cranium is relatively large, while the face is small; the head accounts for approximately 25 percent of overall body length.

12 weeks
The cartilage model of the skull bones has been laid down.

16 weeks
Bone spreads out from ossification centers in the cartilage.

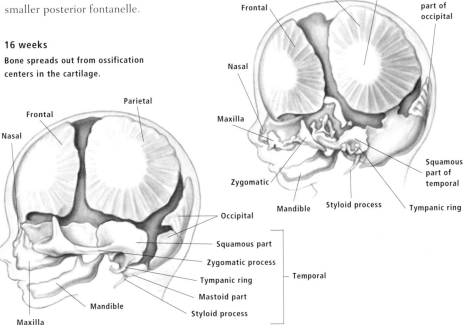

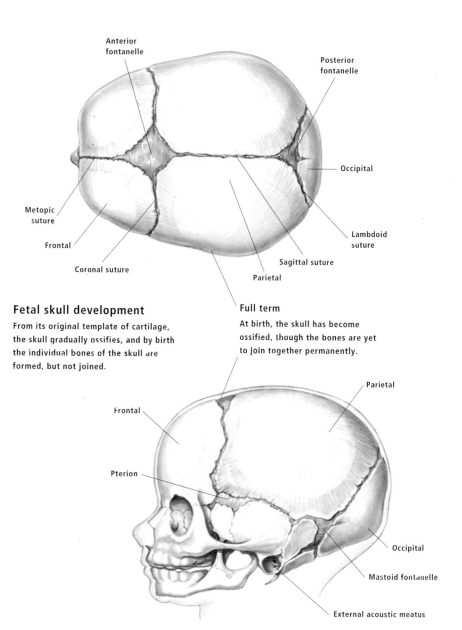

Anterior
fontanelle

Posterior
fontanelle

Occipital

Metopic
suture

Frontal

Lambdoid
suture

Coronal suture

Sagittal suture

Parietal

Fetal skull development

From its original template of cartilage,
the skull gradually ossifies, and by birth
the individual bones of the skull are
formed, but not joined.

Full term

At birth, the skull has become
ossified, though the bones are yet
to join together permanently.

Parietal

Frontal

Pterion

Occipital

Mastoid fontanelle

External acoustic meatus

Rib Cage

The rib cage is designed to protect the
vital organs of the lungs and heart. There
are 12 pairs of ribs, each joined to a corresponding thoracic vertebra. The first 7 pairs
of ribs encircle the chest, joining to the
sternum at the front. The next 3 pairs of
ribs do not extend fully around to the sternum, but instead are connected to each
other by costal cartilage, and are then
connected to the 7th pair of ribs.

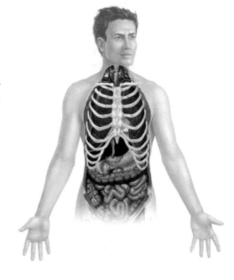

Ribs

**The ribs provide protection to the vital
organs of the chest cavity. The 12 pairs of
ribs each attach to a thoracic vertebra.**

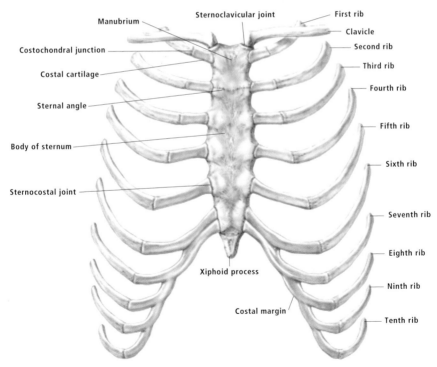

Sternoclavicular joint · First rib

Manubrium · Clavicle

Costochondral junction · Second rib

Costal cartilage · Third rib

· Fourth rib

Sternal angle · Fifth rib

Body of sternum · Sixth rib

Sternocostal joint · Seventh rib

· Eighth rib

Xiphoid process · Ninth rib

Costal margin · Tenth rib

The last 2 pairs of ribs do not extend to the front. The first 7 pairs of ribs are known as the "true ribs," pairs 8 to 10 are known as the "false ribs," while the remaining 2 pairs of ribs are known as "floating ribs." The true ribs are connected to the sternum by costal cartilage.

The ribs play a role in breathing, as the intercostal muscles move the ribs upward and outward to accommodate the expanding lungs.

Clavicle

The two clavicles, or collar bones, are positioned horizontally above the rib cage. These slender bones each articulate with the manubrium of the sternum (breastbone) at one end and acromion of the scapula (shoulder blade) at the other end.

Sternum

The sternum is located in the front wall of the chest, and is divided into three sections: the manubrium at the top, the body of the sternum, and the xiphoid process.

The seven pairs of true ribs are connected to the manubrium and body of the sternum by costal cartilage. The articulation between the manubrium and the clavicle assists in shoulder stability.

Clavicle

These gently curved, slender bones connect to the sternum, forming the sternoclavicular joint; and to the scapula, forming the acromioclavicular joint.

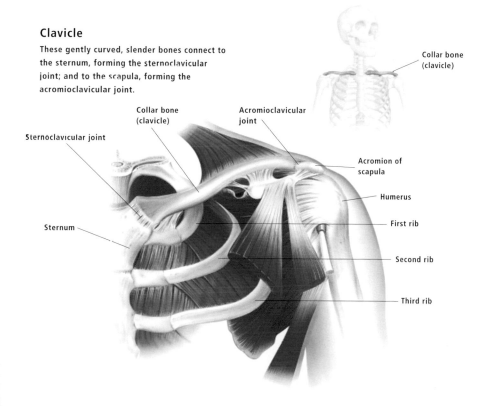

Collar bone (clavicle)

Collar bone (clavicle)

Acromioclavicular joint

Sternoclavicular joint

Acromion of scapula

Humerus

Sternum

First rib

Second rib

Third rib

Spine

Extending from the base of the skull to just above the buttocks, the spine is a column of vertebrae, running down the midline of the back. It is responsible for a variety of functions, including supporting body weight, providing a flexible base for bending and twisting movements, supplying a protective housing for the vulnerable spinal cord, and accommodating all the blood vessels and nerves associated with the spinal cord.

A side view shows the spine to have a gentle S-shaped curvature. This curvature is acquired in childhood, as natural processes firm the final shaping. At birth, the spine is curved in a C-shape—and this shaping remains into adulthood in the thoracic and sacral regions—these are known as primary curvatures.

As an infant begins to lift its head, this prompts the formation of the cervical curvature. Then, as the infant learns to sit, stand, and walk, the formation of the lumbar curvature begins. As these two curvatures occur after birth, they are known as secondary curvatures.

The individual vertebrae are stacked one upon another, creating a tower of bone. Running through the center of almost the entire structure is a hollow canal, the spinal canal, designed to house and protect the spinal cord. The connections between each vertebra create an opening, known as an intervertebral foramen, to accommodate the spinal nerves leading to and from the spinal cord.

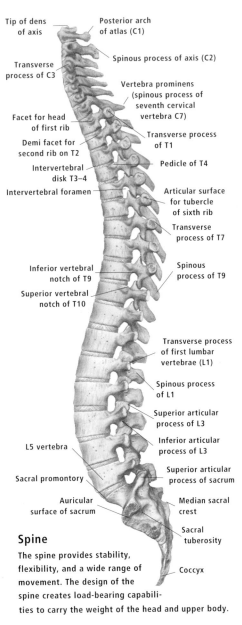

Tip of dens of axis

Posterior arch of atlas (C1)

Spinous process of axis (C2)

Transverse process of C3

Vertebra prominens (spinous process of seventh cervical vertebra C7)

Facet for head of first rib

Transverse process of T1

Demi facet for second rib on T2

Pedicle of T4

Intervertebral disk T3–4

Intervertebral foramen

Articular surface for tubercle of sixth rib

Transverse process of T7

Inferior vertebral notch of T9

Spinous process of T9

Superior vertebral notch of T10

Transverse process of first lumbar vertebrae (L1)

Spinous process of L1

Superior articular process of L3

L5 vertebra

Inferior articular process of L3

Sacral promontory

Superior articular process of sacrum

Auricular surface of sacrum

Median sacral crest

Sacral tuberosity

Coccyx

Spine

The spine provides stability, flexibility, and a wide range of movement. The design of the spine creates load-bearing capabilities to carry the weight of the head and upper body.

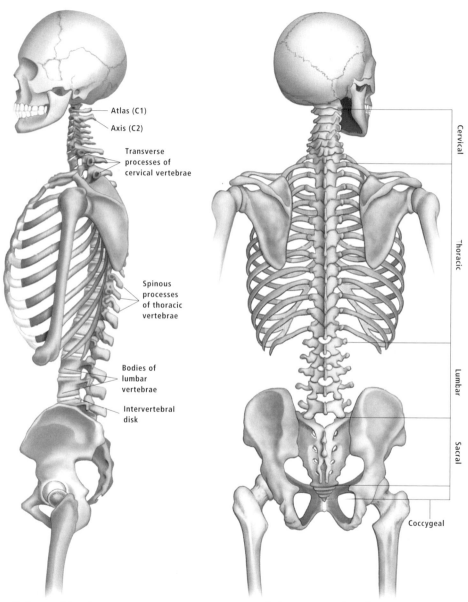

Atlas (C1)

Axis (C2)

Transverse
processes of
cervical vertebrae

Spinous
processes
of thoracic
vertebrae

Bodies of
lumbar
vertebrae

Intervertebral
disk

Cervical

Thoracic

Lumbar

Sacral

Coccygeal

Spine—side view

Spine—posterior view

Vertebrae

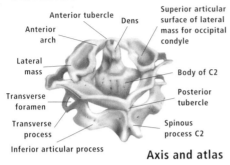

Anterior tubercle

Dens

Superior articular
surface of lateral
mass for occipital
condyle

Anterior
arch

Lateral
mass

Body of C2

Posterior
tubercle

Transverse
foramen

Transverse
process

Spinous
process C2

Inferior articular process

Axis and atlas

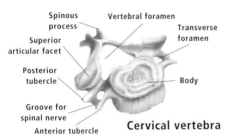

Spinous
process

Vertebral foramen

Transverse
foramen

Superior
articular facet

Posterior
tubercle

Body

Groove for
spinal nerve

Anterior tubercle

Cervical vertebra

Vertebrae

The vertebral column (spine) comprises 7 cervical vertebrae, 12 thoracic vertebrae, 5 lumbar vertebrae, the sacrum, and the coccyx. The individual vertebrae share some uniform structural characteristics, although each type has specific features relevant to its location and purpose in the spine.

Each vertebra has a body at the front, with the lumbar vertebrae having the largest body area to cope with its load bearing requirements. The body joins to the vertebral arch, creating the aperture through which the spinal cord runs.

Extending out from the center of the vertebral arch is the spinous process, a "tail" of bone projecting along the center-line of the spine.

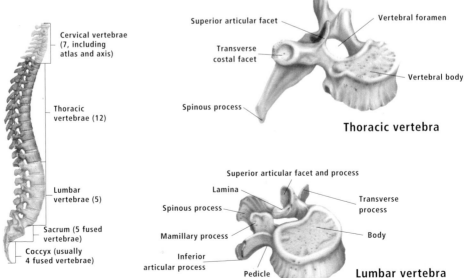

Cervical vertebrae
(7, including
atlas and axis)

Thoracic
vertebrae (12)

Lumbar
vertebrae (5)

Sacrum (5 fused
vertebrae)

Coccyx (usually
4 fused vertebrae)

Superior articular facet

Vertebral foramen

Transverse
costal facet

Vertebral body

Spinous process

Thoracic vertebra

Superior articular facet and process

Lamina

Spinous process

Transverse
process

Mamillary process

Body

Inferior
articular process

Pedicle

Lumbar vertebra

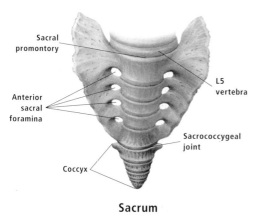

Sacral promontory

L5 vertebra

Anterior sacral foramina

Sacrococcygeal joint

Coccyx

Sacrum

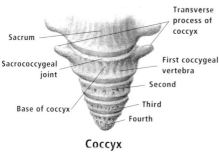

Transverse process of coccyx

Sacrum

First coccygeal vertebra

Sacrococcygeal joint

Second

Base of coccyx

Third

Fourth

Coccyx

Projecting out from either side of the vertebral arch are the transverse processes, while the articular processes extend from the top and bottom surfaces of each side of the arch. The spinous and transverse processes act as levers during movement; the articular processes form pairs of joints between the vertebrae. The anterior intervertebral joints occur between the bodies of adjoining vertebrae and incorporate the intervertebral disks.

The one exception to these structural benchmarks is the first cervical vertebra, C1, also known as the atlas, which does not have a body or spinous process.

Sacrum

The sacrum lies at the lower end of the spine beneath the lumbar vertebrae, and it forms part of the bony pelvis, connected to the hip bones at the sacroiliac joint.

Running through the sacrum is the sacral canal, a continuation of the spinal canal; the spinal cord runs through the sacral canal, with its roots, known as the cauda equina, leading off through holes in the sacrum, the sacral foramina.

Coccyx

The coccyx, or tailbone, is the result of the fusion of three to five vertebrae. Joined to the sacrum, it lies at the base of the spine.

Intervertebral Disks

The intervertebral disks separate each individual vertebra, providing a cushioning layer between neighboring vertebrae. The disks are flexible pads of cartilage, each with a soft center and fibrous outer layer. This pliable structure allows the disk to alter shape when vertebral movements are required.

The intervertebral disks are held in place by ligaments that run along the length of the spine, attaching to the body of each vertebra. These ligaments serve to prevent excessive movement.

Intervertebral disks

Shoulder

Each shoulder comprises the clavicle (collar bone), scapula (shoulder blade), and humerus. The clavicle performs a bracing action on the arm, keeping it to the side of the body. The acromioclavicular joint is formed where the clavicle and the acromion of the scapula meet. This joint plays a role in supporting the arm, with a limited amount of gliding movement to coincide with arm movements. The scapula is a flat plate of bone, with a raised spine; the outer end (glenoid cavity) provides the cup-like structure to hold the ball-shaped end of the humerus, forming the ball-and-socket joint of the shoulder. This is the most mobile joint in the body, with a wide range of movement in most directions.

Movements of the shoulder include flexion, extension, adduction, abduction, and rotation. This wide range of movement is possible because the contact area between the humerus and the scapula is small, creating a high degree of mobility, but this limited contact also creates a less stable joint. The joint is encased in a capsule lined with synovial membrane and filled with synovial fluid, while the ends of the articulating bones are covered in smooth cartilage—all contributing factors to the smooth movements achieved by the shoulder.

Shoulder

The shoulder, comprising the clavicle, scapula, and humerus, offers a wide range of movement. The ball-and-socket joint created by the scapula and humerus is the most mobile joint in the body.

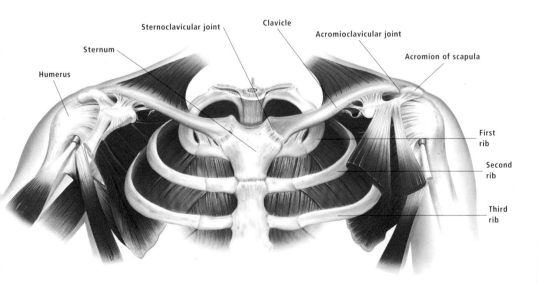

Sternoclavicular joint

Clavicle

Acromioclavicular joint

Sternum

Acromion of scapula

Humerus

First rib

Second rib

Third rib

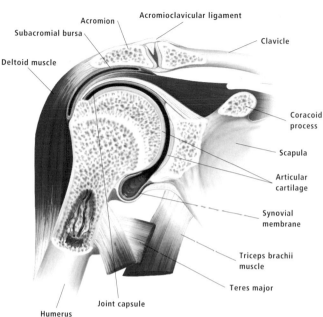

Acromion

Acromioclavicular ligament

Subacromial bursa

Clavicle

Deltoid muscle

Coracoid process

Scapula

Articular cartilage

Synovial membrane

Triceps brachii muscle

Teres major

Joint capsule

Humerus

Shoulder—cross section

The smooth movements achieved by the shoulder are the result of several contributing factors. A small contact area between the articulation surfaces and a generous, well-lubricated joint capsule provide friction-free movement.

Shoulder joint

The scapula, humerus, and clavicle unite to form the shoulder joint. The scapula lies at the back of the shoulder and provides the housing for the head of the humerus. The clavicle acts to stabilize the shoulder joint, while still permitting great range of movement.

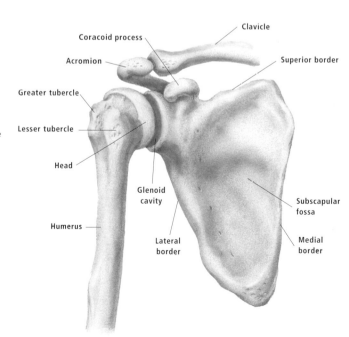

Clavicle

Coracoid process

Superior border

Acromion

Greater tubercle

Lesser tubercle

Head

Glenoid cavity

Humerus

Subscapular fossa

Lateral border

Medial border

Scapula

The scapula, or shoulder blade, is connected to the humerus and the clavicle, articulating with both bones. The acromioclavicular joint is where the clavicle and scapula meet. The ball-and-socket joint of the shoulder is where the scapula and humerus meet.

The scapula is a flat plate of bone, roughly triangular in shape, with a prominent spine. The acromion extends out slightly from the spine of the scapula, forming the high point on the shoulder blade. At the outer end of the scapula is the glenoid cavity, which forms the socket of the ball-and-socket joint of the shoulder.

Arm

The upper limb of the arm is considered to extend from the shoulder to the wrist. It is then divided into two regions: the upper arm, from the shoulder to the elbow, and the forearm, from the elbow to the wrist.

Humerus

The upper arm contains only one bone—the humerus. The humerus is a long bone, with a long cylindrical shaft, a rounded head at its upper end, and paired condyles at its lower end. The upper, rounded head articulates with the scapula at the shoulder joint. The condyles at the lower end articulate with the radius and ulna at the elbow joint.

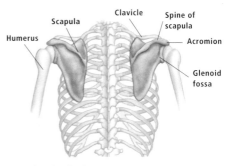

Scapula—back view

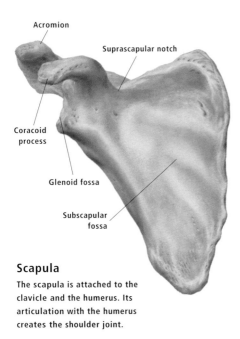

Scapula

The scapula is attached to the clavicle and the humerus. Its articulation with the humerus creates the shoulder joint.

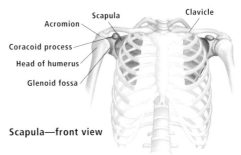

Scapula—front view

Bones of the arm

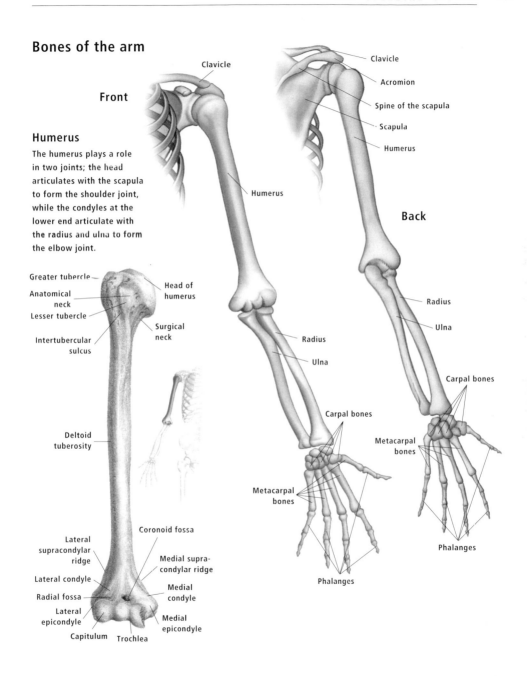

Front

Humerus

The humerus plays a role in two joints; the head articulates with the scapula to form the shoulder joint, while the condyles at the lower end articulate with the radius and ulna to form the elbow joint.

Clavicle

Clavicle

Acromion

Spine of the scapula

Scapula

Humerus

Humerus

Back

Humerus

Greater tubercle

Anatomical neck

Lesser tubercle

Intertubercular sulcus

Head of humerus

Surgical neck

Deltoid tuberosity

Radius

Ulna

Radius

Ulna

Carpal bones

Metacarpal bones

Carpal bones

Metacarpal bones

Lateral supracondylar ridge

Lateral condyle

Radial fossa

Lateral epicondyle

Capitulum   Trochlea

Coronoid fossa

Medial supra-condylar ridge

Medial condyle

Medial epicondyle

Phalanges

Phalanges

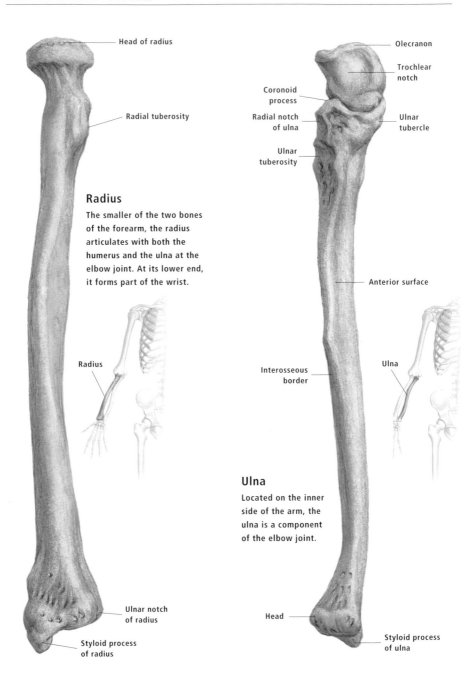

Head of radius

Radial tuberosity

Radius

The smaller of the two bones of the forearm, the radius articulates with both the humerus and the ulna at the elbow joint. At its lower end, it forms part of the wrist.

Radius

Ulnar notch of radius

Styloid process of radius

Olecranon

Trochlear notch

Coronoid process

Radial notch of ulna

Ulnar tubercle

Ulnar tuberosity

Anterior surface

Interosseous border

Ulna

Ulna

Located on the inner side of the arm, the ulna is a component of the elbow joint.

Head

Styloid process of ulna

Radius

The forearm contains two bones—the radius and the ulna—with the radius located on the thumb side of the arm. The upper end is rounded; this upper end articulates with the humerus at the elbow joint. Close to the upper end is the radial tuberosity, a raised area that articulates with the ulna. The rotational movement of the radius around the ulna mobilizes the wrist. Wider at its lower end, this part of the radius articulates with the carpal bones of the wrist.

Ulna

The longer of the two bones of the forearm, the ulna lies on the little finger side of the arm. The upper end of the ulna articulates with the humerus, creating a hinge joint at the elbow. The ulna articulates with the radius at the elbow, and joins with the radius at its lower end, although it does not play a role in the wrist joint.

Elbow

The point where the humerus, radius, and ulna meet is the elbow joint. The elbow joint has two ranges of movement, being both a hinge joint and a pivot joint. The articulation between the radius and ulna forms the pivot joint, and the articulation between the humerus and the radius and ulna forms the hinge joint.

Brachial artery — Median nerve — Ulna (trochlear notch)
Trochlea — Ulnar nerve
Triceps brachii tendon
Olecranon bursa
Capitulum — Olecranon
Basilic vein — Biceps tendon — Head of radius (articular surface) — Brachioradialis muscle
Radial nerve — Olecranon

Elbow joint

The elbow joint is formed by the articulation of the humerus, radius, and ulna. The arrangement of these bones creates two joint types, a hinge joint and a pivot joint.

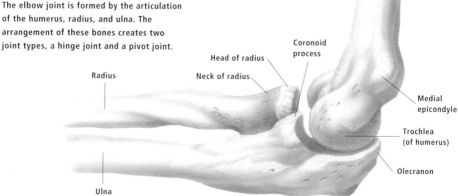

Humerus
Coronoid process
Head of radius
Neck of radius
Radius
Medial epicondyle
Trochlea (of humerus)
Olecranon
Ulna

Wrist

The eight bones of the wrist, known as the carpal bones, connect the bones of the forearm to the bones of the hand. The configuration of the carpal bones creates gliding joints between each of the bones, allowing them to slide against one another, and enabling such hand movements as flexion, extension, and side to side.

Hand

The bones of the hand lie between the carpal bones of the wrist and the phalangeal bones of the fingers. Joints created between the hand bones (metacarpal bones) and the adjoining regions of the

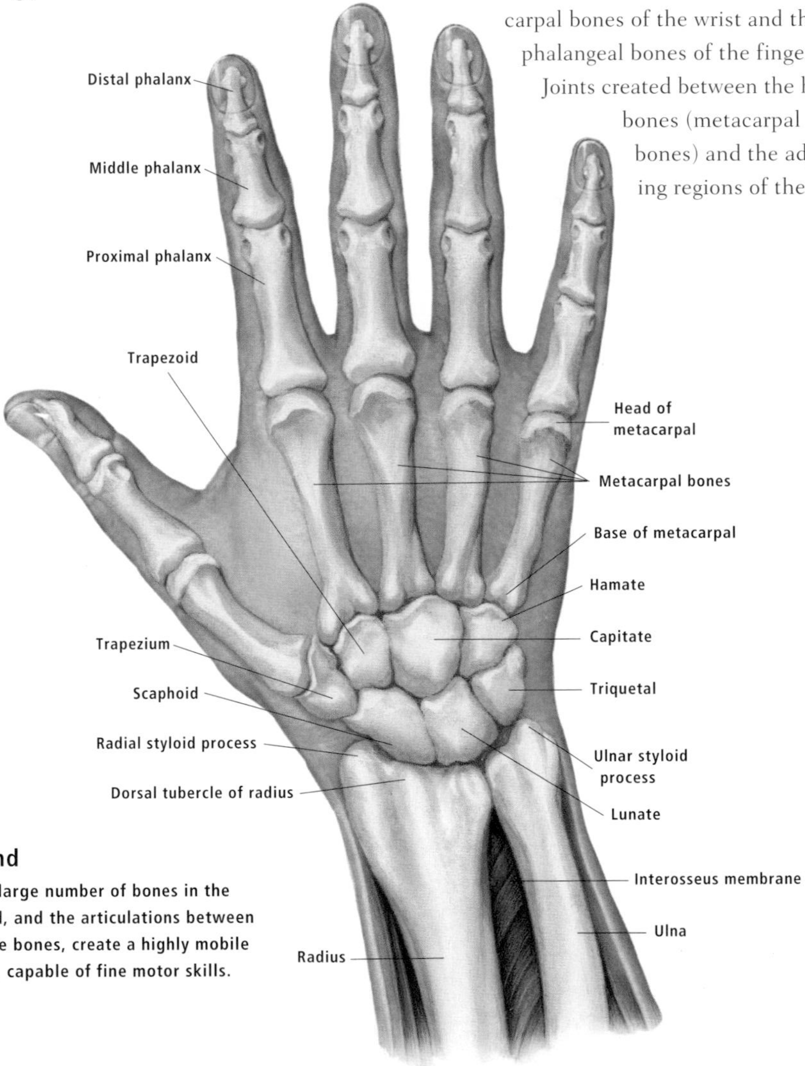

Distal phalanx

Middle phalanx

Proximal phalanx

Trapezoid

Head of metacarpal

Metacarpal bones

Base of metacarpal

Hamate

Trapezium

Capitate

Scaphoid

Triquetal

Radial styloid process

Dorsal tubercle of radius

Ulnar styloid process

Lunate

Hand

The large number of bones in the hand, and the articulations between these bones, create a highly mobile unit, capable of fine motor skills.

Interosseus membrane

Ulna

Radius

wrist and fingers contribute to the overall dexterity achieved by the hand. In particular, where the trapezium of the wrist and the metacarpal bone of the thumb meet, the saddle joint created by the articulation of these two bones allows great mobility, to enable the wide range of movements demanded of the thumb.

Finger

Each of the four fingers contains three bones, the phalanges, while the thumb contains two phalangeal bones.

Finger

The slender bones of the fingers are known as the phalanges (singular, phalanx). The four fingers, index, middle, ring, and little finger, each contain three phalanges: the proximal phalanx, the middle phalanx, and the distal phalanx. The articulation between the phalanges, the interphalangeal joints, forms hinge joints. The thumb contains only two phalanges, the proximal and distal phalanges, forming a single hinge joint.

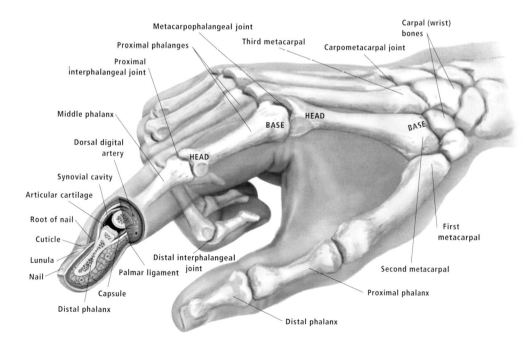

Metacarpophalangeal joint
Proximal phalanges
Proximal interphalangeal joint
Middle phalanx
Dorsal digital artery
Synovial cavity
Articular cartilage
Root of nail
Cuticle
Lunula
Nail
Palmar ligament
Capsule
Distal phalanx
Distal interphalangeal joint
HEAD
BASE
HEAD
Third metacarpal
Carpometacarpal joint
Carpal (wrist) bones
BASE
First metacarpal
Second metacarpal
Proximal phalanx
Distal phalanx

Pelvis

The ring of bones that form the pelvis includes the hip bones of the ilium, ischium, and pubis, and the vertebral bones of the sacrum and coccyx.

The male and female pelvis differ in shape and size, to cater to differing requirements. The male pelvis is comprised of much larger bones and larger joint surfaces than that of the female. The larger size of the male pelvis indicates the need for increased load-bearing requirements due to the generally stronger build and greater weight of men.

While the male pelvis is larger, the heart-shaped pelvic inlet is smaller, the outlet is smaller, and the cavity created by the pelvic bones is elongated and conical.

The shape of the female pelvis, and the dimensions of the pelvic inlet and outlet, are designed to cope with the demands of gestation and childbirth. As a general rule, though there are several variations, the

Pelvis—male

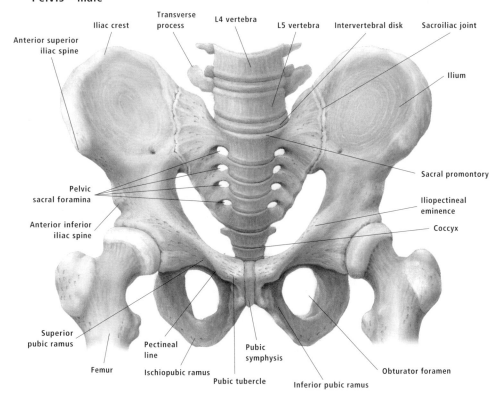

female pelvic inlet is wider than it is deep, thus facilitating the passage of the baby's head during the birthing process. The dimensions of the pelvic outlet allow for the rotation of the baby's head during childbirth. This typical female pelvic shape is called gynecoid. Other shapes include android (heart-shaped, similar to that of the male), anthropoid (deep from front to back, narrow from side to side), and platypelloid (extremely wide, and shortened from front to back).

Pelvis

While the components of the male and female pelvis are the same, the shape and dimensions differ. The male pelvis is generally larger, to meet the demands of a heavier build and weight. The female pelvis is designed to support the fetus during pregnancy and facilitate childbirth; the pelvic inlet is wider to allow the passage of the baby's head during childbirth.

Pelvis—female

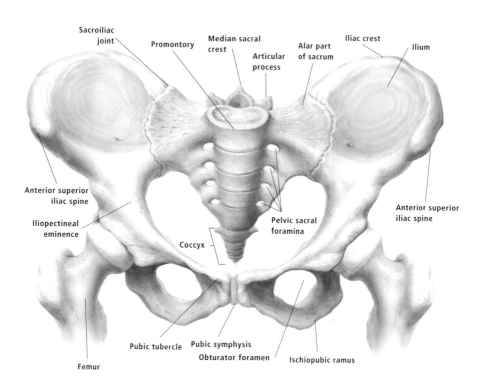

Sacroiliac joint

Promontory

Median sacral crest

Articular process

Alar part of sacrum

Iliac crest

Ilium

Anterior superior iliac spine

Iliopectineal eminence

Coccyx

Pelvic sacral foramina

Anterior superior iliac spine

Pubic tubercle

Pubic symphysis

Obturator foramen

Ischiopubic ramus

Femur

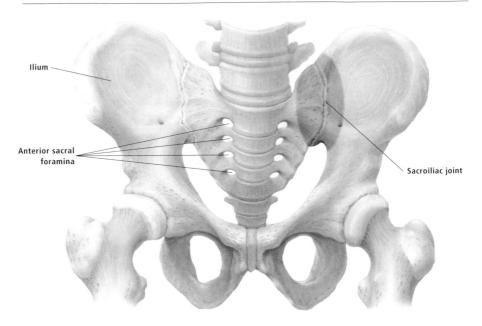

Ilium

Anterior sacral
foramina

Sacroiliac joint

Sacroiliac Joint

The sacroiliac joint occurs where the
sacrum and ilium meet. The entire pelvic
region is placed under stresses as it transfers the weight of the upper body to the
lower limbs of the legs. In order to cope
with the stresses of the pressure, the sacroiliac joint is held firmly in position by
tough, strong ligaments, which permit
extremely limited movement. The configuration of the margins of the two bones is
mutually agreeable to permit a close fit
and thus provide stability.

Part synovial, part fibrous, the sacroiliac
joint must be able to withstand not only
the weight of the upper body, but also
the twisting and bending movements that
the body undergoes.

Sacroiliac joint

**The connection between the sacrum at the base
of the spine and the ilium of the pelvis sees the
formation of the sacroiliac joint.**

Hip

Three bones join to form the hip: the ilium,
ischium and the pubis. At its highest point,
the ilium is located at the waist, while the
pubis is located at the lowest point of the
soft anterior abdominal wall. The tubero-
sities of the ischium are the bony protuberances we sit on.

During adolescence the three hip bones
fuse at the acetabulum, the socket structure of the hip joint. Partnered by the long
bone of the upper leg, the femur, these two
structures articulate at the hip joint. The

Hip bone and joint

Three bones form the hip—the ilium, the ischium, and the pubis.
At the point where all three meet, the bones fuse to create
the acetabulum. The acetabulum is the socket for the head
of the femur—together they form the hip joint.

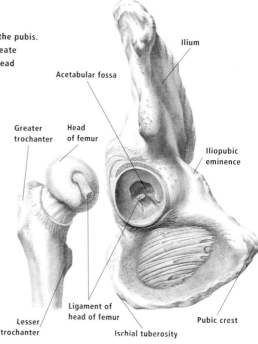

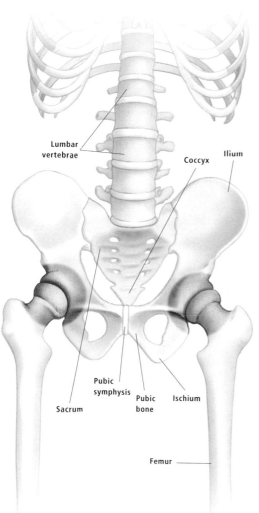

rounded head of the femur sits in its
socket, the acetabulum, creating a mobile
ball-and-socket joint. Necessarily more
stable than the other major ball-and-socket
joint of the shoulder, the hip joint aids in
transferring the weight of the upper body to
the lower limbs. Its stability is achieved by
having a large contact surface between the
femur and the acetabulum, and by the arrangement of strong, supporting ligaments
around the joint. The joint still provides a
wide range of movement; the joint capsule
is lined with synovial membrane and filled
with synovial fluid to lubricate the area,
providing almost friction-free articulation.

Bones of the leg

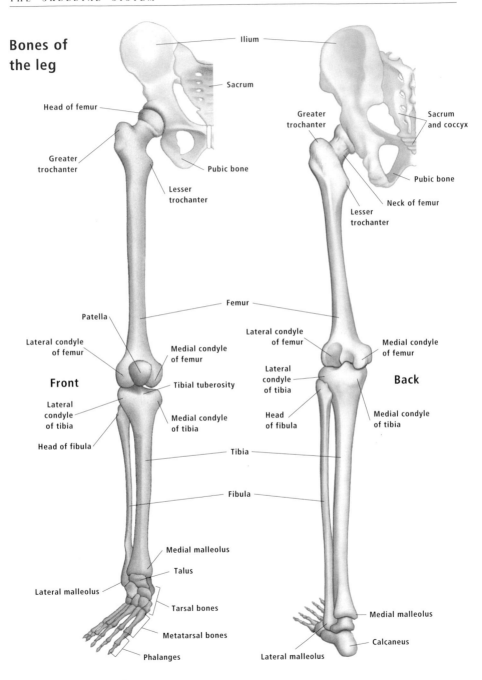

Ilium

Sacrum

Head of femur

Greater trochanter

Pubic bone

Lesser trochanter

Greater trochanter

Sacrum and coccyx

Pubic bone

Neck of femur

Lesser trochanter

Femur

Patella

Lateral condyle of femur

Lateral condyle of femur

Medial condyle of femur

Medial condyle of femur

Front

Tibial tuberosity

Lateral condyle of tibia

Medial condyle of tibia

Head of fibula

Lateral condyle of tibia

Head of fibula

Back

Medial condyle of tibia

Tibia

Fibula

Medial malleolus

Talus

Lateral malleolus

Tarsal bones

Metatarsal bones

Phalanges

Medial malleolus

Calcaneus

Lateral malleolus

Leg

Technically, the leg denotes the area between the knee and the ankle, though it is generally it is regarded as the entire limb with the exception of the foot. The arrangement of the bones is similar to that of the arm, with a large, strong bone in the upper area, and two smaller bones in the lower region. The bones of the leg are necessarily strong to withstand the upper body weight, and to provide bipedal locomotion. Large, strong joints with strong, supporting ligaments are a feature of the leg.

Femur

Extending from the hip joint, where it articulates with the pelvic bones, down to the knee joint, where it articulates with the bones of the lower leg, the femur is the longest and strongest bone in the body. The design of the femur includes a rounded head, a long neck and two enlargements, called trochanters, which provide a point of attachment for the strong muscles of the upper leg. Below these features is the long cylindrical shaft of the femur, and at the lower end are the two condyles. These protuberances articulate with the tibia of the lower leg and the patella (kneecap).

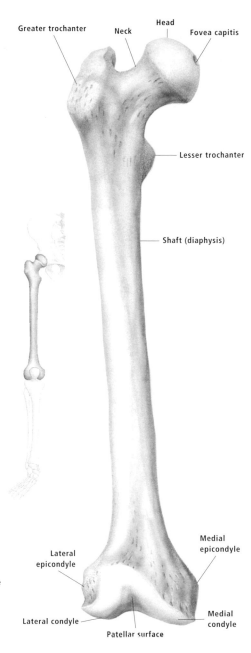

Greater trochanter · Neck · Head · Fovea capitis

Lesser trochanter

Shaft (diaphysis)

Medial epicondyle

Lateral epicondyle

Lateral condyle

Medial condyle

Patellar surface

Femur

The design and integrity of the long bone of the femur are essential to its weight-bearing responsibilities. The femur is the longest and strongest bone in the body.

Tibia

The tibia is the second longest bone in the body. Its design incorporates a widened head, a long, cylindrical shaft, and a wide, lower end, with a protrusion, the medial malleolus, projecting down from the lower edge. The upper surface of the tibia articulates with the femur at the knee joint, while the lower surface articulates with the fibula and the talus to form the ankle joint.

Fibula

The fibula lies on the outer edge of the lower leg. Its upper rounded end sits below the knee joint, while its lower rounded end articulates with the tibia and the talus, with the fibula forming the outer (lateral) side of the ankle joint. The long slender bone of the fibula does not play a major role in weight bearing, but instead serves as a structure for muscle attachment.

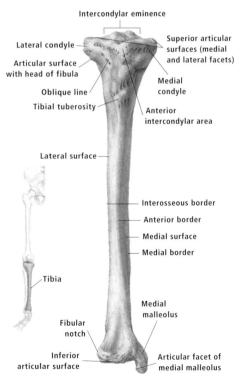

Tibia

The tibia (shin bone) plays a role in two joints. It articulates with the femur at the knee joint and with the talus and fibula at the ankle joint.

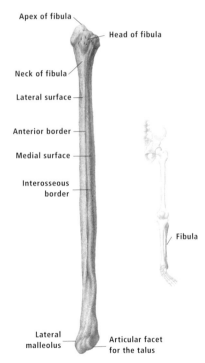

Fibula

The fibula forms the outer edge of the ankle in conjunction with the tibia. The two bones form a bracket-type arrangement around the talus, holding it firm within the ankle joint.

Knee

Connecting the thigh and the lower leg is the knee joint. This complex joint plays a major role in weight bearing and mobility. The knee joint is formed by the femur, the patella, and the tibia. The design of the lower end of the femur includes a concavity to accommodate the gliding movements of the patella. Each of the rounded condyles of the femur fits into the shallow depressions in the corresponding condyle of the tibia. The connection between the bones of the knee joint is not a particularly close fit, with ligaments and muscles responsible for keeping the joint stable.

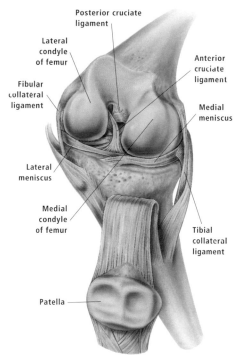

Knee joint

The knee joint is a complex hinge joint, with the most extensive synovial membrane area in the body.

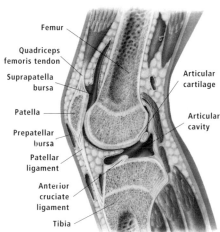

The three bones of the knee are enclosed in a joint capsule. The capsule is lined with synovial membrane, which produces synovial fluid. The lubricating qualities of the capsular components, along with the smooth cartilage coating of the articular surfaces, contribute to provide smooth, friction-free movement.

Strong ligaments maintain knee stability, and allow the actions associated with hinge joint movement, such as bending and straightening. The knee is also capable of some backward and forward gliding movement and a small degree of rotation.

Ligaments and bones of the knee

The bones of the knee joint—the femur, tibia, and patella—are held firmly in place by ligaments and muscles. In this illustration, the patella has been reflected to expose the inner ligaments and bones.

Ankle

The ankle joint is formed by the talus and the bones of the lower leg, the tibia and the fibula. The connection of the three bones forms a fairly stable hinge joint. Protrusions on the side ends of the tibia and fibula, known as malleoli (singular, malleolus), form a socket around the talus. These malleoli are the protuberances than can be seen and felt at the sides of the ankle. The joint is designed to enable raising and lowering of the foot, movements known as dorsiflexion and plantarflexion.

The ankle is strengthened and held firm by ligaments, primarily the deltoid ligament, which connects the medial malleolus

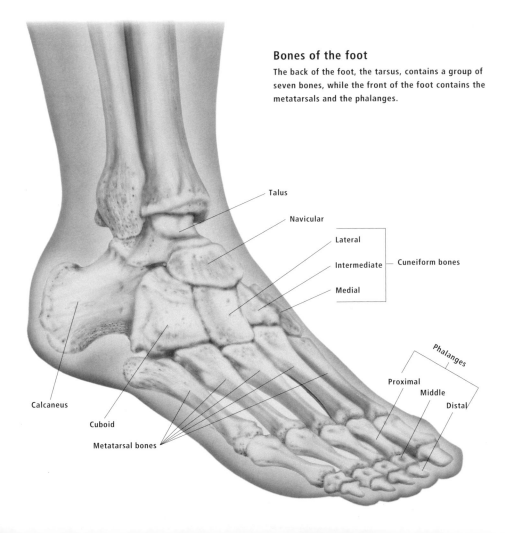

Bones of the foot

The back of the foot, the tarsus, contains a group of seven bones, while the front of the foot contains the metatarsals and the phalanges.

Talus

Navicular

Lateral

Intermediate — Cuneiform bones

Medial

Phalanges

Proximal

Middle

Distal

Calcaneus

Cuboid

Metatarsal bones

of the tibia to the talus, navicular, and calcaneus bones, in conjunction with the ligaments that connect the lateral malleolus of the fibula to the tibia, talus, and calcaneus.

Ankle—posterior view

The ankle is formed by the tibia, fibula, and talus. Ligaments keep the joint stable and prevent excessive movement.

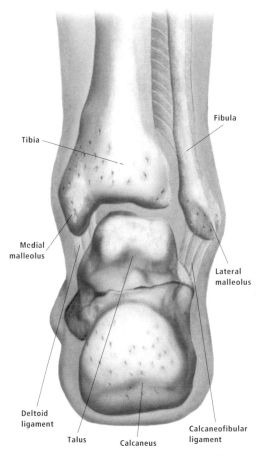

Fibula

Tibia

Medial malleolus

Lateral malleolus

Deltoid ligament

Talus

Calcaneus

Calcaneofibular ligament

Foot

The foot supports the body weight and acts as a lever to propel the body forward during walking. There are 26 bones in total in the foot, comprising: 7 bones in the ankle, 5 bones in the front of the foot, and the 14 bones of the toes.

The back half of the foot, known as the tarsus, contains the seven tarsal bones. The tarsal bones include the talus, which plays a role in the ankle joint; the calcaneus, which forms the heel; the navicular bone; the cuboid bone; and the lateral, intermediate, and medial cuneiform bones.

At the front of the foot are the five metatarsals and the fourteen phalangeal bones that form the toes. The slender metatarsal bones form joints with the tarsus and the phalanges.

Each of the smaller toes contains three phalanges (proximal, middle, and distal), while the big toe contains only two phalangeal bones (proximal and distal).

In order to stabilize the numerous bones of the foot, strong ligaments are arranged around much of the foot and ankle, providing support and stability. However, it is this segmented design of the foot that allows flexibility and adaptability to various surfaces and situations.

The foot is capable of a range of movements including: dorsiflexion (feet point upward), plantarflexion (feet point downward), inversion or supination (sole faces inward), and eversion or pronation (sole faces outward).

Diseases and Disorders of the Skeletal System

Arthritis

Arthritis is the painful inflammation of joints, and can affect any number of joints. Symptoms include redness, swelling, and pain, and it can result in reduction or loss of mobility. While there are many types of arthritis, two of the most common are osteoarthritis and rheumatoid arthritis.

Osteoarthritis is a degenerative disorder, and is the most common form of joint disease. Characterized by the deterioration of cartilage within the joints, and the growth of bony spurs and thickening of bone at the contact surfaces of the joint, osteoathritis can affect all ages. It is more common from middle age onward, and often affects weight-bearing joints, such as the hip and knee joints.

Rheumatoid arthritis is the chronic inflammation of the connective tissue of the body. While it can affect many of the major organs of the body, it more commonly affects the smaller joints of the body, such as those found in

Rheumatoid arthritis
Rheumatoid arthritis most commonly affects the small joints of the hands and feet.

Osteoarthritis
The weight-bearing joints of the hips and knees are the most likely to be affected by osteoarthritis.

the hands, feet, and elbows. Symptoms include pain, swelling, inflammation, and stiffness.

Hip replacement is often required when the joint has become badly deteriorated. This involves replacing the head of the femur with a metal ball, and replacing the socket (acetabulum) with a polyethylene cup. Advances in technology are continually breaking new ground in the components of artificial joints.

Osteoarthritis—knee joint

The gradual breakdown of cartilage within joints affected by osteoarthritis, coupled with the development of nodular bony outcrops in the articular surface, creates painful joints.

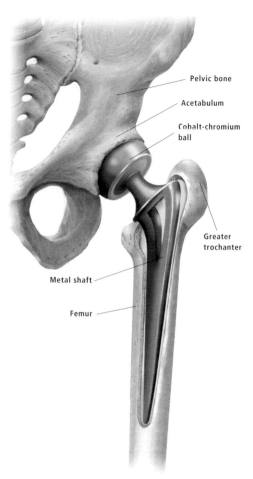

Pelvic bone

Acetabulum

Cobalt-chromium ball

Greater trochanter

Metal shaft

Femur

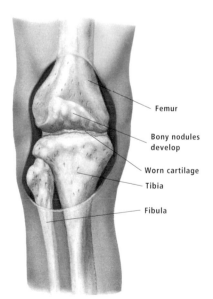

Femur

Bony nodules develop

Worn cartilage

Tibia

Fibula

Hip replacement

The components of a joint cannot regenerate, and once they have deteriorated, joint replacement can be an option in some situations. Hip replacement is becoming an increasingly common procedure for arthritis sufferers.

Bone tumors

Tumors affecting bone can be benign or malignant. Forms of benign bone tumors include osteochondromas and osteomas, neither of which usually requires treatment. Malignant tumors are usually associated with primary cancers of the organs or glands.

Primary malignant bone tumors are rare, and include osteosarcomas, Ewing's sarcoma, fibrosarcoma, and chondrosarcoma. Treatment for these types of tumors includes radiation therapy, chemotherapy, and surgical removal, and in some cases, amputation of the affected limb is required.

Fractures

While any bone in the body can suffer a fracture, some bones are more prone to fracture than others. The long bones of the arms and legs are the most likely to fracture. There are several different types of fractures that can be sustained, and the exact type of fracture can be determined by x-ray of the damaged site.

Treatment of fractures depends on the fracture type. Most often immobilization of the bone is required—usually by use of a plaster cast, or sometimes by traction or surgical insertion of rods, plates, or pins.

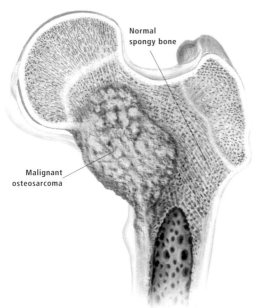

Normal spongy bone

Malignant osteosarcoma

Bone tumor

Most commonly occurring as a result of spread from another organ, primary malignant bone tumors, such as the one pictured here, are quite rare.

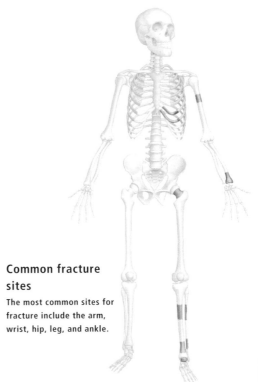

Common fracture sites

The most common sites for fracture include the arm, wrist, hip, leg, and ankle.

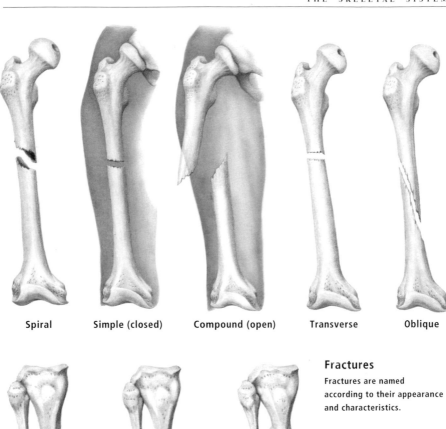

Spiral Simple (closed) Compound (open) Transverse Oblique

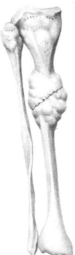

Fractures

Fractures are named according to their appearance and characteristics.

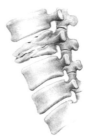

Comminuted Greenstick Pathologic Compression

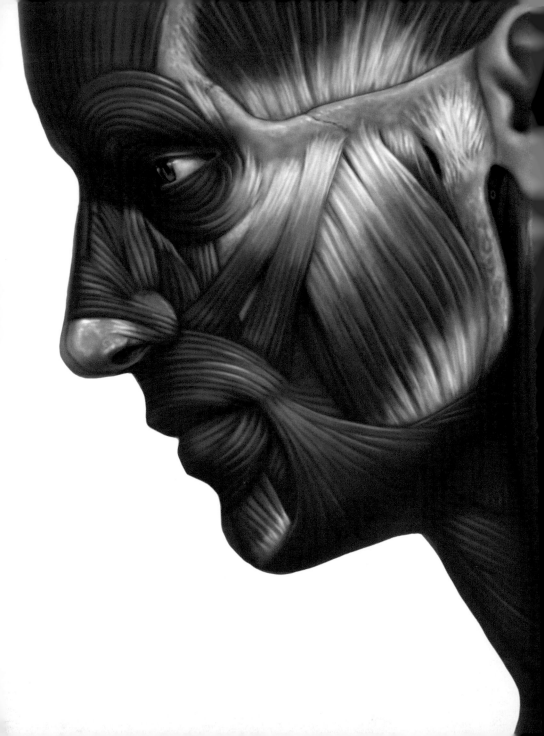

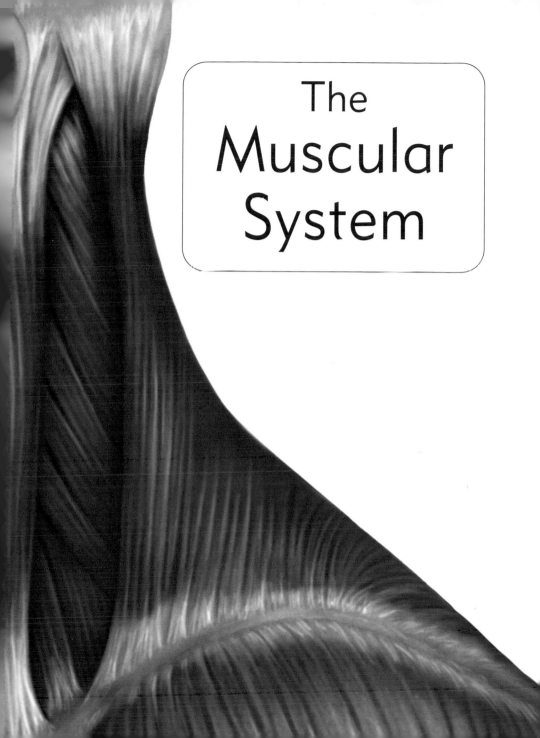

The
Muscular
System

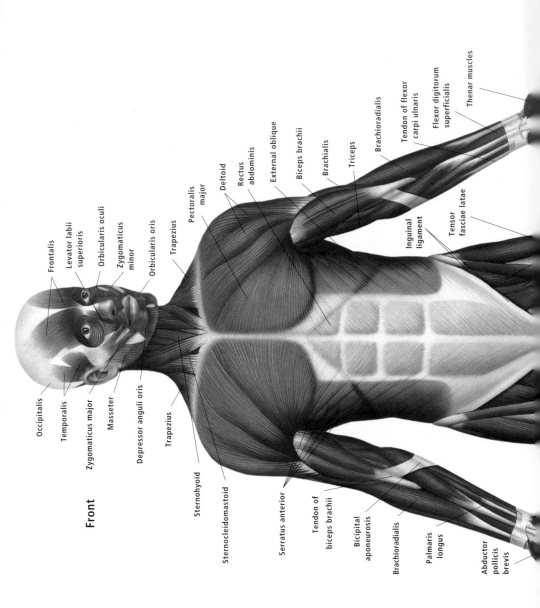

Front

Occipitalis

Temporalis

Zygomaticus major

Masseter

Depressor anguli oris

Trapezius

Sternohyoid

Sternocleidomastoid

Serratus anterior

Tendon of biceps brachii

Bicipital aponeurosis

Brachioradialis

Palmaris longus

Abductor pollicis brevis

Frontalis

Levator labii superioris

Orbicularis oculi

Zygomaticus minor

Orbicularis oris

Trapezius

Pectoralis major

Deltoid

Rectus abdominis

External oblique

Biceps brachii

Brachialis

Triceps

Brachioradialis

Tendon of flexor carpi ulnaris

Flexor digitorum superficialis

Thenar muscles

Inguinal ligament

Tensor fasciae latae

THE MUSCULAR SYSTEM

The voluntary muscles of the body are responsible for body movement. The contours of an individual's physique are shaped by the muscles of the body, which can range in size from the tiny muscles of the forehead that assist in facial expression, to the large, powerful muscles of the thigh.

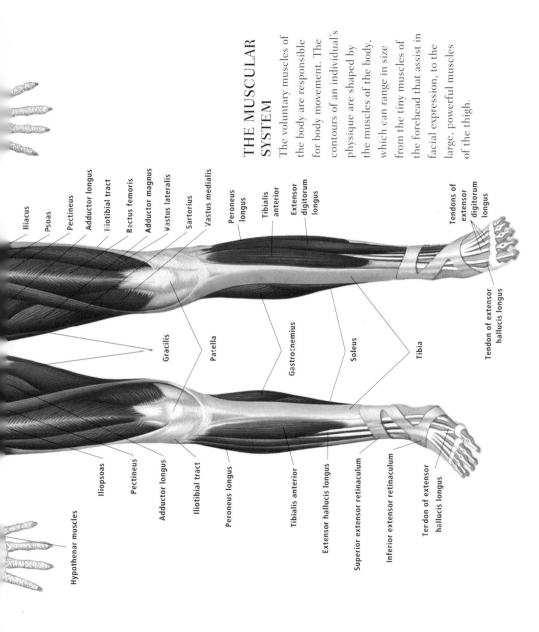

Iliacus

Psoas

Pectineus

Adductor longus

Iliotibial tract

Rectus femoris

Adductor magnus

Vastus lateralis

Sartorius

Vastus medialis

Peroneus longus

Tibialis anterior

Extensor digitorum longus

Tendons of extensor digitorum longus

Gracilis

Patella

Gastrocnemius

Soleus

Tibia

Tendon of extensor hallucis longus

Hypothenar muscles

Iliopsoas

Pectineus

Adductor longus

Iliotibial tract

Peroneus longus

Tibialis anterior

Extensor hallucis longus

Superior extensor retinaculum

Inferior extensor retinaculum

Terdon of extensor hallucis longus

Back

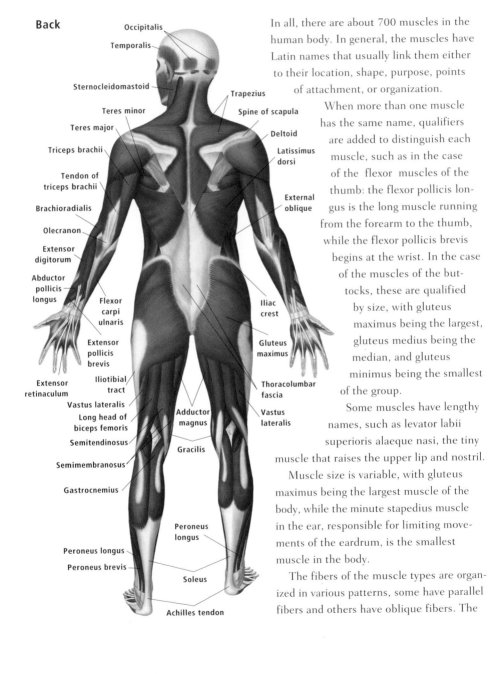

Occipitalis
Temporalis
Sternocleidomastoid
Teres minor
Teres major
Triceps brachii
Tendon of triceps brachii
Brachioradialis
Olecranon
Extensor digitorum
Abductor pollicis longus
Flexor carpi ulnaris
Extensor pollicis brevis
Extensor retinaculum
Iliotibial tract
Vastus lateralis
Long head of biceps femoris
Semitendinosus
Semimembranosus
Gastrocnemius
Peroneus longus
Peroneus brevis

Trapezius
Spine of scapula
Deltoid
Latissimus dorsi
External oblique
Iliac crest
Gluteus maximus
Thoracolumbar fascia
Vastus lateralis
Adductor magnus
Gracilis
Peroneus longus
Soleus
Achilles tendon

In all, there are about 700 muscles in the human body. In general, the muscles have Latin names that usually link them either to their location, shape, purpose, points of attachment, or organization.

When more than one muscle has the same name, qualifiers are added to distinguish each muscle, such as in the case of the flexor muscles of the thumb: the flexor pollicis longus is the long muscle running from the forearm to the thumb, while the flexor pollicis brevis begins at the wrist. In the case of the muscles of the buttocks, these are qualified by size, with gluteus maximus being the largest, gluteus medius being the median, and gluteus minimus being the smallest of the group.

Some muscles have lengthy names, such as levator labii superioris alaeque nasi, the tiny muscle that raises the upper lip and nostril.

Muscle size is variable, with gluteus maximus being the largest muscle of the body, while the minute stapedius muscle in the ear, responsible for limiting movements of the eardrum, is the smallest muscle in the body.

The fibers of the muscle types are organized in various patterns, some have parallel fibers and others have oblique fibers. The arrangement of the fibers denotes the function of the muscle, for instance, if the purpose of the muscle is to move bone, then the fibers will be aligned in the same direction. If the purpose of the muscle is to open and close an entrance, then the fibers will be organized in a circular arrangement. If the purpose of the muscle is to support an organ, the fibers will be interwoven in a criss-cross pattern.

The muscle fibers are attached either to bone or to a tendon that is connected to bone. The tendons give strength to the line of force of the muscle.

Spanning between muscle and bone, tendons are a type of connective tissue. Tendons are tough, fibrous structures comprised of fibrocytes and collagen held within a ground substance.

Tendons are responsible for transfer of the forces created by muscle to the connecting bone, thus moving the bone.

This intermediary role means that the bone and muscle are quite separate entities, with the tendons controlled by muscle movement and completing the task required.

Where tendon meets bone, there is a gradual melding of the tendon fibers into the bone. In some parts of the body the tendons pass through a fibrous sheath, known as a tendon sheath. The protective sheath provides a lubricant of synovial fluid, which aids in the smooth movement of the tendon, isolating it from abrasive moving parts that may otherwise cause damage to the tendon.

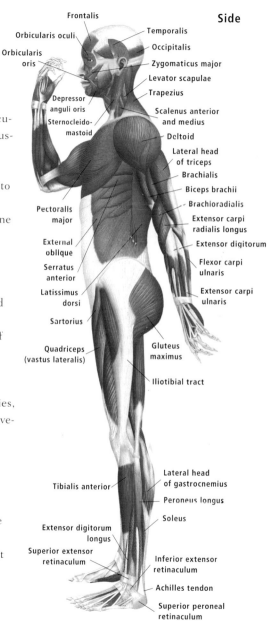

Side

Frontalis
Orbicularis oculi
Orbicularis oris
Temporalis
Occipitalis
Zygomaticus major
Levator scapulae
Depressor anguli oris
Trapezius
Sternocleido-mastoid
Scalenus anterior and medius
Deltoid
Lateral head of triceps
Brachialis
Biceps brachii
Brachioradialis
Pectoralis major
Extensor carpi radialis longus
Extensor digitorum
External oblique
Flexor carpi ulnaris
Serratus anterior
Extensor carpi ulnaris
Latissimus dorsi
Sartorius
Gluteus maximus
Quadriceps (vastus lateralis)
Iliotibial tract
Lateral head of gastrocnemius
Tibialis anterior
Peroneus longus
Soleus
Extensor digitorum longus
Superior extensor retinaculum
Inferior extensor retinaculum
Achilles tendon
Superior peroneal retinaculum

Muscle tissue

Skeletal muscle tissue

Skeletal muscle accounts for a large percentage of body mass. Skeletal muscle provides the muscular movements of the body, and contours our physique.

Smooth muscle tissue

Smooth muscle is involuntary muscle, controlled by the autonomic nervous system.

Cardiac muscle tissue

Cardiac muscle is found only in the heart. Triggered by the sinoatrial node, cardiac muscle contracts and relaxes rhythmically in an involuntary manner.

Muscle Tissue

There are three different types of muscle tissue in the human body: skeletal muscle, smooth muscle, and cardiac muscle.

Skeletal muscle is the most common muscle tissue type, and accounts for around 60 percent of the body's mass. It is referred to as voluntary muscle, indicating that skeletal muscle and its movement is under our conscious control. Nerve impulses trigger a reaction from the muscle which contracts, moving its associated bone or joint.

Smooth muscle is found in the digestive system, reproductive system, major blood vessels, skin, and some internal organs. Triggered by the autonomic nervous system, the operation of smooth muscle is not under our conscious control and is referred to as involuntary. Much of the smooth muscle within the body is in motion continually, performing constant rhythmic movements.

Cardiac muscle is found only in the heart. Like smooth muscle, it is not under our voluntary control. Electrical impulses originating from the sinoatrial node, the

heart's natural pacemaker, initiate heart muscle contraction. Cardiac muscle is also influenced by the autonomic nervous system, with nerves from the system able to regulate heartbeat under certain conditions.

Muscle Fiber Microstructure

Muscles are made up of bundles of fibers known as fascicles. Each fiber is an elongated cell containing thread-like structures called myofibrils. Each myofibril contains thick and thin myofilaments that lie alongside each other. Thick filaments contain myosin, while thin filaments contain actin, troponin, and troposin. When triggered by impulses from the nervous system, the myofilaments slide along each other, causing a chemical reaction as they meet and interlock at crossbridges. This chemical reaction produces a muscular contraction. The myofilaments release and slide back to their original position as the muscle relaxes.

Muscle fiber microstructure

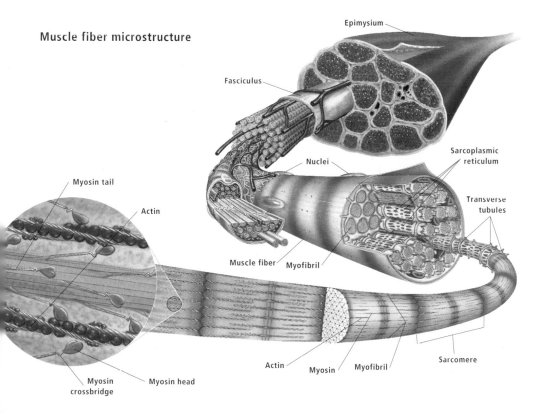

Epimysium

Fasciculus

Nuclei

Sarcoplasmic reticulum

Myosin tail

Actin

Transverse tubules

Muscle fiber Myofibril

Actin Myosin Myofibril

Sarcomere

Myosin crossbridge Myosin head

Muscle Types

The organization of muscle fibers creates various muscle types. Some variations of muscle types have parallel fibers. In pennate muscles, from the Latin penna, meaning feather, the fibers are arranged obliquely down the tendon. As indicated by its name, the fibers in circular muscle are arranged in a circular pattern.

Actions of Muscles

Muscular contractions are controlled by the nervous system. Impulses from the nerves trigger release of calcium in the cells of the muscle fibers, which initiates a contraction.

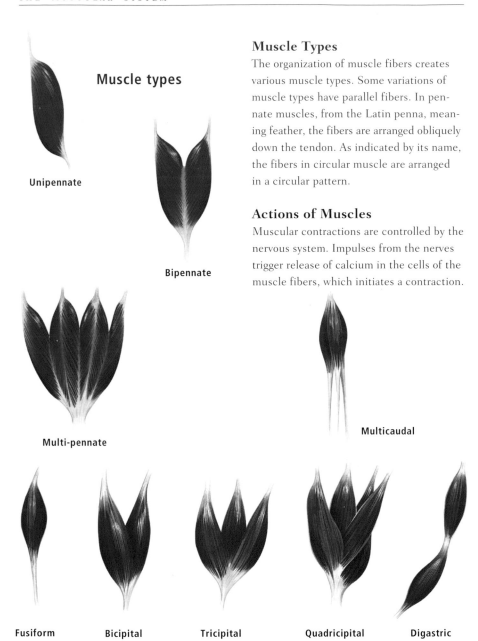

Muscle types

Unipennate

Bipennate

Multi-pennate

Multicaudal

Fusiform

Bicipital

Tricipital

Quadricipital

Digastric

When a muscle contracts, it shortens and exerts a pull on the muscle attachment.

The precise action of a muscle is dependent on its position and purpose—muscles perform many actions including straightening, bending, raising, and lowering. Muscles often work in pairs called agonist and antagonist muscles. Because muscles can only pull in one direction they are arranged in opposing pairs to enable active extension and flexion. When one of these muscles contracts, the other stretches in response, and vice versa, to produce opposing movements. The energy for muscular contraction comes from glycogen, a form of sugar glucose, stored in muscle tissue.

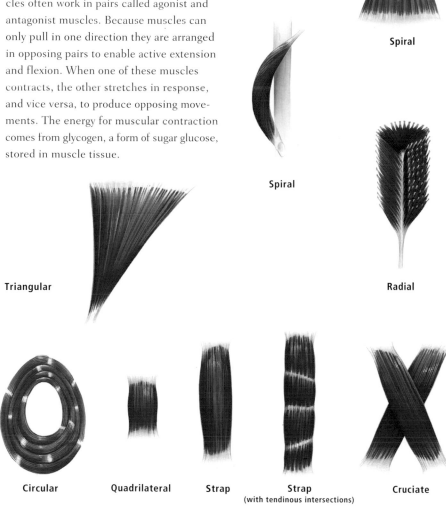

Spiral

Spiral

Radial

Triangular

Circular　　**Quadrilateral**　　**Strap**　　**Strap**
(with tendinous intersections)　　**Cruciate**

Face muscles

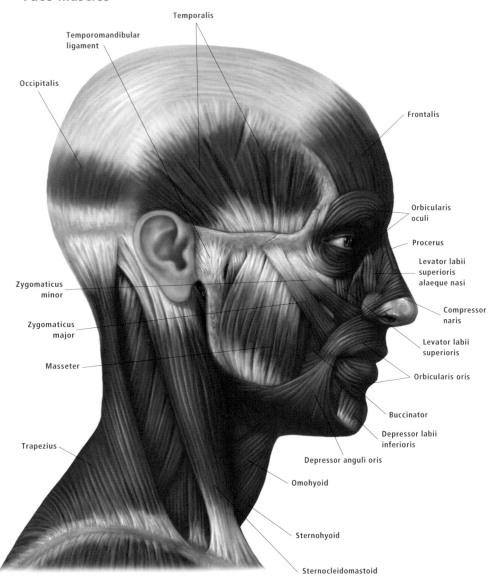

Temporalis

Temporomandibular
ligament

Occipitalis

Frontalis

Orbicularis
oculi

Procerus

Levator labii
superioris
alaeque nasi

Compressor
naris

Levator labii
superioris

Zygomaticus
minor

Zygomaticus
major

Masseter

Orbicularis oris

Buccinator

Depressor labii
inferioris

Trapezius

Depressor anguli oris

Omohyoid

Sternohyoid

Sternocleidomastoid

Facial Muscles

The facial muscles are responsible for the facial expressions that convey much of what we feel and think. A multitude of muscles cover the facial area, from the very small muscles in the forehead, responsible for furrowing of the brow or wrinkling of the forehead, to the circular muscles of the mouth and eye area.

The circular muscles of the lips form the mouth's muscular opening. These muscles participate in speech, the chewing process, and the creation of facial expressions such as smiling.

The muscles of the cheek area also play a part in speech and in holding food in the mouth. The buccinator muscle of the cheek, which is used for blowing air from the mouth, derives its name from the Latin word meaning trumpeter.

Zygomaticus major and zygomaticus minor attach at the corners of the mouth, and stretch across the cheek area, attaching to the zygomatic arch below the eye. These two muscles move the upper lip and lift the corners of the mouth.

Similarly, levator labii superioris and levator labii superioris alaeque nasi both contribute to movements of the upper lip, with the latter also being involved in nostril movements.

The procerus muscle, which runs upward across the bridge of the nose, is responsible for eyebrow movement.

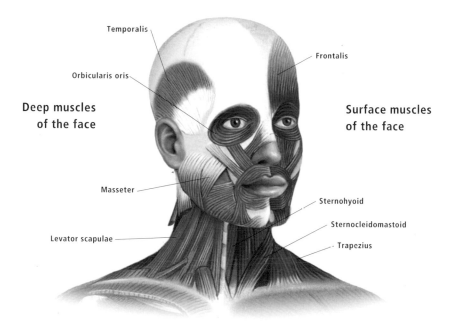

Temporalis

Orbicularis oris

Deep muscles of the face

Frontalis

Surface muscles of the face

Masseter

Levator scapulae

Sternohyoid

Sternocleidomastoid

Trapezius

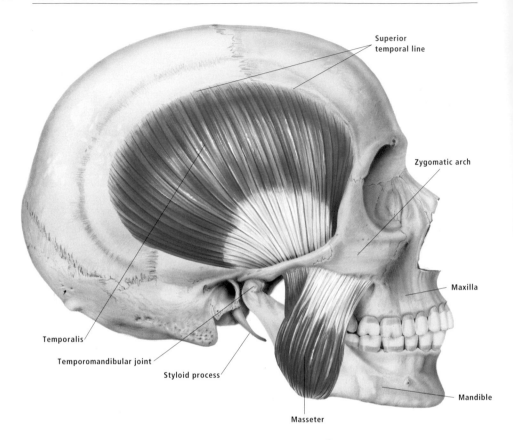

Superior temporal line

Zygomatic arch

Maxilla

Temporalis

Temporomandibular joint

Styloid process

Mandible

Masseter

Surface muscles of the jaw

The muscles of the jaw are used for mastication and in speech. The masseter muscle is regarded as one of the strongest muscles in the body, based on force per unit of mass.

Jaw Muscles

The jaw muscles provide a range of movement to the bones of the jaw—the maxillae and the mandible. The maxillae of the upper jaw are fixed, while the lower jaw, the mandible, can move up, down, sideways, forward, and backward.

The muscles controlling jaw movements are arranged around the temporomandibular joint. This joint is formed where the ramus, a small projection on the upper end of the body of the mandible, links with the base of

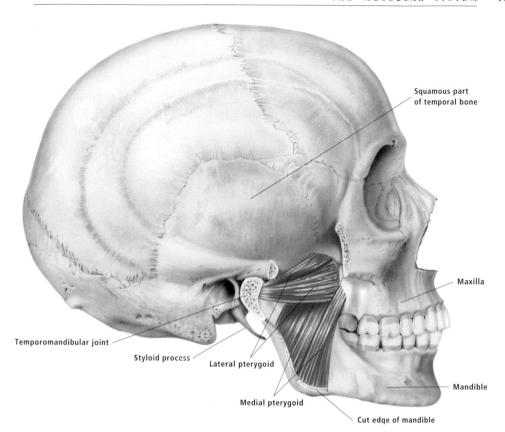

Squamous part
of temporal bone

Maxilla

Temporomandibular joint

Styloid process   Lateral pterygoid

Medial pterygoid

Cut edge of mandible

Mandible

the skull. The jaw muscles are known as the muscles of mastication, and include the medial and lateral pterygoid, masseter, and temporalis muscles.

The medial pterygoid covers the inner, or medial, surface of the ramus; the masseter muscle covers the outer, or lateral, surface; the lateral pterygoid lowers the mandible; and the temporalis muscle raises the mandible.

Based on force per unit of mass, the masseter muscle of the jaw is said to be the strongest muscle in the body.

Deep muscles of the jaw

The muscles around the jaw permit a wide range of movements to the moveable lower jaw (mandible), including up and down movement, backward and forward movement, and sideways movement. Here, the mandible has been cut away to show the pterygoid muscles of the jaw.

Eye Muscles

Six muscles control the movements of the eye: the superior rectus, medial rectus, lateral rectus, inferior rectus, superior oblique, and inferior oblique muscles. The synergy of these muscles allows the eye to look up, down, left, and right, providing a wide field of vision.

Movements of the eye muscles often involve fractions of a degree, and coordinated muscle movement is critical to ensure both eyes are synchronized in order to maintain stereoscopic vision.

The eye is housed in the orbit of the skull. The muscles of the eye hold the eye suspended in the orbit.

The eye muscles work in pairs, though even when one pair of muscles dominates eye movement, the remaining muscles still play a role in achieving the movement required. For instance, in order to move the right eye to look to the right, the lateral rectus on the right side of the eye must contract. Its partner on the left side of the eye, the medial rectus, relaxes to allow the movement. The other muscles of the eye work in a similar way, with the superior rectus and inferior rectus working together in up and down, and medial movements, and the superior oblique and inferior oblique working together to provide up and down, and outward movements.

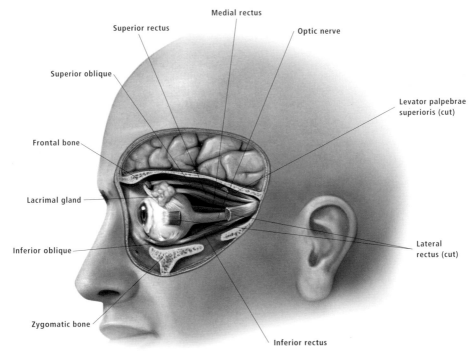

Medial rectus

Superior rectus

Optic nerve

Superior oblique

Levator palpebrae superioris (cut)

Frontal bone

Lacrimal gland

Inferior oblique

Lateral rectus (cut)

Zygomatic bone

Inferior rectus

Eye muscles

The six muscles of the eye work in unison to offer
a wide field of vision, by providing up, down, left,
and right movement.

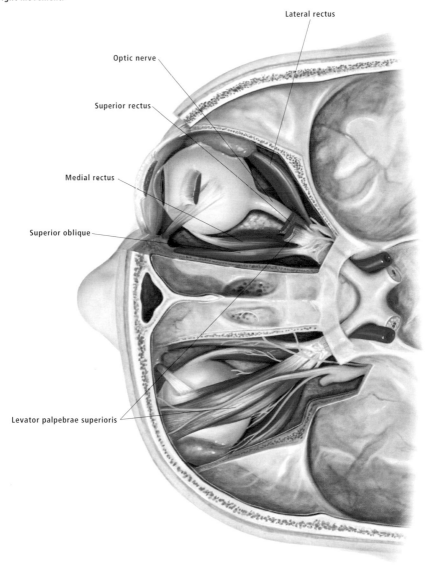

Lateral rectus

Optic nerve

Superior rectus

Medial rectus

Superior oblique

Levator palpebrae superioris

Neck Muscles

Strong muscles surround the neck region, supporting the weight of the head and strengthening the cervical vertebrae at the back of the neck, and the visceral column at the front of the neck, which contains the structures of the larynx and trachea.

The muscles located at the back of the neck include: the splenius capitis, splenius cervicis, semispinalis capitis, semispinalis cervicis, trapezius, levator scapulae, and the scalene muscles. Muscles attach to the front, back, and sides of the vertebrae to produce forward, backward, and sideways movements. Some rotational movement is also provided by the muscles with an oblique arrangement of fibers.

The larger muscles lie to the back of the vertebrae, and provide a variety of movements. Some are involved in moving the head and neck, some are associated with shoulder movement, and some are involved in movement of the upper two ribs.

At the front of the neck, the muscles include: the platysma, sternocleidomastoid, and infrahyoid muscles. The strap-like

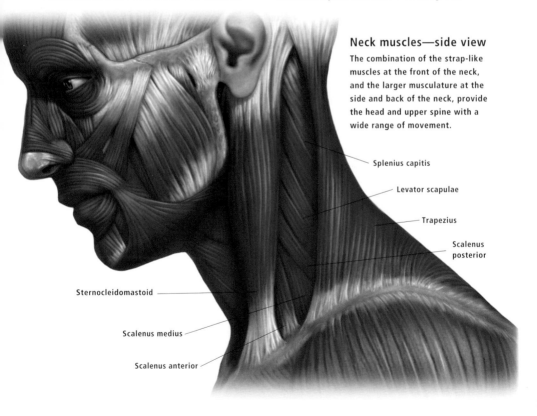

Neck muscles—side view

The combination of the strap-like muscles at the front of the neck, and the larger musculature at the side and back of the neck, provide the head and upper spine with a wide range of movement.

Splenius capitis

Levator scapulae

Trapezius

Scalenus posterior

Sternocleidomastoid

Scalenus medius

Scalenus anterior

muscles of the front of the neck play a role in head movement and facial expression.

The larger muscles of the neck region include the sternocleidomastoid and the trapezius. The sternocleidomastoid muscle, which extends from the clavicle and sternum to the skull, contracts when the head is raised. The trapezius muscle, at the outer margin of the neck, is involved in shoulder movement.

Surface and deep muscles of the neck

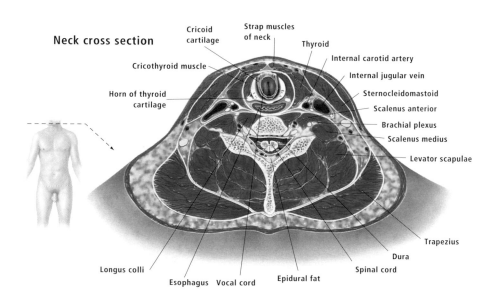

Frontalis

Temporalis

Surface muscles

Deep muscles

Sternohyoid

Masseter

Sternocleido mastoid

Levator scapulae

Trapezius

Neck cross section

Cricoid cartilage

Strap muscles of neck

Thyroid

Cricothyroid muscle

Internal carotid artery

Internal jugular vein

Horn of thyroid cartilage

Sternocleidomastoid

Scalenus anterior

Brachial plexus

Scalenus medius

Levator scapulae

Trapezius

Dura

Spinal cord

Longus colli

Esophagus   Vocal cord

Epidural fat

Muscles of the Throat

The throat is the front portion of the neck. Lying within the throat are the fauces, the opening that leads from the back of the mouth into the pharynx and the pharynx itself. The pharynx is the cavity that connects the nose, mouth, and larynx, and is consequently divided into three regions (in descending order): the naso-pharynx, the oropharynx, and the laryn-gopharynx. The pharynx is the common passageway for air, fluids, and food. As a result of the demands of both the respira-tory and digestive systems governing the function of the pharynx, the muscles of the throat area are involved in a variety of processes and functions.

Muscles control the movement of the epiglottis. During breathing, the epiglottis is held in position by muscles. During swallowing, the muscles bend the epiglottis downward and backward, thus closing off the passageway to the larynx and preventing food from entering. The muscles then re-turn the epiglottis to its original position.

Muscles in the throat, larynx, and mouth are involved in speech production. Modifi-cations brought about by the action of the muscles of the throat can result in a variety of sounds being produced.

Throat—posterior view

The pharynx is one of the most important regions of the throat, comprising structures used for breathing, speech, and swallowing. During swallowing, the muscles of the pharynx lift the larynx and close the epiglottis, to prevent food from entering the airways.

Throat

The throat plays an important role as a passageway of both the respiratory and digestive systems.

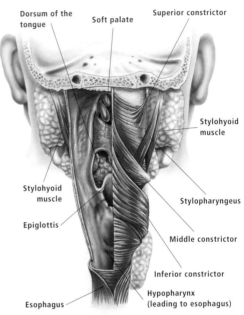

Dorsum of the tongue · Soft palate · Superior constrictor · Stylohyoid muscle · Stylohyoid muscle · Stylopharyngeus · Epiglottis · Middle constrictor · Inferior constrictor · Hypopharynx (leading to esophagus) · Esophagus

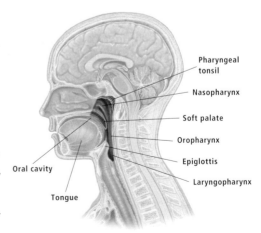

Pharyngeal tonsil · Nasopharynx · Soft palate · Oropharynx · Epiglottis · Laryngopharynx · Oral cavity · Tongue

Epiglottis

Base of
tongue

Epiglottis

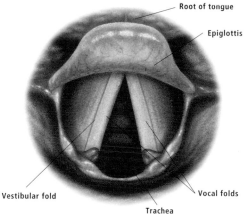

Corniculate
tubercle

Swallowing

Muscles in the throat lower the
epiglottis during swallowing to seal
off the entrance to the airways.

Glottis closed

Root of tongue

Epiglottis

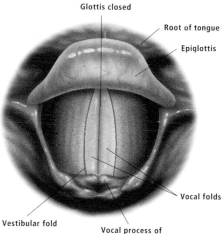

Vocal folds

Vestibular fold

Vocal process of
arytenoid cartilage

Speaking

During speaking, the muscles of the throat hold
the epiglottis in the open position to allow the
passage of air involved in speech production.
The laryngeal muscles adjust the vocal cords,
with the tension and length of the cords
determining the pitch.

Breathing

During breathing, muscles of the
throat hold the epiglottis open, and
muscles of the larynx hold the vocal
folds apart, to allow the inspiration
and expiration of air.

Root of tongue

Epiglottis

Vestibular fold

Vocal folds

Trachea

Back Muscles

The strong muscles of the back stabilize the area, providing support and movement to the spinal column. Beneath the large superficial muscles, such as latissimus dorsi and the trapezius muscle, lie the deeper strap-like muscles, arranged in many directions, and often criss-crossing one another to provide maximum stability and support while also providing the flexibility needed for the movements of the axial skeleton. Many of the muscles of the back are involved in providing movement to the spinal region, but many also are involved

Back muscles

Surface muscles

Deep muscles

- Superior fibers of trapezius
- Spine of scapula
- Middle fibers of trapezius
- Inferior fibers of trapezius
- Latissimus dorsi
- External oblique
- Iliac crest
- Gluteus maximus
- Thoracolumbar fascia
- Cut tendon of semitendinosus

- Semispinalis capitis
- Levator scapulae
- Supraspinatus
- Spine of scapula
- Deltoid
- Teres minor
- Infraspinatus
- Rhomboid minor
- Rhomboid major
- Teres major
- External intercostal muscle
- Erector spinae muscle
- Serratus posterior inferior
- Internal oblique
- Iliac crest
- Posterior superior iliac spine
- Gluteus medius
- Gluteus minimus
- Piriformis
- Lumbar fascia
- Gluteus medius
- Gemellus superior
- Sacrotuberous ligament
- Gemellus inferior
- Quadratus femoris

in the movements of the shoulders and arms, and the neck. For example, the trapezius muscle, which extends across the upper back and up into the neck, attaching to the occipital bone of the skull, acts to steady the shoulder and aid in movements of the scapula.

Abdominal muscles

Abdominal Muscles

The muscles of the abdominal wall provide support to the internal organs and play a role in breathing. The major muscles of the abdomen include rectus abdominis, and the external and internal oblique muscles.

While the diaphragm is the major abdominal muscle involved in the breathing process, the remaining muscles must complement its actions. During breathing, when the diaphragm contracts, the pressure in the abdominal cavity is raised, necessitating the movement of the abdominal muscles.

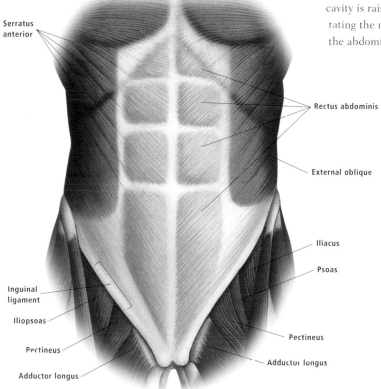

Serratus anterior

Rectus abdominis

External oblique

Iliacus

Psoas

Inguinal ligament

Iliopsoas

Pectineus

Pectineus

Adductor longus

Adductor longus

Pectoral Muscles

Located at the front of the chest, the two pectoral muscles are pectoralis major and pectoralis minor.

From its midline origin, where it is attached to the sternum, collar bone, and rib cage, pectoralis major stretches across the chest to its point of attachment on the humerus. This muscle plays a role in arm movement.

Pectoralis minor extends from the ribs to the coracoid process of the scapula (shoulder blade). It plays a role in shoulder movement and aids in rib movement.

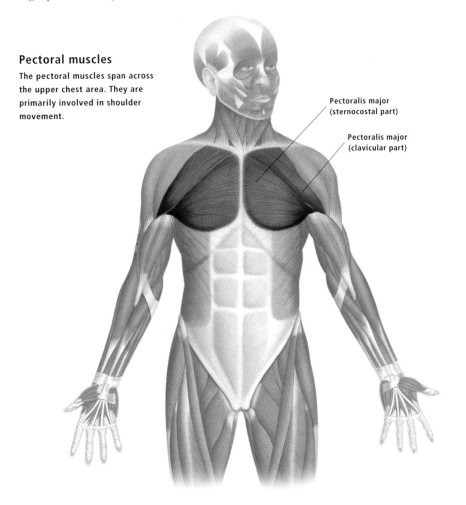

Pectoral muscles

The pectoral muscles span across the upper chest area. They are primarily involved in shoulder movement.

Pectoralis major
(sternocostal part)

Pectoralis major
(clavicular part)

Intercostal Muscles

The intercostal muscles play an important role in breathing. The external intercostal muscles lie over the internal intercostal muscles. These muscles span between the upper and lower borders of neighboring ribs to form a complete unit.

During inspiration, the capacity of the rib cage is increased, enlarging from front to back, side to side, and vertically. With the intake of breath, the intercostal muscles act to raise the ribs. Expiration is a passive process involving the relaxation of tension in the muscles of the ribs, allowing them to return to their original position.

Intercostal muscles

The intercostal muscles support the rib cage and play an important role in breathing.

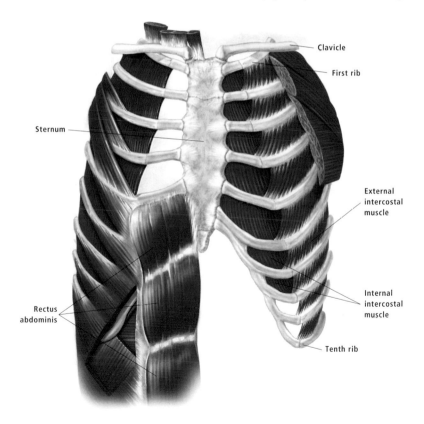

Clavicle

First rib

Sternum

External intercostal muscle

Internal intercostal muscle

Rectus abdominis

Tenth rib

Diaphragm

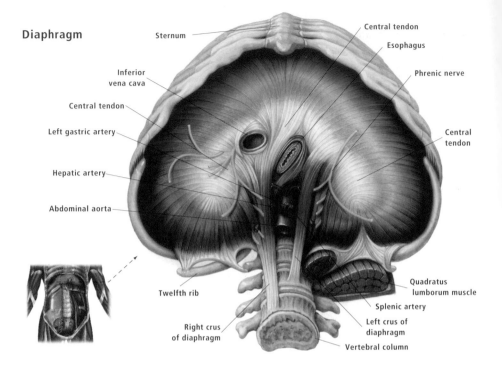

Sternum

Central tendon

Esophagus

Phrenic nerve

Inferior vena cava

Central tendon

Left gastric artery

Central tendon

Hepatic artery

Abdominal aorta

Twelfth rib

Quadratus lumborum muscle

Splenic artery

Right crus of diaphragm

Left crus of diaphragm

Vertebral column

Diaphragm

The muscular layer of the diaphragm is attached to the vertebral column at the back, to the ribs at the side of the chest, and to the sternum at the front; it forms the floor of the chest cavity and the roof of the abdominal cavity. Several structures penetrate through the diaphragm, including the major structures of the esophagus, the aorta, and the inferior vena cava. The central part of the diaphragm is called the central tendon. It is fibrous rather than muscular and has the pericardial sac, which surrounds the heart, firmly attached to its upper surface.

The diaphragm is a vital component of the breathing process, with its muscular actions crucial to inspiration. When at rest, the diaphragm forms a high arched dome. When it contracts, the dome descends, increasing the height of the chest cavity, resulting in air being drawn into the lungs. This process is known as inspiration. Expiration is a passive process, being the result of relaxation of the contracted muscles of the chest and abdomen. During expiration, or breathing out, the diaphragm returns to its at rest position, but does not play an active role in the process.

Pelvic Floor Muscles

The pelvic floor muscles are the coccygeus and levator ani muscles. The pelvic floor muscles straddle the pelvis, forming a muscular cradle for the pelvic organs. These muscles contract during the everyday actions of sneezing, laughing, and coughing. They control the orifices of the rectum, vagina, and urethra using sphincteric or constrictive actions, thus maintaining fecal and urinary continence. These muscles are also involved in assisting in increasing intra-abdominal pressure.

Groin

The region where the abdomen meets the thigh is known as the groin or inguinal region.

Groin

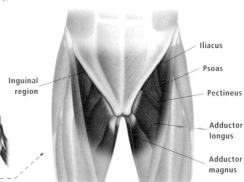

Inguinal region

Iliacus

Psoas

Pectineus

Adductor longus

Adductor magnus

Pelvic floor muscles

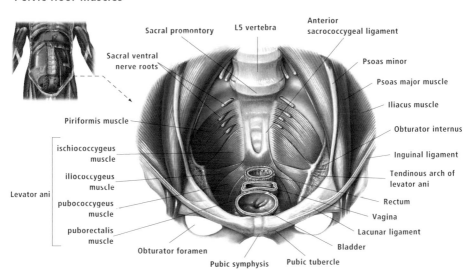

Sacral promontory

L5 vertebra

Anterior sacrococcygeal ligament

Sacral ventral nerve roots

Psoas minor

Psoas major muscle

Iliacus muscle

Piriformis muscle

Obturator internus

Levator ani

ischiococcygeus muscle

iliococcygeus muscle

pubococcygeus muscle

puborectalis muscle

Inguinal ligament

Tendinous arch of levator ani

Rectum

Vagina

Lacunar ligament

Bladder

Obturator foramen

Pubic symphysis

Pubic tubercle

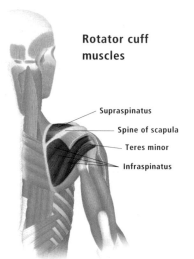

Rotator cuff muscles

Supraspinatus

Spine of scapula

Teres minor

Infraspinatus

Shoulder Muscles

The muscles associated with the shoulder fall into two groups: those attaching the humerus of the upper arm to the shoulder girdle, and those attaching the shoulder girdle to the trunk.

Muscles attaching the humerus to the shoulder girdle include the deltoid, pectoralis major, latissimus dorsi, teres major, and the rotator cuff muscles. The deltoid muscle is positioned on the outer side of the shoulder joint, gives the shoulder its contours, and is involved in most movements of the shoulder.

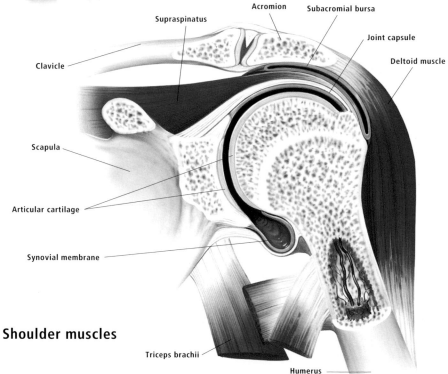

Supraspinatus

Acromion

Subacromial bursa

Joint capsule

Deltoid muscle

Clavicle

Scapula

Articular cartilage

Synovial membrane

Triceps brachii

Humerus

Shoulder muscles

Pectoral girdle

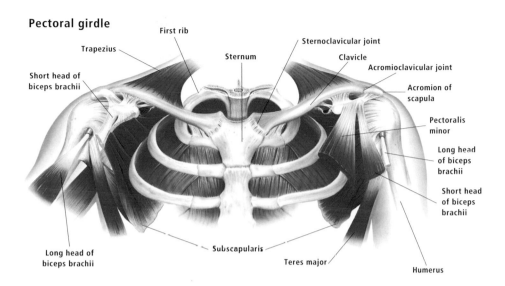

The rotator cuff muscles are a group of muscles comprising subscapularis, infraspinatus, supraspinatus, and teres minor. These muscles originate at the scapula (shoulder blade), running across its outer face, and attaching around the head of the humerus.

Muscles attaching the shoulder girdle to the trunk include the rhomboids and levator scapulae, the trapezius, latissimus dorsi, serratus anterior, and pectoralis minor.

The shoulder, the most mobile joint in the body, is capable of a wide range of movements. Generally, muscles that pass in front of the shoulder joint act to flex or rotate the humerus, those passing behind the shoulder extend and/or rotate the humerus, and those passing over the shoulder abduct the humerus.

Pectoralis major and the deltoid muscle are involved in flexion of the shoulder joint; teres major, latissimus dorsi, and the deltoid muscles are involved in extension; abduction is achieved by the movements of the deltoid, in conjunction with supraspinatus; adduction is primarily brought about by latissimus dorsi at the back and pectoralis major at the front; medial rotation involves the actions of subscapularis, latissimus dorsi, teres major, pectoralis major, and part of the deltoid muscle; and lateral rotation involves infraspinatus and teres minor. The trapezius muscle plays a role in bringing about such actions as shrugging of the shoulders, and in raising the arm above the head, requiring upward rotation of the scapula. Scapular rotation is also dependent on the involvement of serratus anterior.

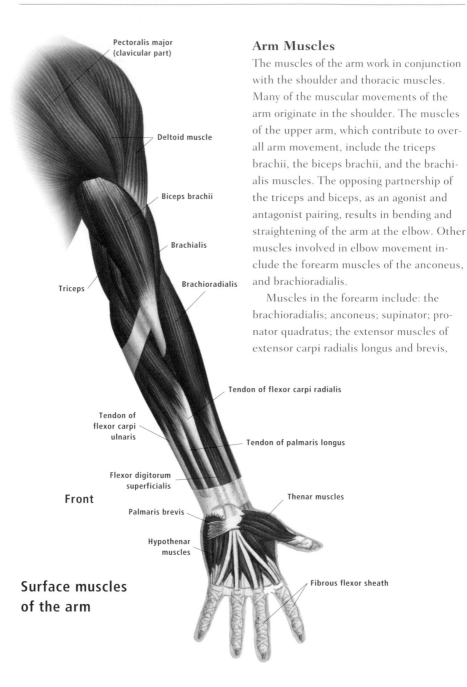

Pectoralis major
(clavicular part)

Deltoid muscle

Biceps brachii

Brachialis

Triceps

Brachioradialis

Tendon of flexor carpi radialis

Tendon of
flexor carpi
ulnaris

Tendon of palmaris longus

Flexor digitorum
superficialis

Front

Thenar muscles

Palmaris brevis

Hypothenar
muscles

**Surface muscles
of the arm**

Fibrous flexor sheath

Arm Muscles

The muscles of the arm work in conjunction
with the shoulder and thoracic muscles.
Many of the muscular movements of the
arm originate in the shoulder. The muscles
of the upper arm, which contribute to overall arm movement, include the triceps
brachii, the biceps brachii, and the brachialis muscles. The opposing partnership of
the triceps and biceps, as an agonist and
antagonist pairing, results in bending and
straightening of the arm at the elbow. Other
muscles involved in elbow movement include the forearm muscles of the anconeus,
and brachioradialis.

Muscles in the forearm include: the
brachioradialis; anconeus; supinator; pronator quadratus; the extensor muscles of
extensor carpi radialis longus and brevis,

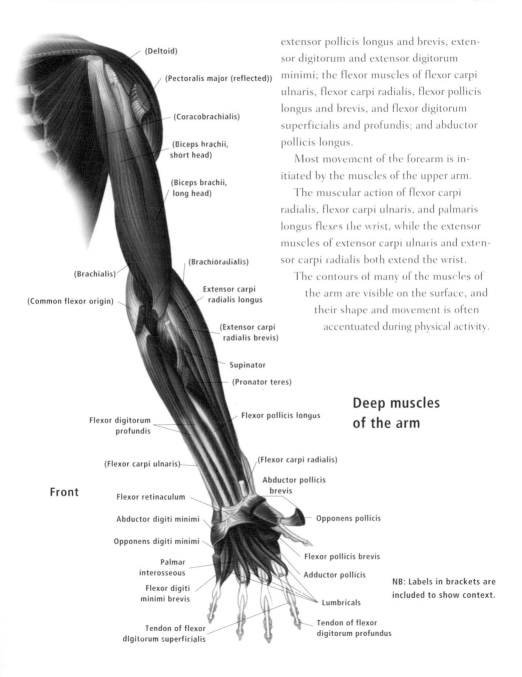

(Deltoid)

(Pectoralis major (reflected))

(Coracobrachialis)

(Biceps brachii, short head)

(Biceps brachii, long head)

(Brachioradialis)

(Brachialis)

Extensor carpi radialis longus

(Common flexor origin)

(Extensor carpi radialis brevis)

Supinator

(Pronator teres)

Flexor pollicis longus

Flexor digitorum profundis

(Flexor carpi ulnaris)

(Flexor carpi radialis)

Abductor pollicis brevis

Front

Flexor retinaculum

Abductor digiti minimi

Opponens pollicis

Opponens digiti minimi

Flexor pollicis brevis

Palmar interosseous

Adductor pollicis

Flexor digiti minimi brevis

Lumbricals

Tendon of flexor digitorum superficialis

Tendon of flexor digitorum profundus

extensor pollicis longus and brevis, extensor digitorum and extensor digitorum minimi; the flexor muscles of flexor carpi ulnaris, flexor carpi radialis, flexor pollicis longus and brevis, and flexor digitorum superficialis and profundis; and abductor pollicis longus.

Most movement of the forearm is initiated by the muscles of the upper arm.

The muscular action of flexor carpi radialis, flexor carpi ulnaris, and palmaris longus flexes the wrist, while the extensor muscles of extensor carpi ulnaris and extensor carpi radialis both extend the wrist.

The contours of many of the muscles of the arm are visible on the surface, and their shape and movement is often accentuated during physical activity.

Deep muscles of the arm

NB: Labels in brackets are included to show context.

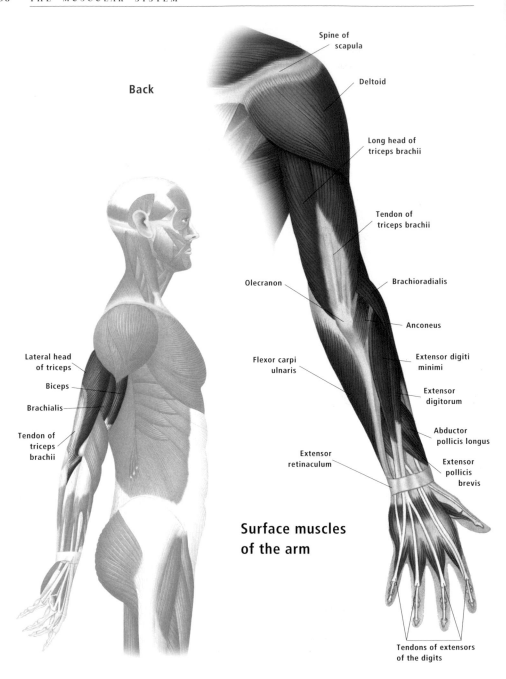

Back

Spine of scapula

Deltoid

Long head of triceps brachii

Tendon of triceps brachii

Brachioradialis

Olecranon

Anconeus

Extensor digiti minimi

Flexor carpi ulnaris

Extensor digitorum

Abductor pollicis longus

Extensor retinaculum

Extensor pollicis brevis

Lateral head of triceps

Biceps

Brachialis

Tendon of triceps brachii

Surface muscles of the arm

Tendons of extensors of the digits

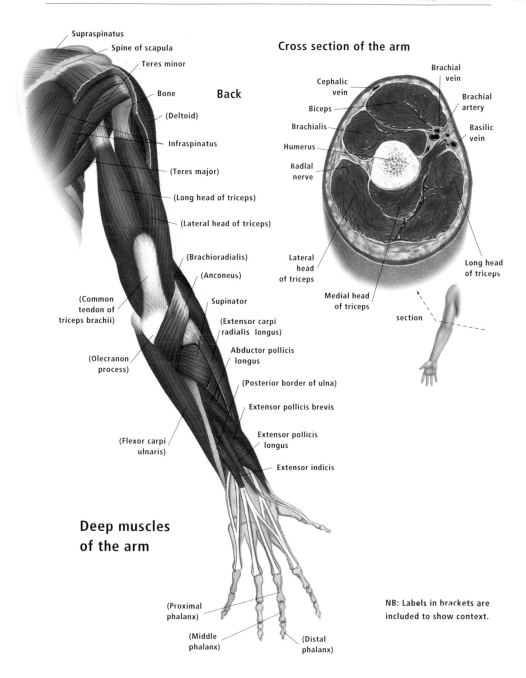

Supraspinatus

Spine of scapula

Teres minor

Bone

Back

(Deltoid)

Infraspinatus

(Teres major)

(Long head of triceps)

(Lateral head of triceps)

(Brachioradialis)

(Anconeus)

(Common tendon of triceps brachii)

Supinator

(Extensor carpi radialis longus)

(Olecranon process)

Abductor pollicis longus

(Posterior border of ulna)

Extensor pollicis brevis

Extensor pollicis longus

(Flexor carpi ulnaris)

Extensor indicis

Deep muscles of the arm

(Proximal phalanx)

(Middle phalanx)

(Distal phalanx)

Cross section of the arm

Brachial vein

Cephalic vein

Brachial artery

Biceps

Brachialis

Basilic vein

Humerus

Radial nerve

Lateral head of triceps

Long head of triceps

Medial head of triceps

section

NB: Labels in brackets are included to show context.

Elbow

The muscles of the upper arm work together to produce elbow movement. Triceps brachii and biceps brachii work together, with triceps brachii initiating elbow extension, while biceps brachii initiates elbow flexion and also supinates the forearm, that is, with the palm facing up.

The brachialis and brachioradialis muscles also contribute to elbow flexion, while the anconeus muscle contributes to elbow extension.

Contraction of the biceps muscle results in the arm bending at the elbow. In a reciprocal movement, when the biceps contracts, the triceps relaxes to allow ease of movement. In returning to its original position, the triceps contracts and the biceps relaxes. This partnership of opposing muscles, is known as agonist and antagonist.

Elbow

Elbow—front

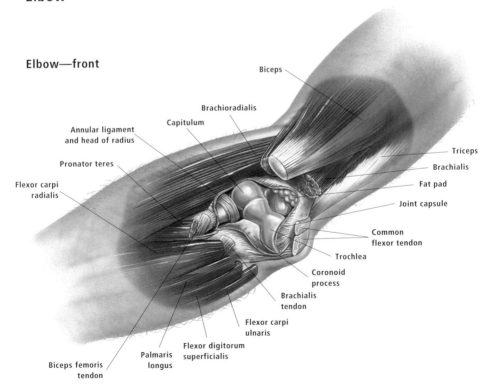

Biceps

Brachioradialis

Capitulum

Annular ligament
and head of radius

Pronator teres

Flexor carpi
radialis

Triceps

Brachialis

Fat pad

Joint capsule

Common
flexor tendon

Trochlea

Coronoid
process

Brachialis
tendon

Flexor carpi
ulnaris

Flexor digitorum
superficialis

Palmaris
longus

Biceps femoris
tendon

Elbow muscles

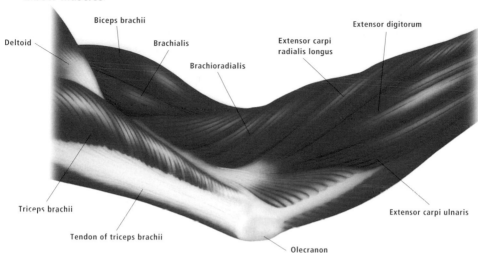

Deltoid
Biceps brachii
Brachialis
Brachioradialis
Extensor carpi radialis longus
Extensor digitorum
Triceps brachii
Tendon of triceps brachii
Olecranon
Extensor carpi ulnaris

Elbow—back

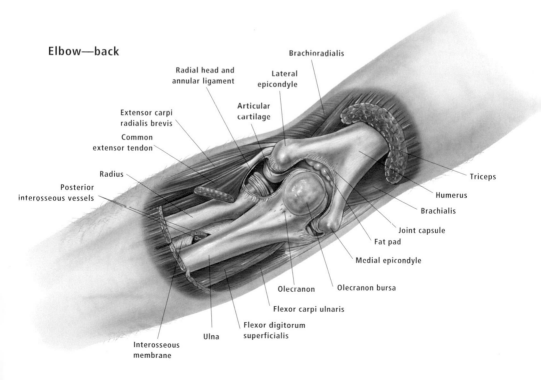

Radial head and annular ligament
Brachioradialis
Lateral epicondyle
Articular cartilage
Extensor carpi radialis brevis
Common extensor tendon
Radius
Posterior interosseous vessels
Triceps
Humerus
Brachialis
Joint capsule
Fat pad
Medial epicondyle
Olecranon bursa
Olecranon
Flexor carpi ulnaris
Interosseous membrane
Ulna
Flexor digitorum superficialis

Wrist and Hand Muscles

While some of the muscles of the forearm have points of attachment at the lower end of the forearm bones of the radius and ulna, others converge, with their tendons extending over the wrist area, to enable movements of the hand and fingers. The long flexor tendons to the fingers and thumb pass over the front of the wrist through tendon sheaths, which are designed to protect the tendons and reduce friction as they pass over the carpal bones. Passing over the back of the wrist, also through lubricating sheaths, are the tendons associated with extension of the fingers and thumb. The dexterity of the hand is controlled to a large extent by the muscles of the forearm, but there are also many muscles in the hand itself that contribute to the fine motor abilities of the hand.

The palm of the hand has a slightly concave surface, created in part by the thenar muscles that form the fleshy mound between the wrist and the thumb. These muscles play a part in thumb movement. At the

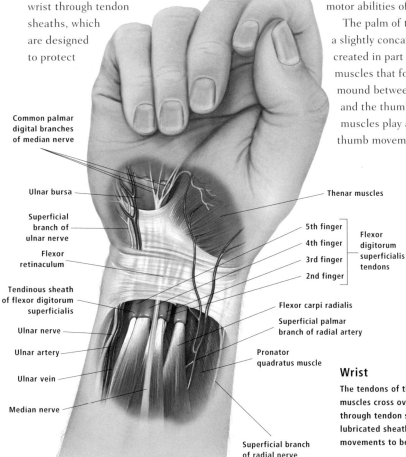

Common palmar digital branches of median nerve

Ulnar bursa

Superficial branch of ulnar nerve

Flexor retinaculum

Tendinous sheath of flexor digitorum superficialis

Ulnar nerve

Ulnar artery

Ulnar vein

Median nerve

Thenar muscles

5th finger
4th finger
3rd finger
2nd finger

Flexor digitorum superficialis tendons

Flexor carpi radialis

Superficial palmar branch of radial artery

Pronator quadratus muscle

Superficial branch of radial nerve

Wrist

The tendons of the forearm muscles cross over the wrist area through tendon sheaths. These lubricated sheaths allow smooth movements to be achieved.

side of the palm, between the wrist and the little finger, are the hypothenar muscles. Between each of the metacarpal bones of the palm of the hand are the interosseus muscles. These muscles are involved in moving the fingers apart and bringing them together, and also play a role in flexion and extension of the fingers. The lumbricals are a group of thin muscles that connect the flexor tendons and the phalanges, enabling finger movement. There are no muscles in the fingers themselves, with movement of the finger tendons being initiated by the muscles of the palm or forearm.

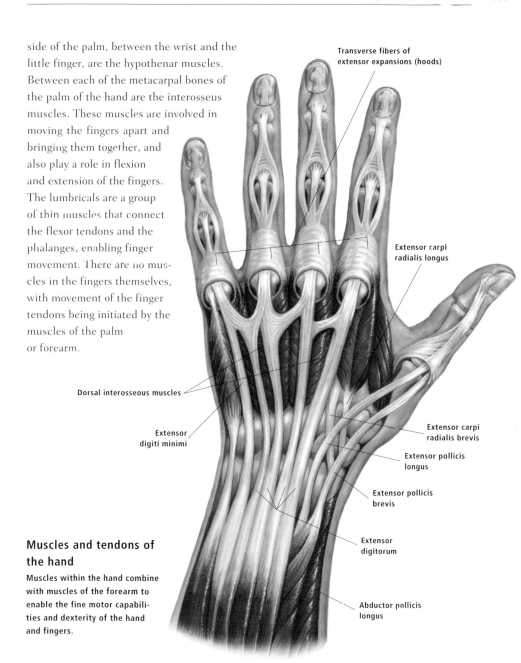

Transverse fibers of extensor expansions (hoods)

Extensor carpi radialis longus

Dorsal interosseous muscles

Extensor digiti minimi

Extensor carpi radialis brevis

Extensor pollicis longus

Extensor pollicis brevis

Extensor digitorum

Abductor pollicis longus

Muscles and tendons of the hand

Muscles within the hand combine with muscles of the forearm to enable the fine motor capabilities and dexterity of the hand and fingers.

Muscles of the leg—front

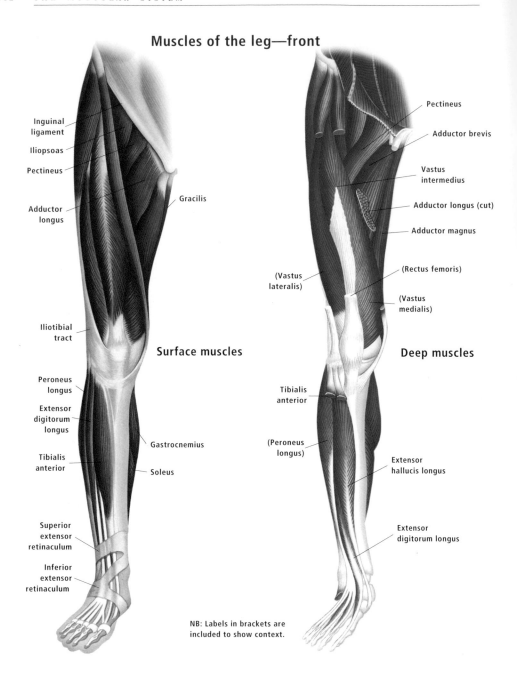

Inguinal
ligament

Iliopsoas

Pectineus

Adductor
longus

Gracilis

Iliotibial
tract

Surface muscles

Peroneus
longus

Extensor
digitorum
longus

Tibialis
anterior

Gastrocnemius

Soleus

Superior
extensor
retinaculum

Inferior
extensor
retinaculum

Pectineus

Adductor brevis

Vastus
intermedius

Adductor longus (cut)

Adductor magnus

(Vastus
lateralis)

(Rectus femoris)

(Vastus
medialis)

Deep muscles

Tibialis
anterior

(Peroneus
longus)

Extensor
hallucis longus

Extensor
digitorum longus

NB: Labels in brackets are
included to show context.

Leg Muscles

Weight transference from the pelvis to the legs means that the legs must carry the weight of the upper body—added to this is the responsibility for locomotion. As a result, the muscles of the legs are particularly powerful in order to fulfill these requirements.

The hip region contains powerful muscles, including the largest muscle in the body, gluteus maximus, which gives the buttocks their familiar shape. Gluteus maximus extends the thigh, while the other gluteal muscles, gluteus medius and minimus, work to keep the pelvis level and swing the opposite side forward during walking.

The thigh muscles include the quadriceps muscle at the front, and the hamstring muscles and adductor muscles at the back.

The quadriceps muscle and hamstrings extend and flex the knee respectively, while the adductor muscles hold the leg toward the midline.

The muscles of the lower leg are separated by deep fascia into three compartments, with each compartment contributing to the various movements of the foot. The foot is moved upward by the muscles of the front compartment; the sole is moved outward by the muscles of the lateral compartment. In the rear compartment the muscles are divided into a superficial group and a deep group. The foot is moved downward by the muscles of the superficial group, while the muscles of the deep group contribute to the action of pushing off from the big toe during walking.

Leg cross section

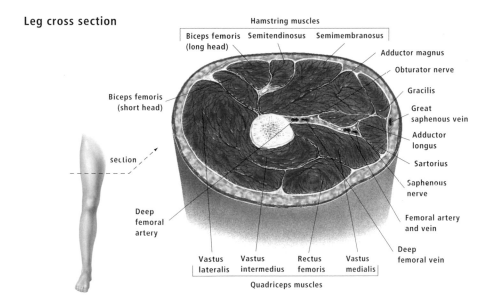

Hamstring muscles

Biceps femoris (long head)   Semitendinosus   Semimembranosus

Adductor magnus

Obturator nerve

Gracilis

Biceps femoris (short head)

Great saphenous vein

Adductor longus

Sartorius

section

Saphenous nerve

Femoral artery and vein

Deep femoral artery

Deep femoral vein

Vastus lateralis   Vastus intermedius   Rectus femoris   Vastus medialis

Quadriceps muscles

Muscles of the leg—back

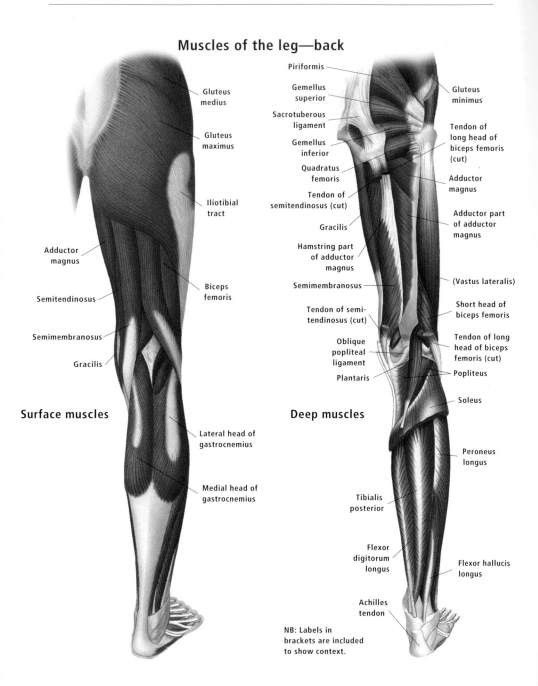

Gluteus medius

Gluteus maximus

Iliotibial tract

Adductor magnus

Semitendinosus

Semimembranosus

Gracilis

Biceps femoris

Surface muscles

Lateral head of gastrocnemius

Medial head of gastrocnemius

Piriformis

Gemellus superior

Sacrotuberous ligament

Gemellus inferior

Quadratus femoris

Tendon of semitendinosus (cut)

Gracilis

Hamstring part of adductor magnus

Semimembranosus

Tendon of semi-tendinosus (cut)

Oblique popliteal ligament

Plantaris

Gluteus minimus

Tendon of long head of biceps femoris (cut)

Adductor magnus

Adductor part of adductor magnus

(Vastus lateralis)

Short head of biceps femoris

Tendon of long head of biceps femoris (cut)

Popliteus

Soleus

Deep muscles

Peroneus longus

Tibialis posterior

Flexor digitorum longus

Flexor hallucis longus

Achilles tendon

NB: Labels in brackets are included to show context.

Quadriceps muscle

The thigh muscles provide strength for locomotion, and give stability to the important knee joint. The muscles of the thigh region are separated into compartments. The quadriceps muscle lies in the front compartment, forms the mass of the front and side of the region, and is responsible for extending or straightening the knee.

The quadriceps muscle gets its name from the Latin, meaning "four heads," and it is indeed comprised of four parts, namely: rectus femoris, vastus lateralis, vastus medialis, and vastus intermedius.

The quadriceps muscle originates at the upper end of the femur, widening to form the bulk of the front and side of the thigh, before the four parts taper and merge to

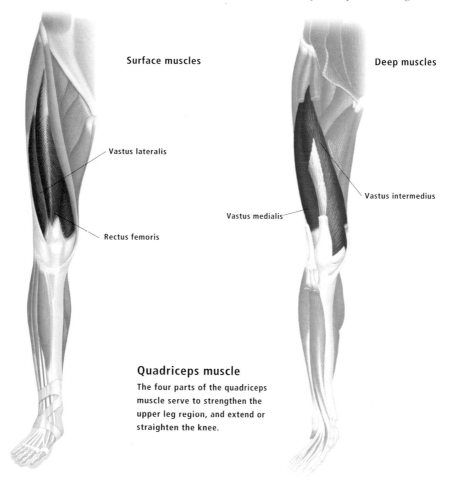

Surface muscles

Vastus lateralis

Rectus femoris

Deep muscles

Vastus intermedius

Vastus medialis

Quadriceps muscle

The four parts of the quadriceps muscle serve to strengthen the upper leg region, and extend or straighten the knee.

become a single tendon. The tendon extends across the knee joint, down into the lower leg, joining to the tibia. The patella (kneecap) is attached to the tendon at the knee joint, with the movements of the quadriceps muscle group facilitating the gliding movement of the patella over the femur in movement of the knee joint.

Hamstring muscles

The group of muscles known as the hamstring muscles is located at the back of the upper leg, above the knee. This group of muscles is involved in straightening the hip joint and bending the knee joint.

The hamstrings are comprised of three muscles: the semimembranosus, the semitendinosus, and biceps femoris. These three muscles originate at the pelvis—attaching at the ischium, they pass behind the hip joint and serve to straighten the hip. As they extend down from the hip, the muscles diverge, with the tendon of biceps femoris passing behind the outside of the knee and attaching to the fibula of the lower leg. The tendons of semimembranosus and semitendinosus pass behind the inside of the knee and attach to the tibia.

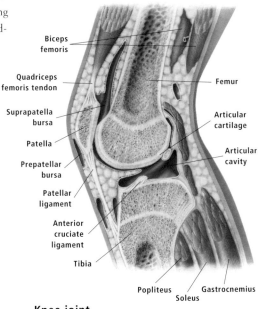

Biceps femoris

Quadriceps femoris tendon

Suprapatella bursa

Patella

Prepatellar bursa

Patellar ligament

Anterior cruciate ligament

Tibia

Biceps femoris

Semitendinosus

Semimembranosus

Femur

Articular cartilage

Articular cavity

Popliteus

Soleus

Gastrocnemius

Hamstring muscles

The group of three muscles known as the hamstrings work to straighten the hip joint and bend the knee joint.

Knee joint

The strong muscles of the upper leg are primarily responsible for initiating knee joint movements.

Muscles of the lower leg

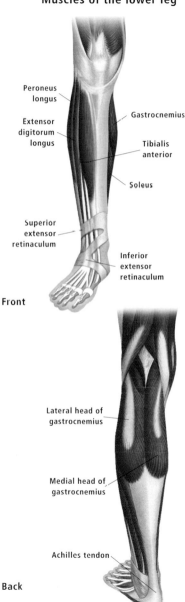

Peroneus longus

Gastrocnemius

Extensor digitorum longus

Tibialis anterior

Soleus

Superior extensor retinaculum

Inferior extensor retinaculum

Front

Lateral head of gastrocnemius

Medial head of gastrocnemius

Achilles tendon

Back

Muscles of the lower leg

The muscles of the lower leg are compartmentalized by sheets of connective tissue. The muscles of this region are thus divided into three groups.

The anterior compartment, at the front of the leg, contains the muscles of tibialis anterior, extensor hallucis longus, and extensor digitorum longus. These muscles play a role in upward movement (dorsiflexion) of the foot, raising the foot at the ankle when walking.

The lateral compartment, on the outer side of the leg, contains the muscles of peroneus longus and peroneus brevis. These muscles attach to the fibula at the top, and run down the leg, passing through tendon sheaths at the ankle, and join to the fifth metatarsal of the foot. Movements provided by these muscles include plantarflexion and eversion of the foot.

At the back of the leg, the posterior compartment is divided into two sections: the superficial posterior compartment and the deep posterior compartment. Within the superficial posterior compartment are the muscles of gastrocnemius, soleus, and plantaris, while the deep posterior compartment contains the muscles of flexor hallucis longus, flexor digitorum longus, tibialis posterior, and popliteus. The muscles of the superficial posterior compartment enable plantarflexion of the ankle, lowering the foot at the ankle joint, while the muscles of the deep posterior compartment enable the foot to push off from the ground.

Muscles of the foot

Foot—front

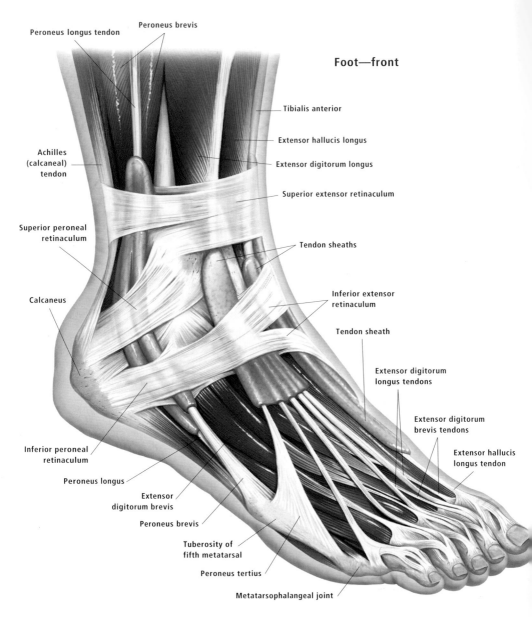

Peroneus longus tendon

Peroneus brevis

Achilles (calcaneal) tendon

Superior peroneal retinaculum

Calcaneus

Inferior peroneal retinaculum

Peroneus longus

Extensor digitorum brevis

Peroneus brevis

Tuberosity of fifth metatarsal

Peroneus tertius

Metatarsophalangeal joint

Tibialis anterior

Extensor hallucis longus

Extensor digitorum longus

Superior extensor retinaculum

Tendon sheaths

Inferior extensor retinaculum

Tendon sheath

Extensor digitorum longus tendons

Extensor digitorum brevis tendons

Extensor hallucis longus tendon

Muscles of the Ankle and Foot

The ankle joins the lower leg to the foot. Many of the muscles of the lower leg converge at the ankle, thinning into tendons, which run across the front, back, and sides of the ankle joint. These tendons are carried across the joint in tendons sheaths. The sheaths produce synovial fluid to lubricate the passage to reduce friction, allow smooth movements, and protect the tendons. The tendons then extend to their points of attachment throughout the foot. Running behind the ankle is perhaps the best-known tendon of the lower leg area, the Achilles tendon. Extending from the gastrocnemius muscle in the calf region of the lower leg, this tendon attaches to the calcaneus (heel).

The movements of the foot include dorsiflexion (feet point upward), plantarflexion (feet point down), inversion or supination (sole faces inward), and eversion or pronation (sole faces outward). The main muscles involved in dorsiflexion of the ankle lie at the front of the lower leg, and the muscles producing plantarflexion lie at the back of the lower leg.

Interosseus muscles, which serve to flex the toes, lie between each of the metatarsal bones of the foot. There are no muscles in the phalanges of the toes, with movement provided by tendons that attach to the toe bones. The tendons of the flexor muscles provide flexion of the toes, while the tendons of the extensor muscles extend the toes.

Flexor hallucis longus muscle

Tibialis posterior muscle

Flexor digitorum longus muscle

Tibia

Flexor digitorum longus tendon

Tibialis posterior tendon

Posterior tibial artery

Tibial nerve

Flexor retinaculum

First metatarsal

Calcaneal tuberosity

Fibula

Peroneus longus tendon

Flexor hallicus longus tendon

Achilles (calcaneal) tendon

Foot—back

Muscles of the soles of the feet

Long tendons extend along the sole of the foot, which also contains a number of small muscles that are arranged in four layers.

Originating at the calcaneus, and fanning out over the sole of the foot at the superficial level, is flexor digitorum brevis. This muscle separates into tendons, which extend to the middle phalanges of the four

Muscles of the soles of the feet

Surface muscles

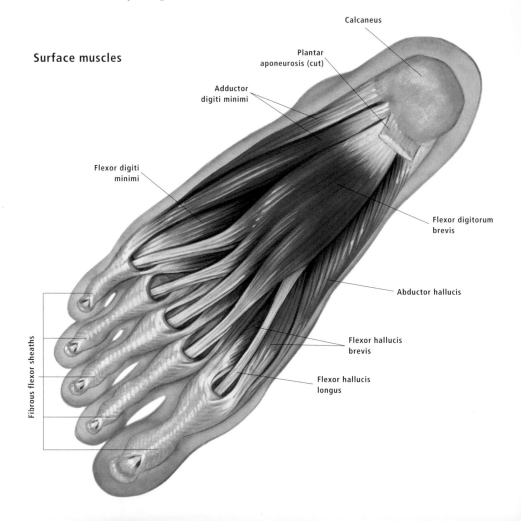

Calcaneus

Plantar aponeurosis (cut)

Adductor digiti minimi

Flexor digiti minimi

Flexor digitorum brevis

Abductor hallucis

Flexor hallucis brevis

Flexor hallucis longus

Fibrous flexor sheaths

small toes. Underlying these tendons are the tendons of flexor digitorum longus. This muscle originates at the lower part of the tibia, and the tendons join to the distal phalanges of the four small toes. The tendons of flexor hallucis longus and brevis attach to the big toe, allowing independent flexion of the big toe.

The small muscles of the lumbricals assist in joint movement between the metatarsal bones and the phalanges of the toes. The abductor muscles of abductor hallucis and abductor digiti minimi abduct the big and small toe respectively. The adductor hallucis muscle provides flexion and adduction to the big toe.

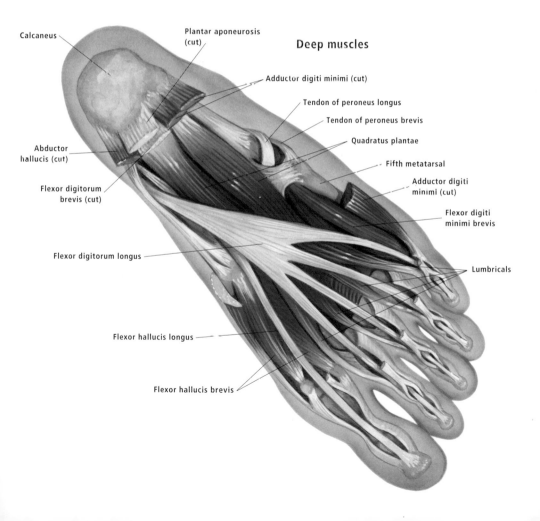

Calcaneus

Plantar aponeurosis (cut)

Deep muscles

Adductor digiti minimi (cut)

Tendon of peroneus longus

Tendon of peroneus brevis

Quadratus plantae

Abductor hallucis (cut)

Fifth metatarsal

Adductor digiti minimi (cut)

Flexor digitorum brevis (cut)

Flexor digiti minimi brevis

Flexor digitorum longus

Lumbricals

Flexor hallucis longus

Flexor hallucis brevis

Diseases and Disorders of the Muscular System

Problems of the muscular system can range from a simple, though painful, pulled muscle, to muscle wasting diseases with serious complications. Several of the more common diseases and disorders of the muscular system are discussed below.

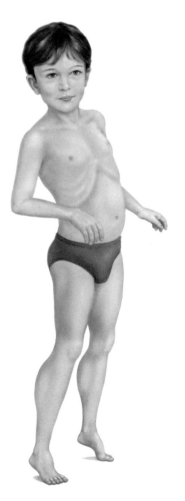

Muscular dystrophy

A group of inherited disorders, muscular dystrophy affects the muscular system, causing gradual loss of muscle bulk and weakening of muscles.

Painless in itself, muscular dystrophy is characterized by the gradual deterioration of muscle function, eventually presenting problems in performing simple everyday tasks.

Different types of muscular dystrophy affect different areas of the body, such as the shoulders, hips, and face. Some affect the organs such as the heart, causing complications such as cardiomyopathy.

The disease is more common to men than women, and a particularly serious form is Duchenne muscular dystrophy. Generally affecting young boys, commencing in infancy or early childhood, this form of the disease causes muscle degeneration and wasting, often affecting the weight-bearing limbs of the body, and can also affect the heart muscles. It is caused by a defective gene carried by the mother. While the mother is unaffected by the disease, there is an even chance that the disease may affect any sons she may produce.

Signs of the disease include bulging muscles, which indicate that protein has been lost from muscle and replaced by fat cells.

There is no known cure for muscular dystrophy, and life expectancy is reduced. Physical therapy can help to maintain some muscle function. Aids such as wheelchairs and braces may be needed to improve mobility.

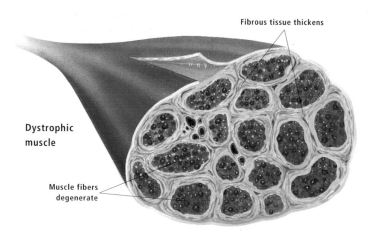

Fibrous tissue thickens

Dystrophic
muscle

Muscle fibers
degenerate

Muscular dystrophy

Muscular dystrophy is a group of inherited diseases,
affecting men more often than women. The various
types of muscular dystrophy can affect different
areas of the body, causing muscle degeneration
and wasting.

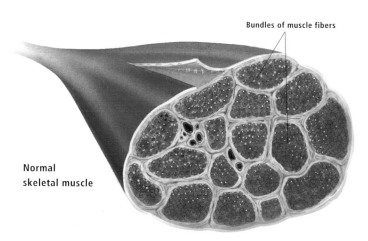

Bundles of muscle fibers

Normal
skeletal muscle

Problems of the Achilles tendon

Rupture of the Achilles tendon is a common problem encountered by athletes and middle-aged men. When placed under stress, the tendon can snap, causing immediate pain.

If the tendon is ruptured, the usual procedure is to surgically repair the tendon. The leg requires a plaster cast following surgery.

An alternative is to place the foot in a cast, immobilizing the area, and allowing the two ends of the tendon to heal. Whichever option is selected, physical therapy is usually advised after removal of the plaster cast, to restore full mobility and use of the tendon. However, full recovery can take up to 6 months.

Another problem that can affect the Achilles tendon is Achilles tendinitis. Overuse of the tendon, or footwear that places the tendon under abnormal stress, can both contribute to Achilles tendinitis, a painful inflammation of the tendon. The inflammation can then render the tendon more susceptible to rupture.

The condition is relieved with rest, painkillers, and anti-inflammatory drugs. The sufferer is, however, likely to have the condition recur.

Prevention is the best solution, with warm-up exercises recommended before undertaking sporting activities, followed by simple stretching exercises afterward. Sensible, supportive footwear is also advisable to avoid recurrent problems.

Tendinitis

As with Achilles tendinitis, tendinitis is the inflammation of a tendon, usually affecting those used in repetitive actions or under strenuous physical activity. Tendinitis causes pain and swelling and can limit movement in the affected area.

Apart from the Achilles tendon, the tendons of the shoulders, elbows, and knees are most often affected, causing conditions such as painful arc syndrome, tennis or golfer's elbow, and jumper's knee respectively. Rest, often in conjunction with temporary immobilization of the affected area; anti-inflammatory drugs; and the use of heat or cold can reduce the pain and inflammation.

Inflammation affecting the tendon sheath is called tenosynovitis. Carpal tunnel syndrome is an example of this condition.

Double vision

When the muscles of the eye malfunction, this can cause double vision. In order for the brain to receive a clear image, both eyes must move in unison.

Double vision occurs when the muscles of the eye are weakened or unable to move due to nerve damage, resulting in lack of coordinated eye movement. When this happens, the good eye focuses on a seen object, but the affected eye is focused elsewhere. As a result, the brain receives two conflicting images, which it is unable to merge into one clear image, thus causing the dual images experienced in double vision.

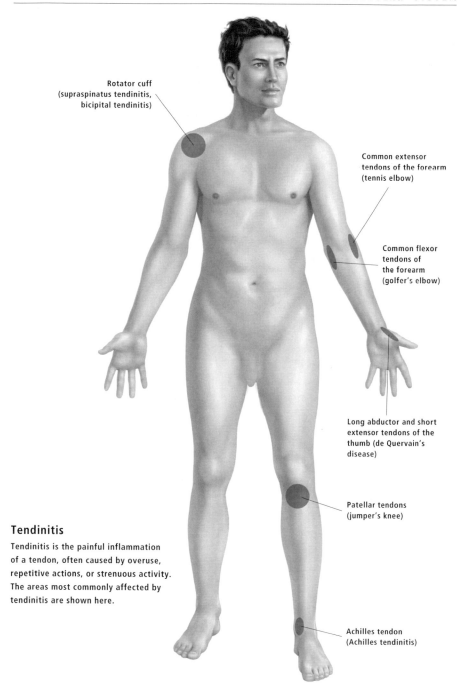

Rotator cuff
(supraspinatus tendinitis,
bicipital tendinitis)

Common extensor
tendons of the forearm
(tennis elbow)

Common flexor
tendons of
the forearm
(golfer's elbow)

Long abductor and short
extensor tendons of the
thumb (de Quervain's
disease)

Patellar tendons
(jumper's knee)

Achilles tendon
(Achilles tendinitis)

Tendinitis

Tendinitis is the painful inflammation
of a tendon, often caused by overuse,
repetitive actions, or strenuous activity.
The areas most commonly affected by
tendinitis are shown here.

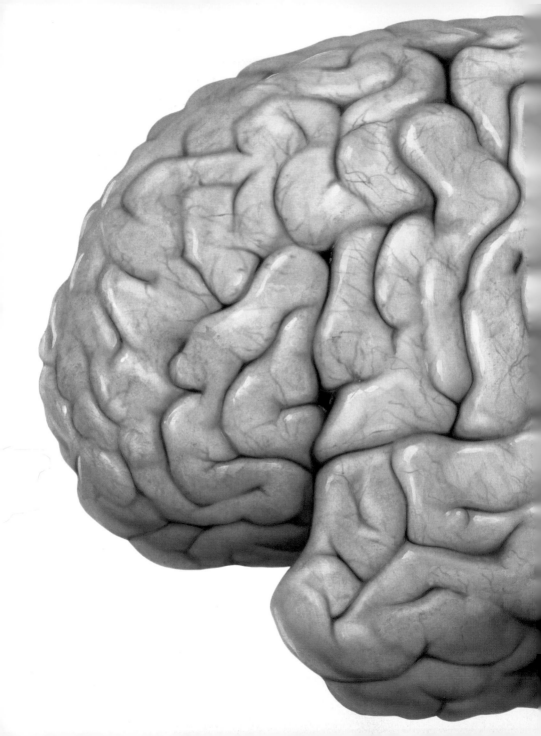

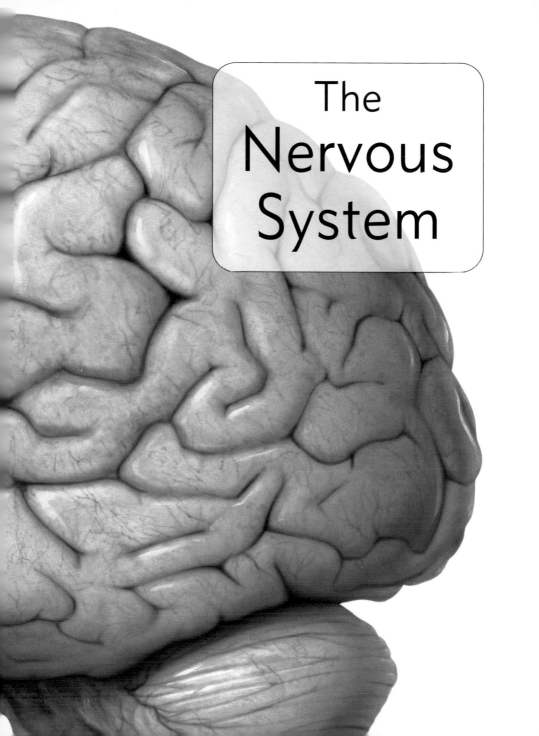

The
Nervous
System

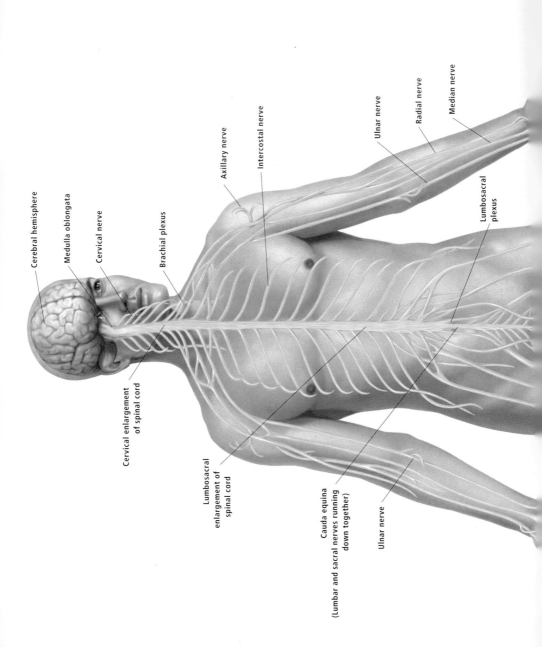

Cerebral hemisphere

Medulla oblongata

Cervical nerve

Brachial plexus

Cervical enlargement of spinal cord

Lumbosacral enlargement of spinal cord

Cauda equina (Lumbar and sacral nerves running down together)

Ulnar nerve

Axillary nerve

Intercostal nerve

Ulnar nerve

Radial nerve

Median nerve

Lumbosacral plexus

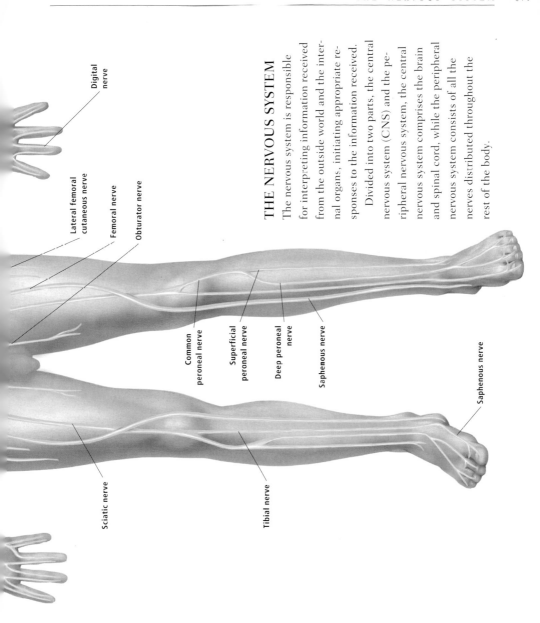

Digital nerve

Lateral femoral cutaneous nerve

Femoral nerve

Obturator nerve

Common peroneal nerve

Superficial peroneal nerve

Deep peroneal nerve

Saphenous nerve

Saphenous nerve

Sciatic nerve

Tibial nerve

THE NERVOUS SYSTEM

The nervous system is responsible for interpreting information received from the outside world and the internal organs, initiating appropriate responses to the information received.

Divided into two parts, the central nervous system (CNS) and the peripheral nervous system, the central nervous system comprises the brain and spinal cord, while the peripheral nervous system consists of all the nerves distributed throughout the rest of the body.

The brain initiates rapid response to information received from the nerve cells. Information is sent from the nerves using electrical signals called action potentials. These signals travel along nerve fibers to the spinal cord and are relayed to the brain. The signals travel at rapid speeds of up to 320 feet (100 meters) per second.

Information is sent to the cerebral cortex, where the brain disseminates the incoming information. Depending on the information received, different areas of the brain are responsible for information translation and response, and interpretation of motor sensory information.

Some parts of the peripheral nervous system govern those body functions that are not under our voluntary control. These nerves belong to the autonomic nervous system, which, like the nervous system, is under the control of the brain.

Nerves

Nerves carry signals regarding sensations or muscle movements from the peripheral nervous system to the brain via the spinal cord. The peripheral nerves are distributed throughout the body, while the cranial nerves arise in the brain and control the nerves of the sense organs. The spinal nerves, which arise from the spinal cord, serve the trunk and limbs.

Neurons

Nerve cells, or neurons, are found in the brain, spinal cord, and nerves. These cells control information processing, muscle and gland activity, and sensation interpretation.

Neuron

Each neuron has a cell body, branching projections (dendrites) that carry impulses to the cell body, and a single elongated projection called an axon that carries impulses away from the cell body.

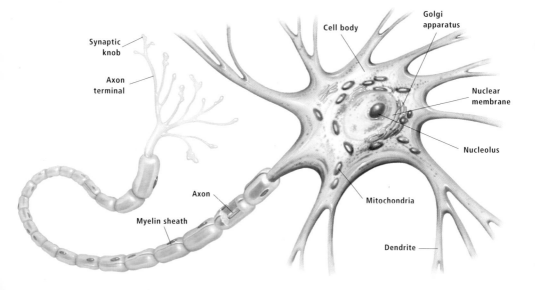

Synaptic knob

Axon terminal

Cell body

Golgi apparatus

Nuclear membrane

Nucleolus

Mitochondria

Axon

Myelin sheath

Dendrite

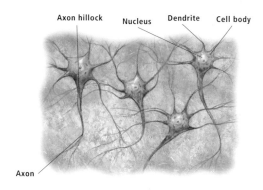

Axon hillock • Nucleus • Dendrite • Cell body

Axon

Neural tissue

Neural tissue processes neural data and conducts electrical impulses from one part of the body to another part. Ninety-eight percent of neural tissue is located in the brain and spinal cord.

The structure of the nerve cell includes: a cell body, within which are the cell nucleus and the organelles, and one or more projections, known as processes, extending out from the cell body. These processes include dendrites—responsible for conveying incoming information to the cell body, and axons—responsible for conveying outward responses from the cell body. In general, each neuron has a single elongated axon and a variable number of dendrites, depending on the cell's function. Many axons have a myelin coating, which improves the conductivity of nerve impulses.

Synapses

As axons convey information from their cell body, and dendrites receive incoming information, the point where two cells meet

is where the information is relayed. This junction between the two processes—the dendrite and the axon—is called a synapse. Information is sent from the sending cell to the receiving cell by the transmission of chemical molecules called neurotransmitters across the synaptic cleft between the two cells.

Synapse

The junction where two nerve cells meet is called a synapse. Nerve signals are passed from one cell to another by the release of chemical molecules called neurotransmitters. These neurotransmitters cross the synaptic cleft and lock into specific receptor sites, thus allowing an electrical charge to take place, which results in the relay of the nerve signal.

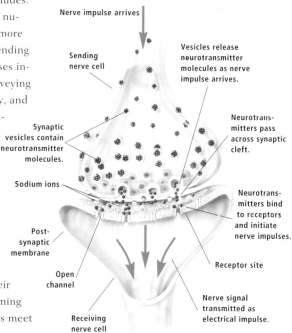

Nerve impulse arrives

Sending nerve cell

Vesicles release neurotransmitter molecules as nerve impulse arrives.

Synaptic vesicles contain neurotransmitter molecules.

Neurotransmitters pass across synaptic cleft.

Sodium ions

Neurotransmitters bind to receptors and initiate nerve impulses.

Post-synaptic membrane

Open channel

Receptor site

Receiving nerve cell

Nerve signal transmitted as electrical impulse.

Reflexes

Reflex actions take place outside our conscious control. When prompted by specific stimuli, our body replies with an immediate response. For instance, when we touch a sharp or hot object, there is an immediate response to move away from the object. This happens at rapid speed, as the information on the stimulus is relayed from the peripheral nerves to a nerve center in the spinal cord and on to the brain.

These nerve impulses can reach speeds of up to 320 feet (100 meters) per second, so a quick response can be initiated auto-matically. The four components of a reflex are: reception, conduction, transmission, and response.

Proprioception

Proprioception is one's own awareness of body position and movement. Special nerve cells called proprioceptors convey information to the brain regarding the overall body state. Sensors in the joints and muscles relay information on status and position, while the sense of balance, initiated in the inner ear by the organs of the labyrinth, conveys information on body position.

Reflexes

Reflexes are rapid and automatic responses to stimuli. Impulses are sent from the stimulus point to a nerve center in the spinal cord. A response to the impulse is initiated by the nerve center, causing a muscle movement. These reactions occur without the person being consciously aware of them.

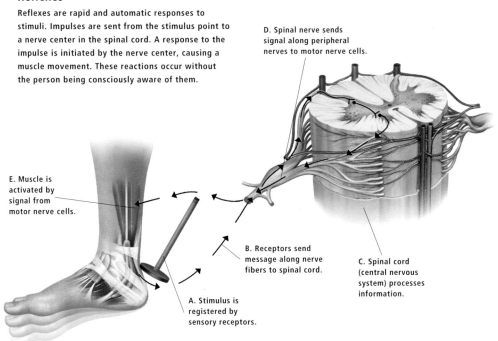

D. Spinal nerve sends signal along peripheral nerves to motor nerve cells.

E. Muscle is activated by signal from motor nerve cells.

B. Receptors send message along nerve fibers to spinal cord.

C. Spinal cord (central nervous system) processes information.

A. Stimulus is registered by sensory receptors.

Pain

Pain is the body's signal to the brain, alerting it of real, impending, or potential injury or damage.

Specialized nerve cells called nociceptors are positioned in the body, primarily in the skin, muscles, joints, and internal organs. These nociceptors have varying degrees of sensitivity, depending on their location. Nociceptors relay information on pain to the spinal nerves, which in turn relay the information on to the brain via the thalamus. Information on pain is registered in the sensory cortex, while emotional responses are regulated by the limbic system.

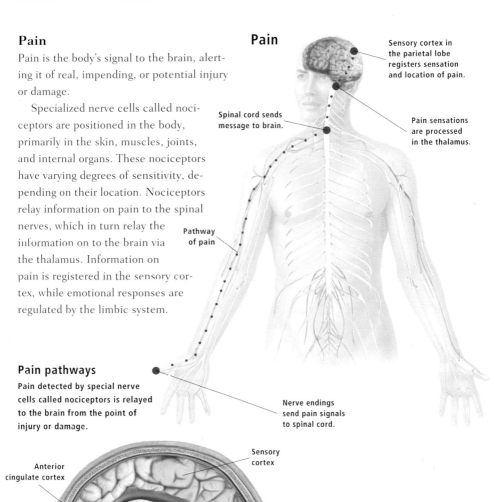

Pain

Sensory cortex in the parietal lobe registers sensation and location of pain.

Pain sensations are processed in the thalamus.

Spinal cord sends message to brain.

Pathway of pain

Pain pathways

Pain detected by special nerve cells called nociceptors is relayed to the brain from the point of injury or damage.

Nerve endings send pain signals to spinal cord.

Anterior cingulate cortex

Sensory cortex

Limbic system

Thalamus

Reticular activating system

Pain processing

The brain controls our perception of pain in a number of different areas. Pain sensation and location is registered in the sensory cortex and emotional responses are governed by the limbic system.

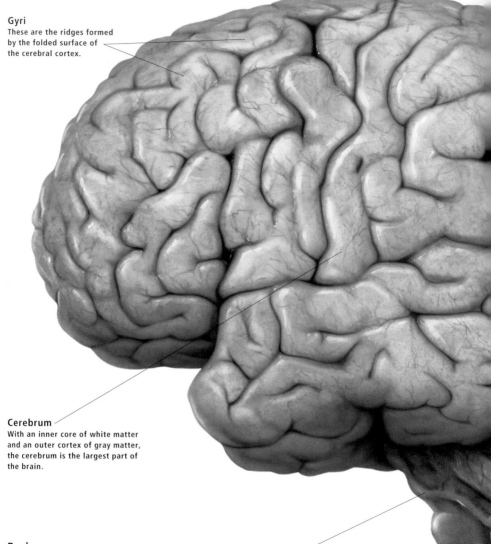

Gyri
These are the ridges formed
by the folded surface of
the cerebral cortex.

Cerebrum
With an inner core of white matter
and an outer cortex of gray matter,
the cerebrum is the largest part of
the brain.

Brain

Compactly packaged within the protective shell
of the skull, the brain is a powerhouse of activity
where billions of brain cells interact to maintain
all our vital body functions.

Brain stem
Consisting of three parts, the midbrain, pons,
and medulla, the brain stem is continuous
with the spinal cord below. The brain stem is
involved in regulation of such vital functions
as breathing, heartbeat, and blood pressure.

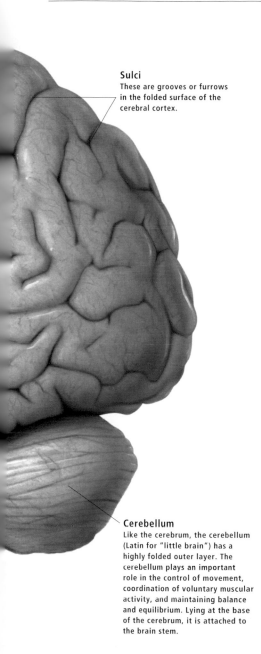

Sulci
These are grooves or furrows in the folded surface of the cerebral cortex.

Cerebellum
Like the cerebrum, the cerebellum (Latin for "little brain") has a highly folded outer layer. The cerebellum plays an important role in the control of movement, coordination of voluntary muscular activity, and maintaining balance and equilibrium. Lying at the base of the cerebrum, it is attached to the brain stem.

Brain

The brain is the center of activity for the nervous system. It is here that nerve signals from throughout the body are received, processed, and acted upon with appropriate responses. The brain controls sensory and motor activities. It is the center of thinking, memory, auditory and visual association, and emotion, and governs our muscular actions, and therefore, our movement. Information from the special sense organs related to sight, hearing, taste, smell, and balance is interpreted here. The brain controls all the vital body functions, from the rhythmic beating of the heart to the blinking of the eye. Like a 24-hour surveillance watchdog, the brain is always on duty, monitoring all of our body systems and functions, maintaining maximum efficiency, preempting potential problems, and acknowledging and counterstriking real dangers, damage, and injury.

The folded surface of the brain creates a massive surface area for neural activity, with billions of nerve cells (neurons) and supporting cells (glia) making up the substance of the brain.

Much of the neural activity takes place in the gray matter, which is found primarily in the cerebral cortex. The gray matter is composed of the cell bodies of neurons, while the white matter is made up of the processes of the nerve cells.

The human brain is generally viewed in four parts: the cerebrum, diencephalon, brain stem, and cerebellum.

Fetal Brain Development

During the third week of gestation, development of the brain and spinal cord begins. At this time, a shallow groove forms along the back of the embryonic disk, gradually becoming deeper until the edges meet and fuse together to form a tube. This neural tube forms into the prosencephalon, the mesencephalon, and the rhombencephalon. The fusion of the neural tube starts in the neck region and extends forward and backward.

By the end of week 4 the tube is fully enclosed, and the prosencephalon has developed into the telencephalon and diencephalon, while the rhombencephalon has become the myelencephalon and metencephalon. The telencephalon further develops into the two cerebral hemispheres.

Once the neural tube has formed, there is further extensive growth and development of the CNS, which continues during the first two years of postnatal life.

Some nerves grow out from the CNS to innervate muscles and organs of the body. Other nerves, including those carrying sensory information, develop separately and later make communication with the appropriate parts of the CNS.

By the end of the gestation period, the brain has all the surface features of the adult brain.

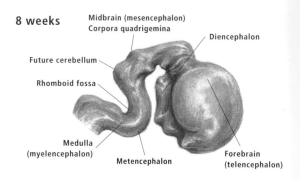

8 weeks

Midbrain (mesencephalon)
Corpora quadrigemina
Diencephalon
Future cerebellum
Rhomboid fossa
Medulla (myelencephalon)
Metencephalon
Forebrain (telencephalon)

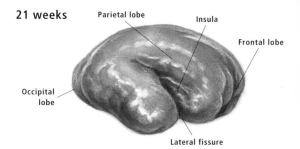

21 weeks

Parietal lobe
Insula
Frontal lobe
Occipital lobe
Lateral fissure

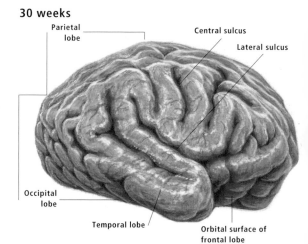

30 weeks

Parietal lobe
Central sulcus
Lateral sulcus
Occipital lobe
Temporal lobe
Orbital surface of frontal lobe

11 weeks

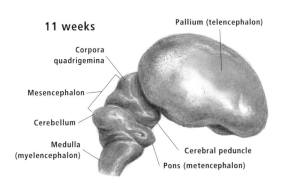

Pallium (telencephalon)

Corpora quadrigemina

Mesencephalon

Cerebellum

Medulla (myelencephalon)

Cerebral peduncle

Pons (metencephalon)

Fetal brain development

Over the gestation period, the brain develops from a simple groove in the embryonic disk into a functioning center for the central nervous system with all the surface features of the adult brain.

26 weeks

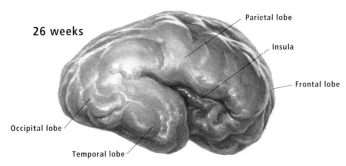

Parietal lobe

Insula

Frontal lobe

Occipital lobe

Temporal lobe

40 weeks

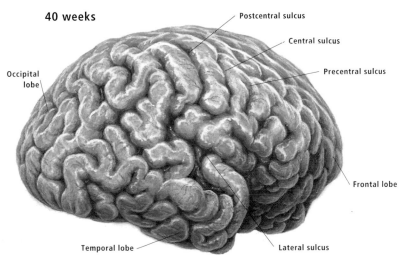

Postcentral sulcus

Central sulcus

Precentral sulcus

Occipital lobe

Frontal lobe

Temporal lobe

Lateral sulcus

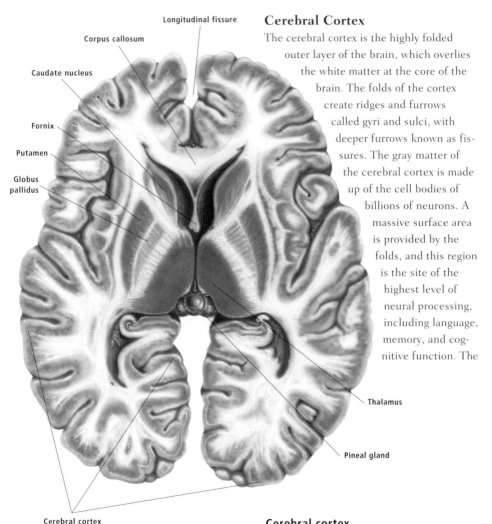

Longitudinal fissure

Corpus callosum

Caudate nucleus

Fornix

Putamen

Globus pallidus

Thalamus

Pineal gland

Cerebral cortex

Cerebral Cortex

The cerebral cortex is the highly folded outer layer of the brain, which overlies the white matter at the core of the brain. The folds of the cortex create ridges and furrows called gyri and sulci, with deeper furrows known as fissures. The gray matter of the cerebral cortex is made up of the cell bodies of billions of neurons. A massive surface area is provided by the folds, and this region is the site of the highest level of neural processing, including language, memory, and cognitive function. The

Cerebral cortex

The cerebral cortex is the outer layer of the brain, and accounts for around 40 percent of the brain mass. Composed of nerve cell bodies, it is the point of origin of most messages from the brain, and is the site of higher learning functions, including thought, hearing, and sight.

white matter is composed of nerve fibers that communicate information between different parts of the cortex and other parts of the brain.

The precentral and postcentral gyri of the cerebral cortex are each involved in motor and sensory activity respectively. The degree of gyri involvement in activity in a particular body region is directly proportionate to the amount of sensory receptors in the area. For example, an area densely covered with sensory receptors, such as the tongue, requires a greater degree of participation by the postcentral gyrus than the arms or legs, where sensory receptors are not as heavily distributed.

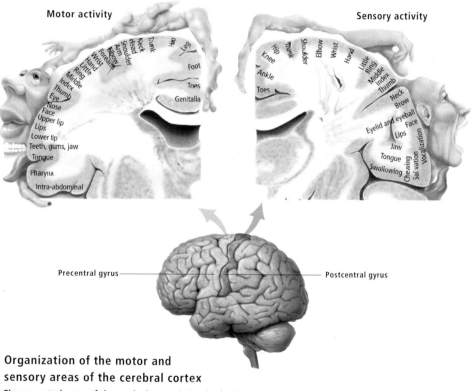

Motor activity

Sensory activity

Precentral gyrus ——————————————— Postcentral gyrus

Organization of the motor and sensory areas of the cerebral cortex

The precentral gyrus of the cerebral cortex is involved with motor activity, and the postcentral gyrus is involved with sensory activity. The illustration uses proportionate sizing of the parts of the body to represent the degree of involvement of each gyrus over that particular region.

Lobes of the Brain

The brain is divided longitudinally into two hemispheres by the longitudinal fissure, with each hemisphere then divided into lobes. Sulci and fissures in the folds of the surface of the cerebral cortex mark the areas of the different lobes of the brain. The central sulcus separates the frontal lobe, at the front of the brain, from the parietal lobe behind it. Located at the back of the brain is the occipital lobe. The temporal lobe lies beneath the frontal and parietal lobes. Each of the lobes is named after the cranial bone which overlies it.

Each of the lobes is strongly linked, though not limited, to a particular function. For example, the occipital lobe is involved in perception of vision; the temporal lobe in

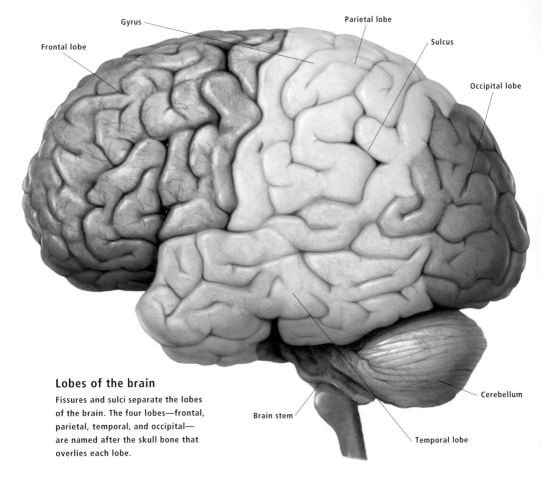

Gyrus

Frontal lobe

Parietal lobe

Sulcus

Occipital lobe

Cerebellum

Brain stem

Temporal lobe

Lobes of the brain

Fissures and sulci separate the lobes of the brain. The four lobes—frontal, parietal, temporal, and occipital—are named after the skull bone that overlies each lobe.

memory; the parietal lobe in the perception of touch and comprehension of speech; and the frontal lobe in movement, thinking, behavior, and personality.

Functional Areas of the Brain

Areas within the lobes of the brain are designated to, though not exclusive to, certain functions. For instance, the visual cortex and visual association cortex are located in the occipital lobe. The cranial nerves of the sensory organs relay information to their allied areas for interpretation. The precentral gyrus is the primary motor cortex of the brain, while the postcentral gyrus is the primary sensory cortex.

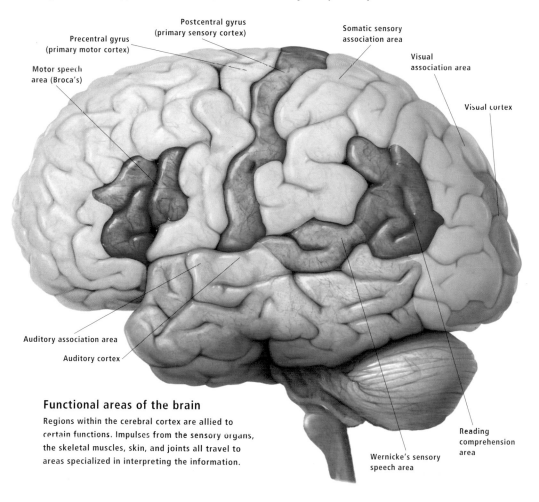

Postcentral gyrus
(primary sensory cortex)

Precentral gyrus
(primary motor cortex)

Motor speech
area (Broca's)

Somatic sensory
association area

Visual
association area

Visual cortex

Auditory association area

Auditory cortex

Wernicke's sensory
speech area

Reading
comprehension
area

Functional areas of the brain

Regions within the cerebral cortex are allied to certain functions. Impulses from the sensory organs, the skeletal muscles, skin, and joints all travel to areas specialized in interpreting the information.

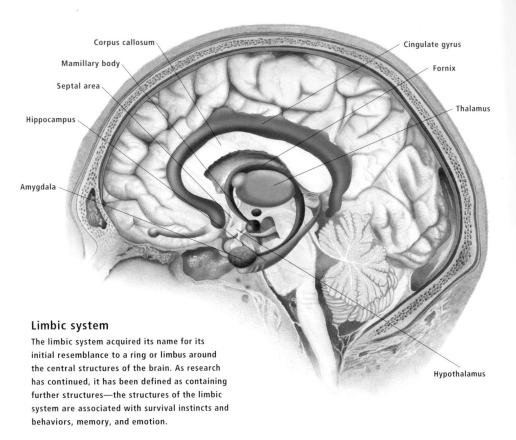

Corpus callosum

Mamillary body

Septal area

Hippocampus

Amygdala

Cingulate gyrus

Fornix

Thalamus

Hypothalamus

Limbic system

The limbic system acquired its name for its initial resemblance to a ring or limbus around the central structures of the brain. As research has continued, it has been defined as containing further structures—the structures of the limbic system are associated with survival instincts and behaviors, memory, and emotion.

Limbic System

The essential elements of the limbic system are the hippocampus, amygdala, septal area, and hypothalamus. These structures are involved in survival behaviors including expression of emotion, feeding, drinking, defense, and reproduction, as well as the formation of memory.

The hippocampus, located on the medial side of the temporal lobe, is particularly im-

portant in the formation of new memories and their recall. The amygdala is located in front of the hypothalamus in the temporal lobe. It plays a role in the olfactory system, and expression of emotion, and is linked to the hippocampus, cerebral cortex, and hypothalamus. The septal area is thought to be the pleasure or reward center. The hypothalamus is interconnected with all parts of the limbic system and is responsible

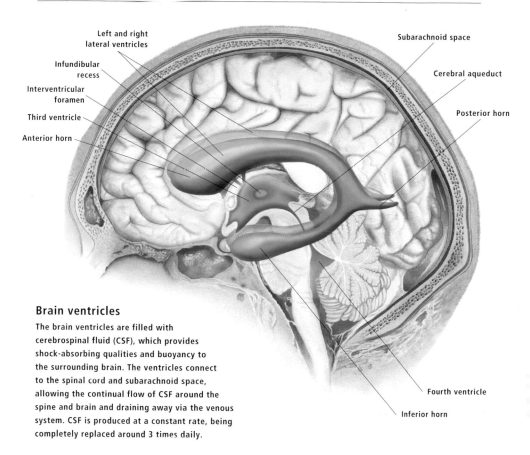

Left and right
lateral ventricles

Infundibular
recess

Interventricular
foramen

Third ventricle

Anterior horn

Subarachnoid space

Cerebral aqueduct

Posterior horn

Fourth ventricle

Inferior horn

Brain ventricles

The brain ventricles are filled with
cerebrospinal fluid (CSF), which provides
shock-absorbing qualities and buoyancy to
the surrounding brain. The ventricles connect
to the spinal cord and subarachnoid space,
allowing the continual flow of CSF around the
spine and brain and draining away via the venous
system. CSF is produced at a constant rate, being
completely replaced around 3 times daily.

for regulation of the body's organs and regulation of hormone production. It is linked to changes associated with emotion—for example when we are anxious or afraid, the hypothalamus causes our blood pressure, heart and breathing rate to increase.

Brain Ventricles

The four ventricles of the brain are chambers of fluid connected to each other; they are also connected to the central canal of the spinal cord, and the subarachnoid space around the outside of the brain.

These structures are linked by passageways that create a continuous route for the flow of cerebrospinal fluid, which serves to protect and cushion the brain. The cerebrospinal fluid is secreted into the ventricles by specialized capillaries called choroid plexuses.

Brain Stem

The brain stem consists of three regions: the midbrain, or mesencephalon; the pons; and the medulla oblongata. The overall structure of the brain stem is continuous with the spinal cord below and meets the cerebrum above. Many of the cranial nerves arise in the brain stem, and processing of information on touch on the face, taste, and hearing occurs in the brain stem. The brain stem is also involved in control of breathing, blood pressure, and heart rate.

Ascending pathways carry sensory nerve impulses from the spine; these impulses pass through the brain stem on their way to the brain. The descending pathways carry the brain's motor responses via the brain stem to the spinal cord.

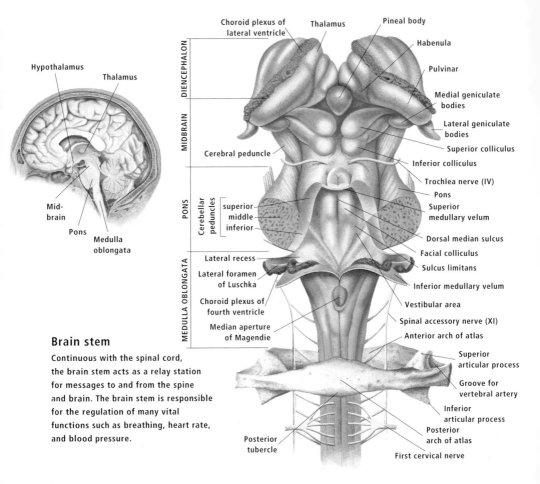

Choroid plexus of lateral ventricle
Thalamus
Pineal body
Habenula
Pulvinar
Medial geniculate bodies
Lateral geniculate bodies
Superior colliculus
Inferior colliculus
Trochlea nerve (IV)
Pons
Superior medullary velum
Dorsal median sulcus
Facial colliculus
Sulcus limitans
Inferior medullary velum
Vestibular area
Spinal accessory nerve (XI)
Anterior arch of atlas
Superior articular process
Groove for vertebral artery
Inferior articular process
Posterior arch of atlas
First cervical nerve
Posterior tubercle
Median aperture of Magendie
Choroid plexus of fourth ventricle
Lateral foramen of Luschka
Lateral recess
Cerebellar peduncles superior / middle / inferior
Cerebral peduncle

DIENCEPHALON
MIDBRAIN
PONS
MEDULLA OBLONGATA

Hypothalamus
Thalamus
Mid-brain
Pons
Medulla oblongata

Brain stem

Continuous with the spinal cord, the brain stem acts as a relay station for messages to and from the spine and brain. The brain stem is responsible for the regulation of many vital functions such as breathing, heart rate, and blood pressure.

Diencephalon

The two main components of the diencephalon are the thalamus and the hypothalamus.

The thalamus acts to relay incoming messages from the spine to the brain, and convey outgoing messages from the brain to the spine. Some of the incoming information is processed at the thalamus before being relayed to its destination in the cerebral cortex. The ovoid structure of the thalamus is comprised of two thalami, made up of nerve cells, which lie on either side of the third ventricle.

The hypothalamus lies at the base of the brain. This small region is essential to life, monitoring body functions such as hormone levels and temperature. The hypothalamus regulates hormone production in the neighboring pituitary gland, and also exerts control over the autonomic nervous system.

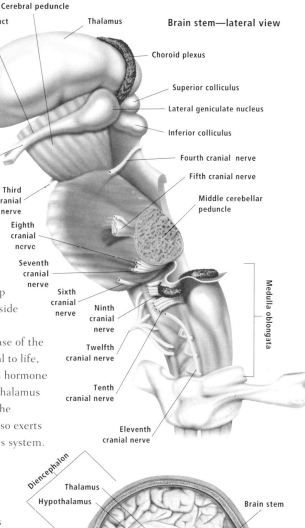

Brain stem—lateral view

Cerebral peduncle
Optic tract
Thalamus
Choroid plexus
Superior colliculus
Lateral geniculate nucleus
Inferior colliculus
Second cranial nerve
Fourth cranial nerve
Fifth cranial nerve
Third cranial nerve
Middle cerebellar peduncle
Eighth cranial nerve
Seventh cranial nerve
Sixth cranial nerve
Ninth cranial nerve
Twelfth cranial nerve
Tenth cranial nerve
Eleventh cranial nerve
Medulla oblongata

Diencephalon
Thalamus
Hypothalamus
Brain stem
Pituitary gland
Cerebellum

Diencephalon

The diencephalon comprises the thalamus and hypothalamus. These two structures each have important functions. The thalamus not only relays nerve impulses to various areas of the brain, but is also responsible for processing some of the incoming information. The hypothalamus plays a pivotal role in homeostasis.

Meninges

Surrounding the brain and spine are the meninges, a series of three layers of membrane, which provide a protective, shock-absorbent cushion. The three layers comprise: the outer layer, the dura mater; the middle layer, the arachnoid mater; and the inner layer, the pia mater.

The outer layer, the dura mater, is comprised of tough, fibrous membrane. The middle layer, the arachnoid mater, is comprised of collagen and elastin, and forms a lacy, web-like structure. This web-like structure aids its connection to the inner membrane layer. The inner layer, the pia

Meninges in the brain

The three layers of membrane that comprise the meninges—the dura mater, the arachnoid mater, and the pia mater—provide protection to the major elements of the central nervous system: the brain and spinal cord.

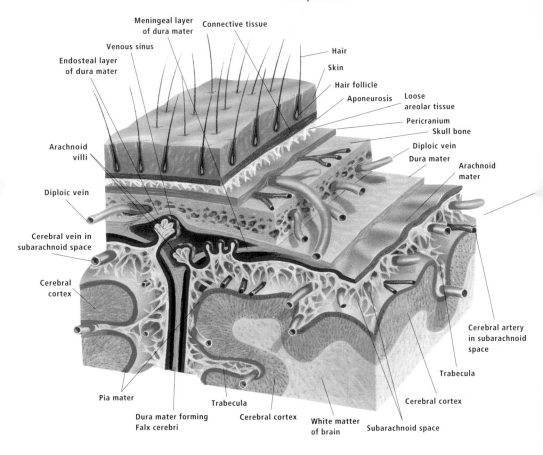

Meningeal layer of dura mater

Connective tissue

Venous sinus

Endosteal layer of dura mater

Hair

Skin

Hair follicle

Aponeurosis

Loose areolar tissue

Pericranium

Skull bone

Arachnoid villi

Diploic vein

Dura mater

Arachnoid mater

Diploic vein

Cerebral vein in subarachnoid space

Cerebral cortex

Cerebral artery in subarachnoid space

Trabecula

Pia mater

Dura mater forming Falx cerebri

Trabecula

Cerebral cortex

White matter of brain

Subarachnoid space

Cerebral cortex

mater, is also comprised of collagen and elastin, and has numerous blood vessels running through its structure.

Between the arachnoid and the pia mater is the subarachnoid space. Flowing through this space is the cerebrospinal fluid (CSF) produced in the choroid plexuses of the brain ventricles.

The purpose of the cerebrospinal fluid is to complement the cushioning qualities of the meninges, supply nutrients to the CNS, and carry away waste products. The composition of cerebrospinal fluid includes water, glucose, and chemical ions. These nutrients are essential to cell stability and regeneration. The cerebrospinal fluid also transports waste products away from the area.

Produced at a steady rate, and replaced about three times daily, the cerebrospinal fluid is absorbed by arachnoid villi and drains into the venous system.

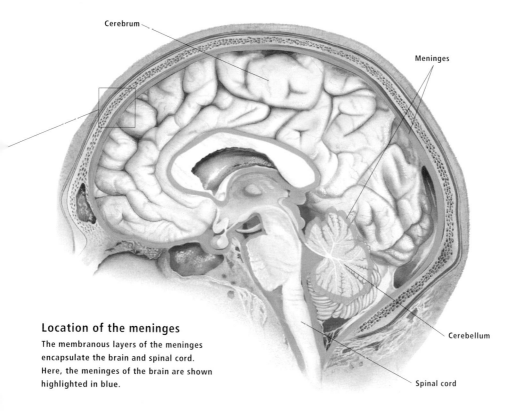

Cerebrum

Meninges

Cerebellum

Spinal cord

Location of the meninges

The membranous layers of the meninges encapsulate the brain and spinal cord. Here, the meninges of the brain are shown highlighted in blue.

Cranial Nerves

There are 12 pairs of cranial nerves arising at the base of the brain and the brain stem. The nerves carry motor and sensory messages, and are also allied to the special sense organs. The cranial nerves innervate the muscles and sensory structures of the head and neck, including skin, membranes, eyes, and ears. Information regarding the special senses of sight, hearing, balance, taste, and smell is relayed to specific functional areas in the lobes of the brain designated to interpret and respond to these nerve impulses.

Cranial nerves—location

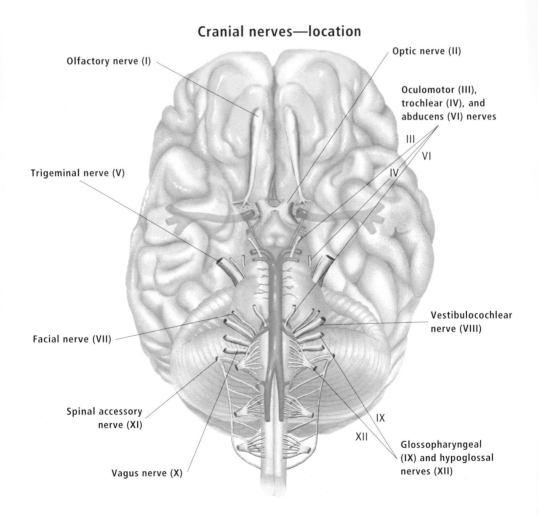

Olfactory nerve (I)

Optic nerve (II)

Oculomotor (III), trochlear (IV), and abducens (VI) nerves

III

VI

IV

Trigeminal nerve (V)

Vestibulocochlear nerve (VIII)

Facial nerve (VII)

Spinal accessory nerve (XI)

IX

XII

Glossopharyngeal (IX) and hypoglossal nerves (XII)

Vagus nerve (X)

Cranial nerves

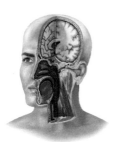

Olfactory nerve (I)

This supplies the chemoreceptor cells of the nose.

Optic nerve (II)

This supplies the photoreceptor cells of the retina.

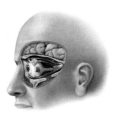

Oculomotor (III), trochlear (IV), and abducens (VI) nerves

These supply the muscles of the eyeball and eyelid.

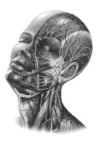

Trigeminal nerve (V)

This supplies the forehead, cheek, and muscles of mastication.

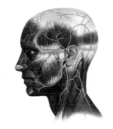

Facial nerve (VII)

This supplies the facial area and the front of the tongue.

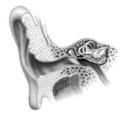

Vestibulocochlear nerve (VIII)

This supplies the balance organs and organs of hearing.

Glossopharyngeal (IX) and hypoglossal (XII) nerves

These supply the back of tongue, soft palate, reflex control of heart (IX), and tongue movement (XII).

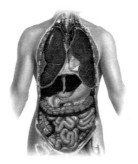

Vagus nerve (X)

This supplies parts of the abdominal cavity.

Spinal accessory nerve (XI)

This supplies the head, neck column, and associated structures.

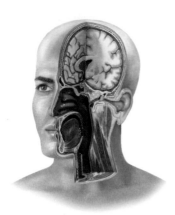

Olfactory nerve (I)

Olfactory nerve (I)

The olfactory nerve (I) is the first cranial nerve, and is associated with the sense of smell. Nerve fibers originating in the mucous membranes of the nose transmit sensory information to the region in the cerebral cortex responsible for processing and responding to the information received.

Optic nerve (II)

Images focused on the retina of the eye are transmitted by the rod and cone cells of the retinal surface to the optic nerve. The optic nerves from each of the eyes meet at

Optic nerve (II)

the optic chiasm, before continuing on to the thalamus and then to the visual cortex for processing.

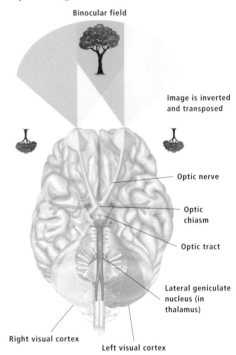

Binocular field

Image is inverted and transposed

Optic nerve

Optic chiasm

Optic tract

Lateral geniculate nucleus (in thalamus)

Right visual cortex

Left visual cortex

Optic path

Each eye has a somewhat different, yet overlapping visual field. The difference between the images in the binocular field enables us to determine an object's distance and 3-D construction. The images that are translated into nerve impulses move along nerve fibers to the optic nerves, the optic chiasm, and the lateral geniculate nuclei of the thalamus. The nuclei process information, transmitting it to the visual cortex, which then combines and interprets images. Although the images received on the retina are upside down, the brain automatically corrects these images so we see them the right way up.

While at the meeting point of the optic chiasm some impulses cross over from left to right and vice versa. This results in each hemisphere of the brain receiving information from both eyes. From the optic chiasm, the impulses continue on to the left and right thalamus and then on to the visual cortex.

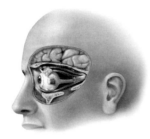

Oculomotor (III), trochlear (IV), and abducens (VI) nerves

Oculomotor (III), Trochlear (IV), and Abducens (VI) nerves

These three cranial nerves are all associated with the eye and eye movements. Together they are responsible for controlling the six muscles of each eye, moving the eyeball and eyelids and enabling focusing.

The oculomotor nerve (III) controls the inferior oblique, the superior, inferior, and medial rectus muscles of the eye, and the levator palpebrae superioris muscle of the eyelid. The trochlear nerve (IV) controls the superior oblique muscle of the eye, while the abducens nerve (VI) controls the lateral rectus muscle of the eye.

Trigeminal nerve (V)

The trigeminal nerve arises in the brain stem. It is responsible for innervation of the skin on the face and scalp; the mucous membranes in the nose, mouth, and eye; and the jaw muscles.

After leaving the brain stem, the trigeminal nerve branches into three nerves, each supplying a separate area.

The ophthalmic branch of the trigeminal nerve innervates the skin of the front scalp area, the forehead and the skin of the upper eyelid, some of the nose, and the cornea. The maxillary branch innervates the skin on the temple and the cheek area, from the lower eyelid to the upper lip.

The mandibular branch innervates the muscles and skin of the jaw, the temple and ear, the lower teeth, the mucous membrane of the floor of the mouth, and middle and front of the tongue.

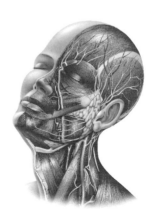

Trigeminal nerve (V)

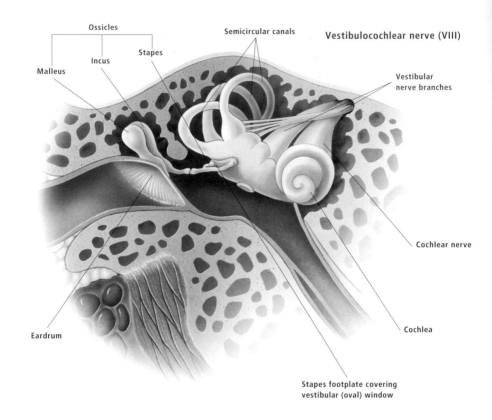

Ossicles

Malleus

Incus

Stapes

Semicircular canals

Vestibulocochlear nerve (VIII)

Vestibular
nerve branches

Cochlear nerve

Cochlea

Eardrum

Stapes footplate covering
vestibular (oval) window

Facial nerve (VII)

The facial nerve is associated with the sensation of taste at the front of the tongue. The extensive coverage of the face and head muscles achieved by the facial nerve and its many branches, result in a wide range of facial expressions.

Vestibulocochlear nerve (VIII)

The vestibulocochlear nerve divides into two branches as it enters the acoustic canal. The vestibular branch is associated with the sense of balance and the organs of the vestibular labyrinth in the inner ear. Change in body position is registered by the tiny hair-like receptor cells located in the ampulla, which shift in unison with the body. The movement of these hair cells activates the nerve cells of the vestibular branch. The cochlear branch of the vestibulocochlear nerve is associated with the sense of hearing. Working on a similar principle to the balance organs, tiny hair cells fire off nerve impulses when vibrated by cochlear fluid.

Glossopharyngeal (IX) and Hypoglossal (XII) nerves

The ninth cranial nerve, the glossopharyngeal nerve, is responsible for supplying the carotid sinus, the back part of the tongue and the soft palate, the parotid gland, and the reflex control of the heart. It also controls the muscles of the pharynx responsible for the swallowing action.

The twelfth cranial nerve, the hypoglossal nerve, controls the muscles of the tongue.

Vagus nerve (X)

Extending from the brain stem, the vagus nerve runs through the neck and thorax to the abdominal cavity, and is involved in a diverse range of functions including breathing, swallowing, speaking, heartbeat, and digestion. It plays a key role in digestion, being responsible for stimulating stomach movement, the secretion of stomach acids, and controlling the movements of smooth muscle in the small and large intestines during the peristaltic movements of the digestive process.

The extensive coverage achieved by the vagus nerve sees it providing information from the external ear, pharynx, esophagus, and thoracic and abdominal organs. It also exerts influence over the constriction of blood vessels, and some of the respiratory passageways.

The vagus nerve is associated with the autonomic nervous system, and many of its wide-ranging functions take place without our conscious knowledge.

Spinal accessory nerve (XI)

The spinal accessory nerve divides into two branches—the cranial branch and the spinal branch.

The spinal branch is responsible for muscle movement of the upper shoulders, head, and neck.

The cranial branch controls the muscles of the larynx, pharynx, and palate. Exerting control over these muscles, the cranial branch of the spinal accessory nerve contributes to the swallowing action in the upper portion of the digestive tract.

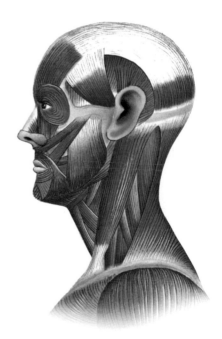

Spinal accessory nerve (XI)

Spinal cord

Connecting the peripheral nerves to centers in the brain, the spinal cord and spinal nerves convey incoming and outgoing messages between the two sources. A column of nerve tissue located in the vertebral canal, the spinal cord is comprised of gray and white matter encased in the protective sheath of the meninges.

Spinal Cord

Measuring between 16 ½–17 ¾ inches (42–45 centimeters), the spinal cord extends almost the full length of the vertebral column, tapering at the lumbar vertebrae at about L1 or L2, and fanning out into the spinal rootlets of the cauda equina (Latin for horse's tail).

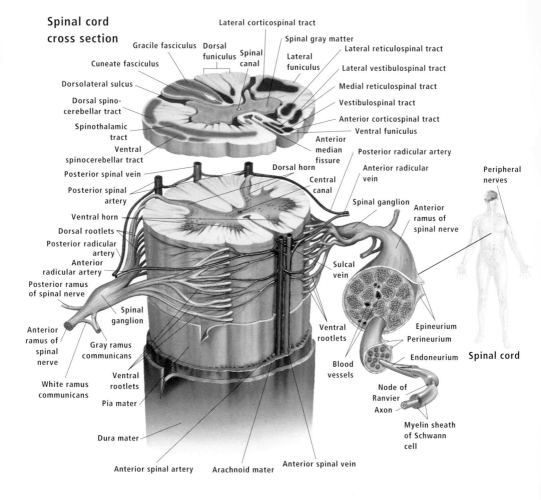

Spinal cord cross section

Lateral corticospinal tract
Gracile fasciculus
Dorsal funiculus
Spinal canal
Cuneate fasciculus
Spinal gray matter
Lateral reticulospinal tract
Lateral funiculus
Dorsolateral sulcus
Lateral vestibulospinal tract
Dorsal spino-cerebellar tract
Medial reticulospinal tract
Spinothalamic tract
Vestibulospinal tract
Anterior corticospinal tract
Ventral spinocerebellar tract
Ventral funiculus
Anterior median fissure
Posterior radicular artery
Posterior spinal vein
Dorsal horn
Anterior radicular vein
Posterior spinal artery
Central canal
Spinal ganglion
Ventral horn
Anterior ramus of spinal nerve
Dorsal rootlets
Posterior radicular artery
Sulcal vein
Anterior radicular artery
Posterior ramus of spinal nerve
Spinal ganglion
Ventral rootlets
Epineurium
Anterior ramus of spinal nerve
Gray ramus communicans
Perineurium
Blood vessels
Endoneurium
White ramus communicans
Ventral rootlets
Node of Ranvier
Pia mater
Axon
Dura mater
Myelin sheath of Schwann cell
Anterior spinal artery
Arachnoid mater
Anterior spinal vein
Peripheral nerves
Spinal cord

The spinal cord is the conductor through which messages from the peripheral nerves are sent to the brain, and messages initiating response are sent from the brain to the peripheral nerves.

Located in a canal formed by the structure of the vertebrae, a column of spinal nerves extends out from the spinal cord from between each pair of vertebrae, connecting the peripheral nerves to the spinal cord. There are 31 pairs of nerves, each allied to a specific body region and a specific skin region called a dermatome.

The internal structure of the spinal cord is made up of an inner core of gray matter and an outer layer of white matter, with a central canal. As in the brain, the gray matter is comprised of cell bodies, while the white matter is comprised of nerve fibers.

The gray matter is structured in an H-shape when viewed in cross section, with the H being formed by pairs of dorsal horns at the back, and ventral horns at the front, with an intermediate zone joining the two pairs of horns.

The dorsal horns are primarily responsible for conveying sensory information to the brain, while the ventral horns are mainly responsible for sending motor responses to the muscles via the spinal nerves. The intermediate zone is concerned with reflex actions, marrying up incoming sensory information and outgoing motor responses that do not involve the brain, thereby effecting rapid response to situations requiring reflex actions.

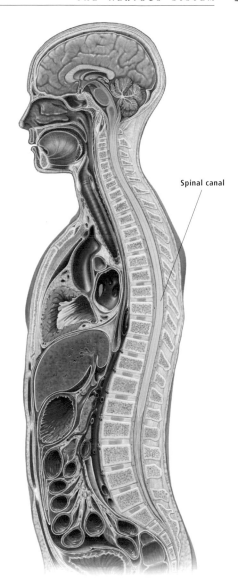

Spinal canal

Spinal canal
Essential protection for the critical structure of the spinal cord is provided by the vertebral column.

Dermatomes

Emerging between each pair of adjacent vertebrae are the spinal nerves of the spinal cord. The spinal nerves emerge from the front and back of the spinal cord as the dorsal and ventral rootlets, merging together, and in so doing, combining the sensory fibers of the dorsal rootlets with the motor fibers of the ventral rootlets into one structure—the spinal nerve. Spaced along the vertebral column are the 31 pairs of spinal nerves—these nerves each serve a specific

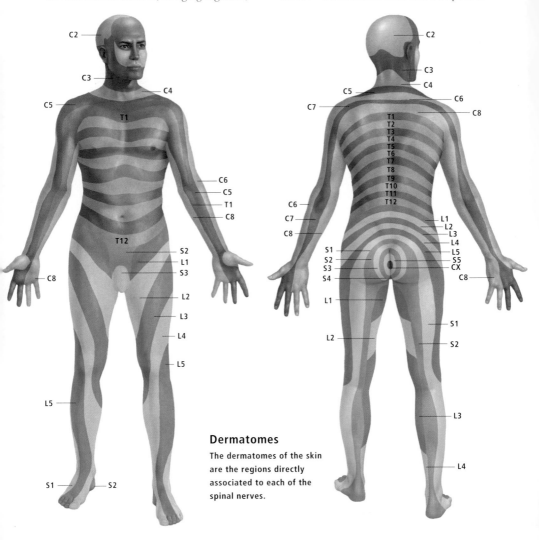

Dermatomes

The dermatomes of the skin are the regions directly associated to each of the spinal nerves.

body region and a specific region of skin, though they often overlap, to have some influence over adjacent areas.

The skin areas controlled by each of the spinal nerves are called dermatomes. Loss of sensation in a particular dermatome can indicate damage to a specific spinal nerve or the spinal cord. For anesthesia purposes, the correct nerve must be identified to numb a particular region.

Nerve-to-skin link

The spinal nerves are numbered and correspond closely to the spinal vertebrae. Each pair of nerves supplies a specific skin area (dermatome) of the body, and a specific group of muscles.

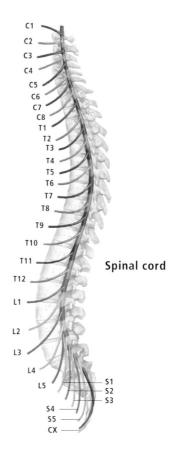

Spinal cord

Facial nerves

Originating at the pons in the brain stem, the
facial nerves spread out over each side of the
face, supplying the muscles of the face, and also
supplying the lacrimal and salivary glands.

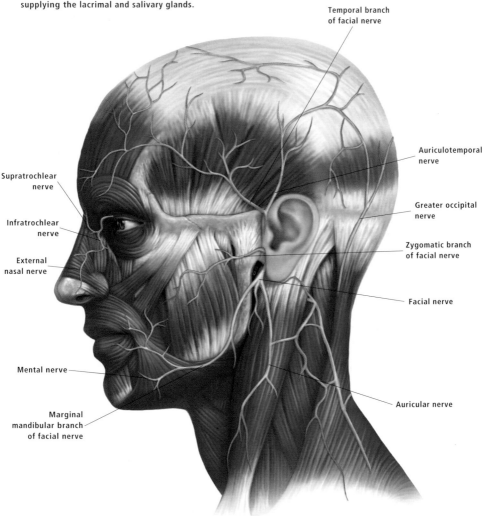

Temporal branch
of facial nerve

Auriculotemporal
nerve

Supratrochlear
nerve

Greater occipital
nerve

Infratrochlear
nerve

Zygomatic branch
of facial nerve

External
nasal nerve

Facial nerve

Mental nerve

Auricular nerve

Marginal
mandibular branch
of facial nerve

Facial Nerves

The seventh cranial nerve, the facial nerve (VII), arises at the pons in the brain stem. Located at either side of the face, the facial nerves emerge through openings in the temporal bone, the stylomastoid foramen, branching out across the face and extending into the head region.

The facial nerve is responsible for the control of the muscles of the face, which enable facial expression, and also supplies the salivary and lacrimal glands.

The various branches of the facial nerve control specific facial muscles. The temporal branch of the facial nerve supplies the temporal muscle at the front of the skull. The zygomatic branch controls the zygomatic muscles that are responsible for lip movements such as smiling. The buccal branch supplies the cheek region, including the skin and the mucous membranes of the area. The cervical branch stretches out across the side of the face and down into the neck, supplying the muscles of this region.

One of the branches of the facial nerve, the stapedius nerve, controls the smallest muscle in the body, the stapedius muscle, located in the ear. The facial nerve is also responsible for the sensation of taste in the front part of the tongue.

While the facial nerve itself is responsible for much of the muscular movement of the face, many of the other cranial nerves contribute to movement and functions in the facial area, particularly those linked to the special sense organs, all of which are found in the head.

Facial nerve (VII)

The facial nerve (VII) is primarily responsible for much of the muscular movement of the face. It is also responsible for taste in the anterior two-thirds of the tongue.

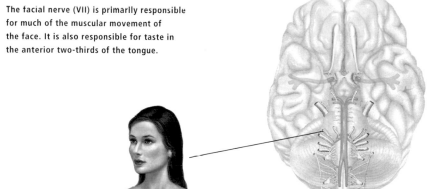

Eye

The eye is our organ of sight. The structure of the eye comprises many components that work together to provide vision. The apparatus of the eye is held in place by six muscles, which coordinate in pairs to provide an extensive field of vision. The upper and lower eyelids protect the outer surface of the eye. Secretions from the lacrimal apparatus lubricate the eye area, draining into the nasolacrimal duct, after washing across the surface of the eye.

The eyeball itself is made up of three layers: the outer layer of the sclera and cornea, the middle layer of the uvea, and the inner layer of the retina. Beneath these layers, the eyeball is divided into two cavities, with the anterior cavity divided again into the anterior and posterior chambers. Each chamber is filled with fluid, which provides

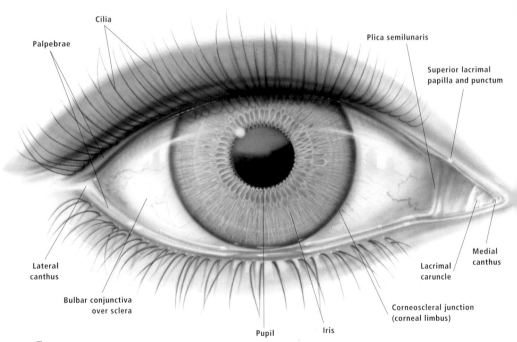

Cilia

Palpebrae

Plica semilunaris

Superior lacrimal papilla and punctum

Lateral canthus

Bulbar conjunctiva over sclera

Pupil

Iris

Lacrimal caruncle

Corneoscleral junction (corneal limbus)

Medial canthus

Eye

The eye is the organ of sight. The combined efforts of the cornea and lens are assisted by muscles, nerves, blood vessels, and lubricating and nourishing fluids, to ensure that clear images are received at the retina.

Optic nerve

The optic nerve (cranial nerve II) is responsible for conducting neural signals from the retinal cells to the visual cortex for interpretation.

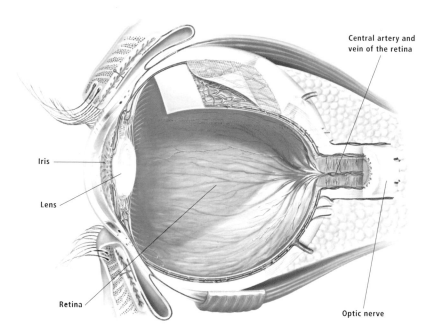

Central artery and vein of the retina

Iris

Lens

Retina

Optic nerve

nutrients to the internal structure. The posterior cavity is filled with a gel-like substance, the vitreous humor, which helps to maintain the shape of the eyeball.

The sclera forms the thick white outer layer—the "white of the eye"—that gives the eyeball its spherical shape. At the front of the sphere, the sclera is replaced by a transparent layer, the cornea. The transparent cornea at the front of the eye is responsible for refracting or bending light.

The sclera maintains a constant distance between the cornea and the retina at the back of the eye. At the back of the eye, the sclera forms a cylindrical tube to carry the optic nerve.

The optic nerve lies behind the retina and conveys seen images—registered as nerve impulses by the cells of the retina, known as rods and cones—along the optic nerve to the part of the brain that interprets the impulses—the visual cortex.

Eyeball

This illustration shows the layers of the eyeball
and the anterior and posterior cavities. Fluid
fills these cavities, providing nourishment
and aiding in maintaining the shape
of the eyeball, to ensure visual
information is not distorted.

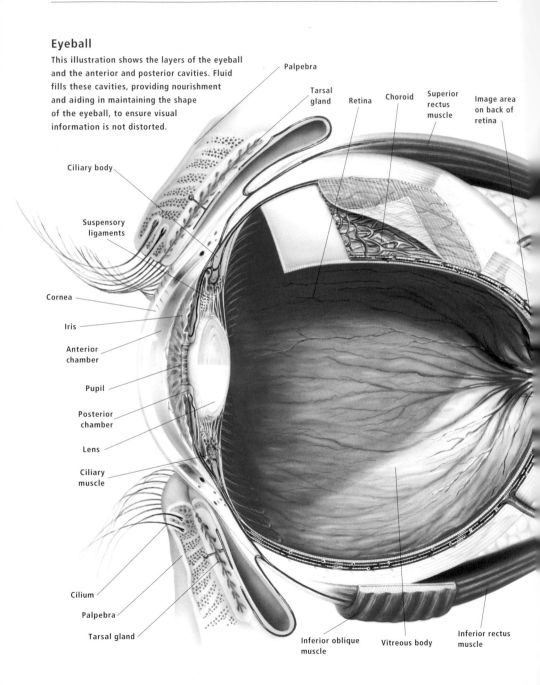

Palpebra

Tarsal
gland

Retina

Choroid

Superior
rectus
muscle

Image area
on back of
retina

Ciliary body

Suspensory
ligaments

Cornea

Iris

Anterior
chamber

Pupil

Posterior
chamber

Lens

Ciliary
muscle

Cilium

Palpebra

Tarsal gland

Inferior oblique
muscle

Vitreous body

Inferior rectus
muscle

The middle layer of the eye, the uvea, is comprised of three parts: the choroid, ciliary body, and iris. The choroid, at the rear of the uvea, is filled with blood vessels and nerves that serve the cornea, ciliary body, and iris. The ciliary body contains the fine fibers of the ciliary muscles, which hold the centrally located lens suspended in position. The ciliary body also secretes a fluid called aqueous humor that provides nourishment to the lens and cornea. The cornea and lens work together to adjust images for projection onto the retina at the back of the eye. The

iris controls the amount of light entering the eye, with its muscles dilating and constricting the pupil depending on the amount of light.

The retina is the inner layer of the eye, and is responsible for receiving images projected through the cornea and lens. Images cast on the retina are converted into neural signals by the rod and cone cells of the retina. Millions of these cells make up the retinal surface, with rod cells being more sensitive to light, and cone cells being more sensitive to color determination, and thus are responsible for providing more detailed visual information.

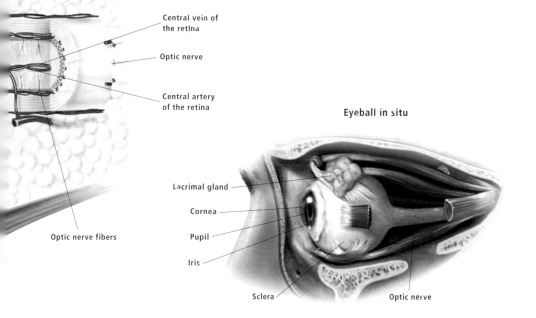

Central vein of the retina

Optic nerve

Central artery of the retina

Optic nerve fibers

Eyeball in situ

Lacrimal gland

Cornea

Pupil

Iris

Sclera

Optic nerve

Ear

Located in a cavity in the temporal bone
of the skull, the ear serves a dual purpose
as the organ of both hearing and balance.
The ear picks up sound waves, converting
them firstly into mechanical vibrations and
then into electrical impulses, which are
carried along the vestibulocochlear nerve
to the brain.

These impulses are interpreted by the
auditory cortex, located in the temporal
lobe. The ear also senses the body's position
relative to gravity, sending information to
the brain that allows the body to maintain
postural equilibrium.

The ear comprises three sections: the
outer ear, the middle ear, and the inner ear,
with each section performing a particular
function. The outer ear comprises the auricle or pinna, which is the visible external
part of the ear; the ear canal, or external
acoustic meatus; and the eardrum. The
auricle channels sound waves through the
ear canal to the eardrum, also known as
the tympanic membrane.

The ear canal contains wax secretions,
which serve to trap foreign particles, thus
protecting the delicate eardrum. The eardrum is a thin, semitransparent membrane
that separates the outer and middle ear.

Sound waves hit the eardrum, causing
vibrations that are transferred to the inner
ear, triggering movement in the auditory
ossicles. The vibrations of the ossicles are
relayed to the membrane that covers the
oval window. The eustachian tube connects

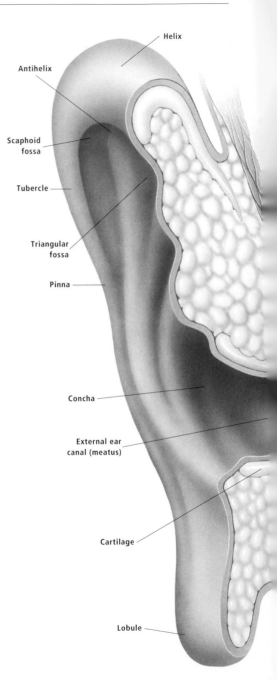

Helix

Antihelix

Scaphoid
fossa

Tubercle

Triangular
fossa

Pinna

Concha

External ear
canal (meatus)

Cartilage

Lobule

Ear

The organ of hearing and balance, the ear contains
many intricate structures that combine to interpret
the sounds we hear, and maintain our sense of
balance and body position.

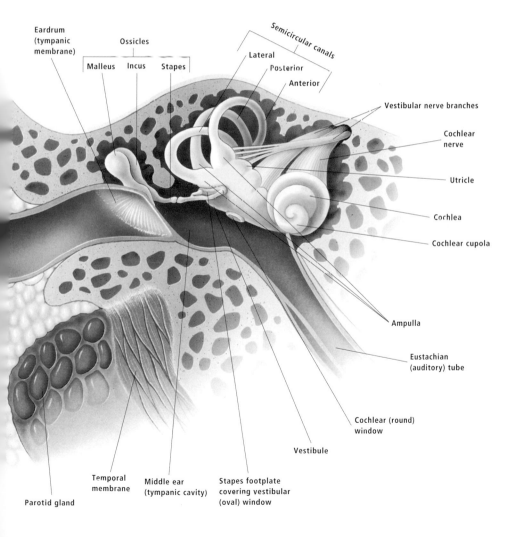

Eardrum
(tympanic
membrane)

Ossicles

Malleus   Incus   Stapes

Semicircular canals

Lateral

Posterior

Anterior

Vestibular nerve branches

Cochlear
nerve

Utricle

Cochlea

Cochlear cupola

Ampulla

Eustachian
(auditory) tube

Cochlear (round)
window

Vestibule

Stapes footplate
covering vestibular
(oval) window

Middle ear
(tympanic cavity)

Temporal
membrane

Parotid gland

the middle ear to the throat and serves to maintain equal air pressure on both sides of the eardrum.

Once the mechanical vibrations have been transmitted to the inner ear, they are converted into electrical impulses to be sent via the vestibulocochlear nerve to the brain. The fluid in the cochlea is moved by the vibrations of the oval window.

This movement instigates activity in the organ of Corti, where the tiny nerve endings pick up movement in the cochlear fluid. When activated, these nerve endings send electrical impulses to the brain, which are interpreted as sound.

The balance component of the ear is comprised of specialized organs, known as the vestibular organs, of the semicircular canals the utricle and saccule.

These tiny organs sense the body's position, and its relationship to gravity. The vestibular branch of the vestibulocochlear nerve conveys messages from the vestibular organs to the brain.

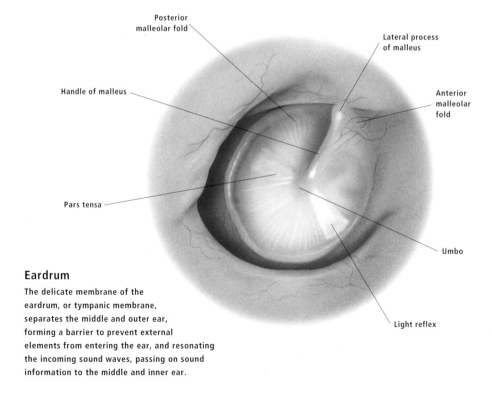

Posterior malleolar fold

Lateral process of malleus

Handle of malleus

Anterior malleolar fold

Pars tensa

Umbo

Light reflex

Eardrum

The delicate membrane of the eardrum, or tympanic membrane, separates the middle and outer ear, forming a barrier to prevent external elements from entering the ear, and resonating the incoming sound waves, passing on sound information to the middle and inner ear.

Cochlea

The cochlea is a small, spiral-shaped structure that contains the hair-like nerve endings of the organ of Corti. Fluid in the cochlea ripples in response to the vibrations of the membrane of the oval window. These ripples trigger the nerve endings, which then send electrical impulses to the brain.

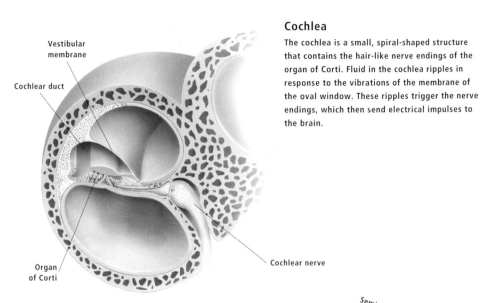

Vestibular membrane

Cochlear duct

Organ of Corti

Cochlear nerve

Ossicles

Malleus

Incus

Stapes

Eardrum

Semicircular canals

Lateral

Posterior

Anterior

Vestibular nerve branches

Cochlea

Stapes footplate covering vestibular (oval) window

Semicircular canals

The fluid in the three semicircular canals registers balance and body position, with each canal registering different movements. Right angles created at the end of each canal contain an ampulla, which has sensory hairs attached. Movement is registered by the fluid, which in turn triggers the nerve endings. These nerve endings send information to the brain regarding body position.

Arm nerves

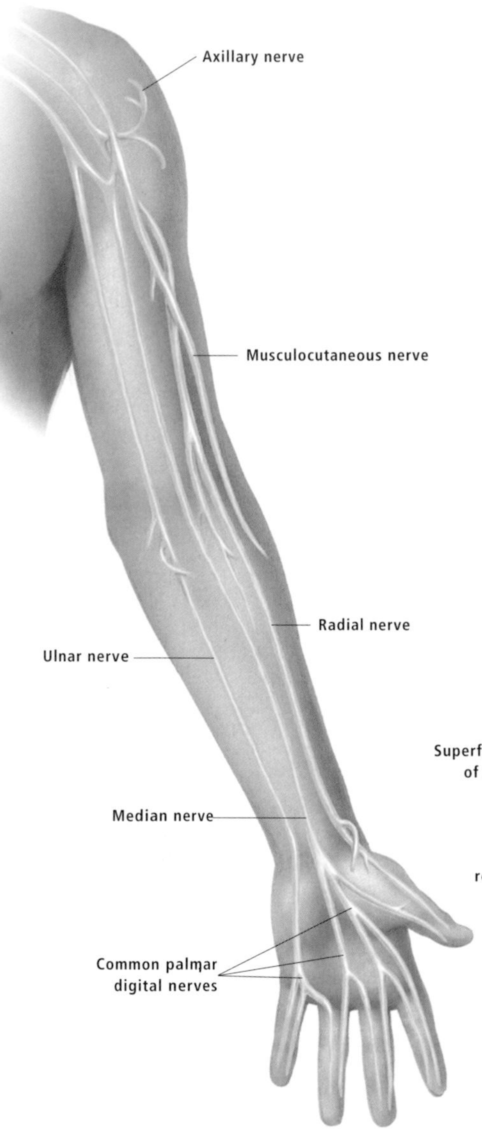

Axillary nerve

Musculocutaneous nerve

Radial nerve

Ulnar nerve

Median nerve

Common palmar digital nerves

Nerves of the Shoulder and Arm

The nerves of the shoulder and arm generally extend from the brachial plexus in the neck, passing through the axillary (armpit) region and then extending down through the arm. The major nerves supplying the shoulder and arm are the ulnar nerve, the radial nerve, and the median nerve.

The ulnar nerve extends down the inner side of the upper arm, elbow, and forearm, with branches of the ulnar nerve extending into the hand. The ulnar nerve supplies the skin on the inner side of the arm, the elbow joint, and some of the flexor muscles of the forearm, including flexor carpi ulnaris, and part of flexor digitorum profundus. It supplies the skin on the little finger side of the hand and some of the small muscles of the hand. When we hit our "funny bone,"

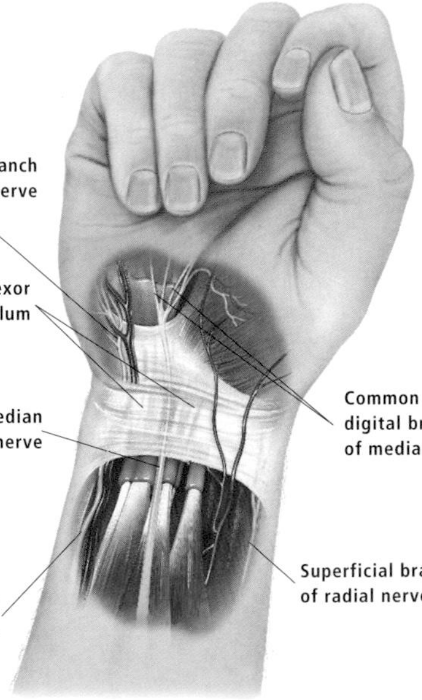

Superficial branch of ulnar nerve

Flexor retinaculum

Median nerve

Common palmar digital branches of median nerve

Ulnar nerve

Superficial branch of radial nerve

what we actually strike is the ulnar nerve, causing a painful sensation.

The radial nerve extends through the axillary (armpit) region and runs down the back of the arm. It supplies the extensor muscles of the arm, such as the triceps; controls the muscles that straighten the elbow and extend the wrist; and supplies the extensor muscles that extend into the wrist, fingers, and thumb. The radial nerve is responsible for supplying skin on the back of the arm and on the thumb side of the hand.

The median nerve extends down the middle of the front of the arm, branching out below the wrist. This nerve supplies some of the flexor muscles of the forearm, including flexor carpi radialis and palmaris longus; muscles of the thumb; and the skin on the front lateral part of the hand.

Nerves of the Leg

The major nerves of the leg are the femoral, obturator, and sciatic nerves. These nerves and their branches are responsible for supplying the muscles of the leg and foot, and much of the skin.

The sciatic nerve and its branches, the tibial and common peroneal nerves, supply the hamstrings and all the muscles of the lower part of the leg and foot. The obturator nerve supplies the gracilis muscle and the adductor muscles of the leg. The femoral nerve, and its major branch, the saphenous nerve, supply the muscles at the front of the thigh and the upper leg joints of the hip and knee.

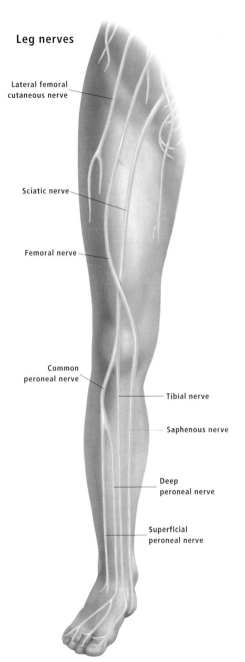

Leg nerves

Lateral femoral cutaneous nerve

Sciatic nerve

Femoral nerve

Common peroneal nerve

Tibial nerve

Saphenous nerve

Deep peroneal nerve

Superficial peroneal nerve

Senses

Changes in our internal and external environments are registered by the senses. The sense organs detect changes such as light, heat, touch, and sound, as well as physical stimuli. The changes are relayed by the sense organs as nerve impulses to the brain. Generally divided into two categories, there are the general senses, and the special senses.

The general senses include touch, pressure, vibration, proprioception, stretch, pain, heat, and cold. Sensory receptors for these senses are distributed throughout the body.

The special senses are sight, hearing, taste, smell, and balance, which each have specialized receptors in their specific location to aid our perception and response to particular sensations.

Touch

The sense of touch is often referred to as our tactile sense. When we come into contact with an object or surface, receptors in the skin are activated. The skin surface is heavily distributed with sensory nerve receptors, known as cutaneous receptors.

These receptors detect pain, pressure, vibration, and temperature change. There are several different types of cutaneous receptors in the skin, including Ruffini's corpuscles, Meissner's corpuscles, and Pacinian corpuscles.

While touch receptors are distributed throughout the entire body, the receptors in some regions have a much higher degree of sensitivity, such as those found in the fingertips, lips, and tongue.

Sensory pathways

Stimuli inside and outside the body activate the sensory receptors of the peripheral nerves. Nerve impulses are transmitted from the point of origin, along the nerve pathways to the central nervous system. These impulses are then transmitted from the spinal cord to the brain for interpretation.

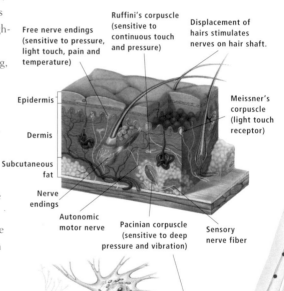

Free nerve endings (sensitive to pressure, light touch, pain and temperature)

Ruffini's corpuscle (sensitive to continuous touch and pressure)

Displacement of hairs stimulates nerves on hair shaft.

Epidermis

Dermis

Subcutaneous fat

Nerve endings

Autonomic motor nerve

Pacinian corpuscle (sensitive to deep pressure and vibration)

Meissner's corpuscle (light touch receptor)

Sensory nerve fiber

Neuron

Nerve endings in the skin, muscles, joints, and internal organs transmit signals of pain, temperature, or pressure along peripheral nerves.

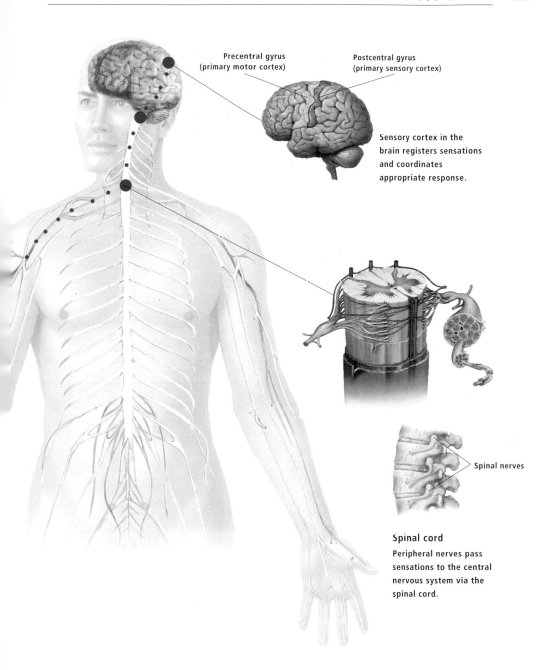

Precentral gyrus
(primary motor cortex)

Postcentral gyrus
(primary sensory cortex)

Sensory cortex in the
brain registers sensations
and coordinates
appropriate response.

Spinal nerves

Spinal cord

Peripheral nerves pass
sensations to the central
nervous system via the
spinal cord.

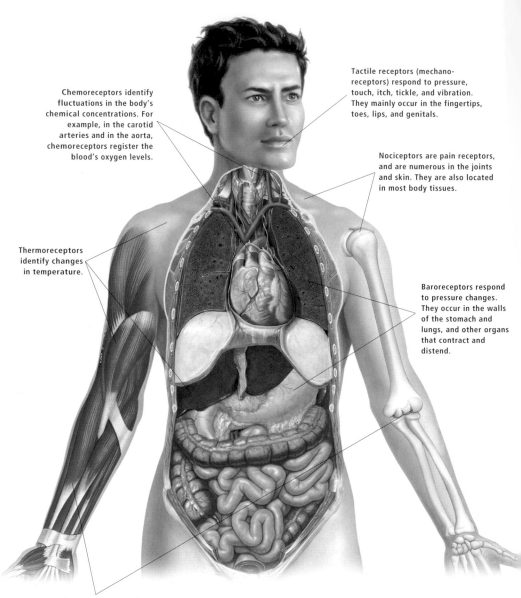

Chemoreceptors identify fluctuations in the body's chemical concentrations. For example, in the carotid arteries and in the aorta, chemoreceptors register the blood's oxygen levels.

Tactile receptors (mechano-receptors) respond to pressure, touch, itch, tickle, and vibration. They mainly occur in the fingertips, toes, lips, and genitals.

Nociceptors are pain receptors, and are numerous in the joints and skin. They are also located in most body tissues.

Thermoreceptors identify changes in temperature.

Baroreceptors respond to pressure changes. They occur in the walls of the stomach and lungs, and other organs that contract and distend.

Proprioceptors monitor the movement and position of muscles, tendons, and joints.

NB: In this illustration the lungs and pleura have been cut to reveal the heart.

General Senses

While many of the body's receptors are found in the skin, there are numerous other receptors that relay specific information to the brain.

The tactile receptors of the skin, which include at least six different types of receptors, are mechanoreceptors. These tactile receptors include: Meissner's corpuscles, Merkel's disks, Krause bulbs, Pacinian corpuscles, Ruffini endings, and free nerve endings. Mechanoreceptors detect sensation such as touch, pressure, and vibration.

Proprioceptors detect general body movement and position, and movement and position of joints and muscles, thus enabling the body to coordinate a response to particular movements, and resolve imbalance.

Thermoreceptors detect temperature variations, with separate receptors for heat and cold. These receptors are particularly sensitive, with the ability to detect temperatures ranging from freezing cold to red hot. They aid in the body's adaptability to fluctuating temperatures, such as when entering a warm indoor environment from the cold outdoor weather. In such instances, temperatures that initially feel too warm or too cold soon feel comfortable as our body adjusts to the new temperature.

Nociceptors are the body's pain receptors, detecting pain from the skin, muscles, joints, and internal organs. Distribution of nociceptors varies, with the surface layers of the skin more densely populated than the muscles, joints, and internal organs. The lighter distribution of nociceptors in the internal organs can often result in pain being perceived in a general, rather than specific, location. There are three types of nociceptors, each specialized to determine pain produced by chemicals, structural damage, and extremes of temperatures.

Baroreceptors are the body's barometric guides, detecting changes in pressure. These receptors are found in organs that extend and contract, such as the lungs and the stomach, and in the blood vessels. Allied to the autonomic nervous system, baroreceptors provide information on bladder fullness, breathing rate, and digestive process status.

Chemoreceptors are responsive to fluctuations in chemical levels in the body. They are active in the aorta and carotid arteries, and detect changes in oxygen levels in the blood. A fall in blood oxygen levels stimulates the chemoreceptors, which then alert the brain to the need for increased oxygen intake in the lungs. Chemoreceptors also provide information on chemical levels in the cerebrospinal fluid (CSF), sensing changes in certain essential constituents.

General senses

The general senses include pain, pressure, proprioception, and vibration; the receptors for these senses are distributed throughout the body. Dependent on the receptor involved, they provide information on both the external and internal environment, alerting the brain to situations that require response, and compensatory movement or action.

Special Senses

The special senses are smell, vision, taste, hearing, and equilibrium (balance), with the organs for the special senses all located in the head. The special sense organs communicate information to the brain via the cranial nerves.

The eyes are the organs of sight, with special photoreceptors called rods and cones. Rod cells provide black and white images in most conditions, while the cone cells are more specialized and require greater light intensity. The cone cells provide color information, with three different types of cone cell receptive to three different colors: blue, red, and green.

These cells on the surface of the retina receive images created by light reflected off a seen object, which they convert into electrical impulses. These impulses are conveyed to the brain by the optic nerve. The visual cortex, responsible for the interpretation of visual information, is located in the occipital cortex.

The nose is the organ of smell, with special receptors in the nasal cavity able to determine thousands of different odors. Odors are detected by chemoreceptors in the nasal cavity that then relay the information to the olfactory centers of the brain via the olfactory nerve. Our sense of smell assists in our sense of taste.

Taste is detected by taste buds located primarily on the tongue, but also distributed on the palate and throat. The taste buds are chemoreceptors, which detect a

Special senses

The special senses are sight, hearing, balance, smell, and taste. The special sense organs are all located in the head, with specialized cells detecting sensory information and relaying impulses to the brain for interpretation.

variety of tastes—sweet, sour, salty, and bitter. The taste buds on the tongue are more sensitive to sweet and salty flavors, while those on the palate and throat are more receptive to sour and bitter flavors. When stimulated, the taste buds transmit information to the brain for interpretation.

The ears are the organs of hearing. Generated sound waves are received at the eardrum, where they are relayed through the structures of the ear to reach the cochlea. Here the mechanical vibrations are converted into nerve impulses by mechanoreceptors in the inner ear. The impulses are relayed to the auditory cortex in the temporal lobe of the brain, where the sounds are interpreted.

The ears are also responsible for our sense of balance, with structures in the inner ear detecting body movement and position. The vestibular organs of the semicircular canals, utricle, and saccule monitor body position, registering changes by means of mechanoreceptors. These tiny receptor cells move in response to body movement, sending nerve impulses to the brain.

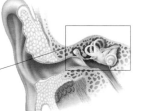

Balance

Receptors in the semicircular canals and otolith organs of the inner ear register position. Movements stimulate the hair cell receptors, which relay information to the brain via the vestibulocochlear nerve.

Hearing

The inner ear contains hair cell mechanoreceptors that relay signals along the vestibulocochlear nerve. The brain thus receives information on sound.

Smell

The tiny hairs (cilia) in the nasal cavity have chemo-receptors, which relay signals to the olfactory centers at the base of the brain, where they are translated to smells.

Sight

Photoreceptors in the eyes are sensitive to light. They send signals along the optic nerve, then to the brain's occipital cortex, which processes visual information.

Taste

The chemoreceptors on the tongue, throat, and palate are commonly known as taste buds. They transmit information on sweet, sour, salty, and bitter tastes to the brain via the cranial nerves.

Sight

Light rays enter the eye, passing through the cornea and lens, which cause the rays to be brought together and cross over, resulting in the image received by the retina to be inverted and transposed. The refraction of the cornea and the shape of the lens adjust, depending on whether the seen object is near or distant. When the light rays reach the retina, they are converted into nerve impulses that are transmitted along the optic nerve to the brain.

Cornea refracts light.

Lens adjusts shape according to distance.

Retina

Light rays

Ciliary muscles

Image on retina is inverted and small.

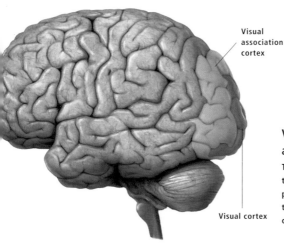

Visual association cortex

Visual cortex and visual association cortex

The visual cortex and visual association cortex lie in the occipital lobe of the brain. The visual cortex processes visual information from the eyes, while the visual association cortex processes the more complex data, such as color and movement.

Visual cortex

Sight

The eyes are responsible for the special sense of sight. Each eye receives light rays reflected from a seen object, which are transferred to the retina at the back of the eye. Before the light rays reach the retina they are adjusted and corrected by the refractive qualities of the cornea and the lens, so that resolution, color, and intensity are maximized. The iris controls the amount of light entering the eye, causing the pupil to contract or dilate according to light intensity.

The light rays then pass through the gel-like vitreous humor, to fall on the layer at the back of the eye, the retina. The curvatures of the cornea and lens cause the image cast on the retina to be inverted and transposed. The retina is an extremely sensitive layer of photoreceptors, specially designed to relay the incoming images.

Millions of these photoreceptors, called rods and cones, make up the layer of the retina, with the area immediately in front of the optic nerve most densely populated with cells. The cells convert the light rays into nerve impulses, which are then relayed along the optic nerve.

Visual pathways

Each eye has a slightly different field of vision, though the fields overlap. This binocular vision allows perception of distance and three dimensionality. Impulses commencing at the retina of each eye are transmitted via the optic nerve to the optic chiasm. At this junction some of the impulses from the left eye cross over to the right optic nerve, and vice versa. The impulses are then sent on to the visual cortex for interpretation.

The image seen by each eye is slightly different—this is binocular vision. The two images become fused into one at the optic chiasm, where the optic nerve from each eye meets, before leading off to the left and right sides of the brain. At the optic chiasm, some impulses from the left eye cross over to the right side, while some impulses from the right eye cross over to the left side. Thus, the nerve impulses sent on to the right and left sides of the brain are a combination of impulses from the left and right eye. The impulses are sent to the visual cortex for interpretation. The brain inverts the image to give us a picture the correct way up, although it is not known how this is done.

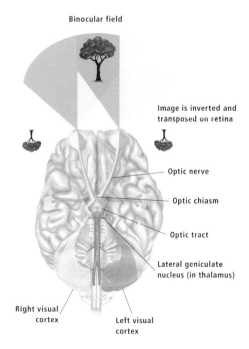

Binocular field

Image is inverted and transposed on retina

Optic nerve

Optic chiasm

Optic tract

Lateral geniculate nucleus (in thalamus)

Right visual cortex

Left visual cortex

Hearing

The ears are the special sense organs of hearing and balance. Hearing is the ability to perceive sound vibrations. These sound vibrations are transmitted through the ears and sent to the brain for processing.

Sound waves enter the ear canal of the outer ear, causing the tympanic membrane, or eardrum, to resonate and vibrate. The vibrations are passed on to the auditory ossicles in the middle ear.

The vibrations ripple through the auditory ossicles of the malleus, incus, and stapes (hammer, anvil, and stirrup). The stapes conveys these vibrations to the oval window of the cochlea.

The mechanical vibrations from the middle ear are converted into electrical impulses in the inner ear, as these vibrations cause the fluid in the cochlea to form waves. These waves stimulate tiny hair-like cells in the organ of Corti.

The organ of Corti is composed of rows of cells with hair-like projections. These projections are stimulated by movement in the cochlear fluid. Dependent on the volume of the sound, either a few hairs move, as in the case of a soft sound, or many hairs move, as in the case of the loud sound.

The organ of Corti sends impulses along the cochlear branch of the vestibulocochlear nerve, which transmits the signals from the ear to the auditory cortex in the brain. Located in the temporal lobe, the auditory cortex processes the nerve impulses and interprets sound.

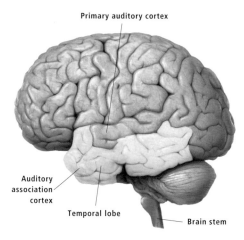

Primary auditory cortex

Auditory association cortex

Temporal lobe

Brain stem

Auditory centers

Nerve impulses from the ear are transmitted to the auditory centers of the brain via the vestibulocochlear nerve. High and low frequency sounds are processed by different parts of the auditory cortex.

How we hear

(a) Sound waves enter ear canal and hit eardrum.

(b) The vibrations of the eardrum are transmitted to the auditory ossicles.

(c) The ossicles intensify the pressure of the sound waves and transmit vibrations to the oval window (a membrane that covers the entrance to the cochlea).

(d) The mechanical vibrations from the middle ear pass into the cochlear spiral, where fluid displaces tiny hair-like receptor cells in the organ of Corti.

(e) These cells send nerve impulses along the cochlear branch of the vestibulocochlear nerve to the auditory centers in the brain.

a

Hearing

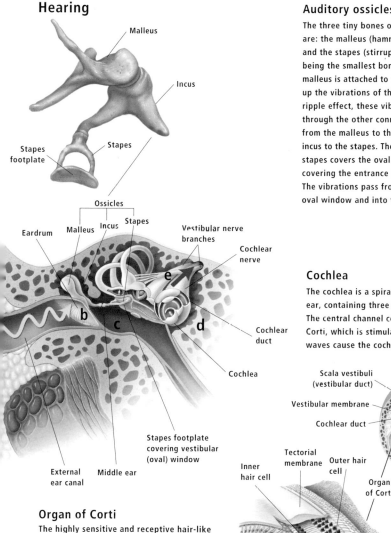

Malleus

Incus

Stapes

Stapes footplate

Ossicles
Malleus Incus Stapes

Eardrum

Vestibular nerve branches

Cochlear nerve

e

b

c

d

Cochlear duct

Cochlea

Stapes footplate covering vestibular (oval) window

External ear canal   Middle ear

Auditory ossicles

The three tiny bones of the auditory ossicles are: the malleus (hammer), the incus (anvil), and the stapes (stirrup), with the stapes being the smallest bone in the body. The malleus is attached to the eardrum, and picks up the vibrations of the membrane. In a ripple effect, these vibrations are passed through the other connecting bones, passing from the malleus to the incus, and from the incus to the stapes. The footplate of the stapes covers the oval window, a membrane covering the entrance to the cochlear spiral. The vibrations pass from the footplate to the oval window and into the cochlear spiral.

Cochlea

The cochlea is a spiral structure in the inner ear, containing three fluid-filled channels. The central channel contains the organ of Corti, which is stimulated when sound waves cause the cochlear fluid to ripple.

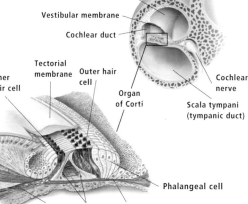

Scala vestibuli (vestibular duct)

Vestibular membrane

Cochlear duct

Tectorial membrane

Inner hair cell

Outer hair cell

Organ of Corti

Cochlear nerve

Scala tympani (tympanic duct)

Phalangeal cell

Nerve fibers   Pillar cell   Basilar membrane

Organ of Corti

The highly sensitive and receptive hair-like cells in the organ of Corti enable us to hear even very faint sounds. The cells are triggered by movement in the cochlear fluid, and when activated, send nerve impulses to the auditory cortex of the brain.

Balance

The ear serves as our organ of balance (equilibrium), although our sense of balance is assisted by the eyes, which give us a visual sense of position, and by the joints and muscles.

Tiny organs in the ear, called the vestibular organs, control our sense of balance. These organs comprise the semicircular canals and the otolith organs of the inner ear—the saccule and the utricle. The vestibular organs alert the brain to changes in body position.

Both the utricle and saccule are hollow sacs, filled with a gelatinous fluid called endolymph. Attached to the inner surface of each structure are tiny hair-like structures, which project into the hollow space. Lying over the hairs are crystals of calcium carbonate, called otoliths. Acting like a spirit level, these otoliths change position with movement, and when they change position, the underlying hair cells are activated. These hair cells fire off nerve impulses to the brain, for reaction and response to correct position.

Connecting with the utricle are the three semicircular canals. Lying at right angles to one another, they contain hair cells in the ampulla. The semicircular canals work on a similar principle to the utricle and saccule. The fluid in the semicircular canals is disturbed when the body moves, triggering hair-like cells in the ampulla that in turn send nerve impulses to the brain, alerting it to change in body position.

Balance

Balance is one of the special senses. Tiny organs in the inner ear are responsible for monitoring balance and alerting the brain to changes in body position. The specialized organs of balance are the semicircular canals and the otolith organs, collectively known as the vestibular organs. Movement in these organs triggers nerve endings, which in turn send nerve impulses to the brain.

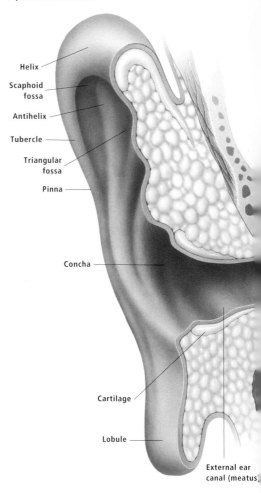

Helix

Scaphoid fossa

Antihelix

Tubercle

Triangular fossa

Pinna

Concha

Cartilage

Lobule

External ear canal (meatus)

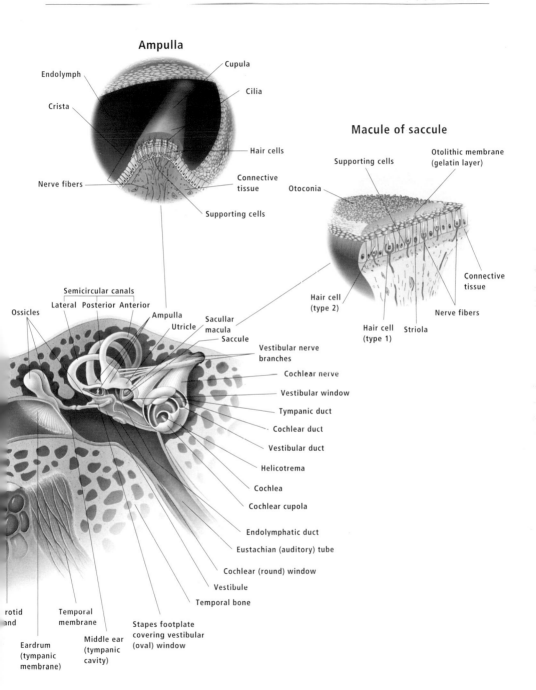

Ampulla

Cupula

Endolymph

Cilia

Crista

Hair cells

Nerve fibers

Connective tissue

Supporting cells

Macule of saccule

Supporting cells

Otolithic membrane (gelatin layer)

Otoconia

Connective tissue

Hair cell (type 2)

Nerve fibers

Hair cell (type 1)

Striola

Semicircular canals

Lateral Posterior Anterior

Ossicles

Ampulla

Utricle

Sacullar macula

Saccule

Vestibular nerve branches

Cochlear nerve

Vestibular window

Tympanic duct

Cochlear duct

Vestibular duct

Helicotrema

Cochlea

Cochlear cupola

Endolymphatic duct

Eustachian (auditory) tube

Cochlear (round) window

Vestibule

Temporal bone

rotid and

Temporal membrane

Stapes footplate covering vestibular (oval) window

Eardrum (tympanic membrane)

Middle ear (tympanic cavity)

Smell

Our sense of smell allows us to detect and recognize airborne substances, and also enhances the sense of taste by recognizing odors related to the food and drink we ingest.

Specialized chemoreceptor nerve cells line the roof of the nasal cavity. Located in the mucous membrane of the nose, the special area for odor detection is known as the olfactory epithelium or olfactory organ. The absorbent properties of the moist surface of the olfactory mucosa trap and dissolve odor molecules from passing air.

Millions of olfactory receptor cells make up the olfactory organ. The tiny cilia of these receptor cells are stimulated by passing odor molecules.

Nerve impulses from the cells are then transmitted through holes in the cribriform plate to the olfactory bulbs. The olfactory

Smell

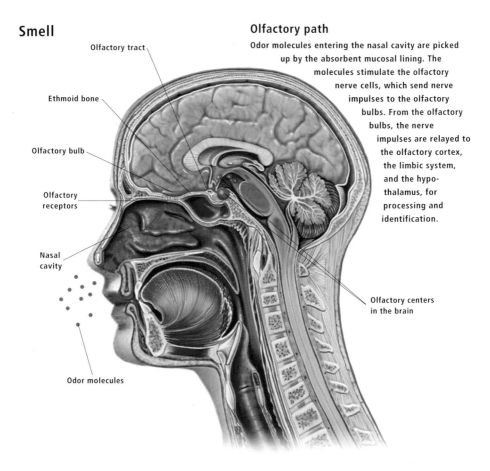

Olfactory tract

Ethmoid bone

Olfactory bulb

Olfactory receptors

Nasal cavity

Odor molecules

Olfactory path

Odor molecules entering the nasal cavity are picked up by the absorbent mucosal lining. The molecules stimulate the olfactory nerve cells, which send nerve impulses to the olfactory bulbs. From the olfactory bulbs, the nerve impulses are relayed to the olfactory cortex, the limbic system, and the hypothalamus, for processing and identification.

Olfactory centers in the brain

bulbs are egg-shaped structures, located in the frontal lobe of the brain, and form the frontal area of the olfactory area of the brain. The medial and lateral olfactory areas are connected to the frontal olfactory area by the olfactory tract. Nerve impulses from the olfactory bulbs are then relayed to the olfactory cortex, the limbic system, and the hypothalamus in the brain, where the information is processed and the smell is determined and recognized.

The medial and lateral olfactory areas have connections with the hypothalamus, hippocampus, and brain stem, with these structures controlling automatic responses to smell, such as salivation.

Anterior nucleus
of thalamus

Cingulate gyrus

Olfactory bulb

Thalamus

Amygdala

Hippocampus

Smell and the limbic system

The limbic system plays a key role in memory and emotion. The connection between the limbic system and the olfactory organs means that nerve impulses received from the olfactory cells often trigger memory and emotion. This is why smells often evoke memories of past places and feelings. The olfactory nerve impulses can activate the hypothalamus and pituitary gland, which triggers the release of hormones associated with appetite and emotional responses.

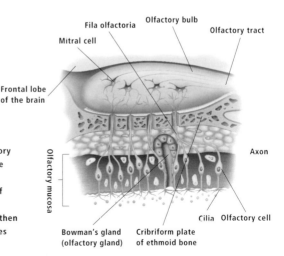

Fila olfactoria

Olfactory bulb

Olfactory tract

Mitral cell

Frontal lobe
of the brain

Olfactory mucosa

Axon

Cilia   Olfactory cell

Bowman's gland
(olfactory gland)

Cribriform plate
of ethmoid bone

Olfactory apparatus

Lying in the roof of the nasal cavity, the olfactory apparatus contains millions of specialized nerve cells dedicated to odor recognition. The odor molecules are picked up by the moist surface of the mucosal lining, stimulating the nerve cells of the olfactory organ. The nerve impulses are then transmitted along the nerve fibers through holes in the cribriform plate to the olfactory bulb.

Taste

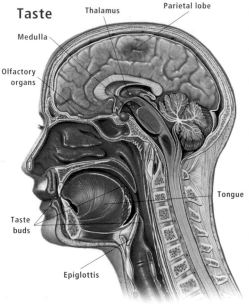

Thalamus

Parietal lobe

Medulla

Olfactory
organs

Tongue

Taste
buds

Epiglottis

Taste pathways

Taste buds on the tongue, palate, and throat send
nerve impulses via the facial nerve, glossopharyngeal nerve, and vagus nerve to the medulla
oblongata in the brain stem. These impulses are
then relayed to the thalamus, and on to centers
in the parietal lobe of the brain for identification.
The olfactory organs play a key role in determining
different tastes.

Taste

Taste buds located on the tongue, palate,
and throat provide our sense of taste.
Located in papillae in these areas, the
taste buds are special chemoreceptor cells,
designed to detect taste sensations.

When food or drink is combined with
saliva, the taste buds receive information
through openings known as taste pores.
This stimulates nerve activity in the taste
bud, with the nerves sending impulses to
the brain for interpretation.

Certain taste buds on the tongue are
more receptive to specific flavors. The front
of the tongue is more sensitive to sweet, the
sides of the tongue register salty and sour,
while the back of the tongue is more sensitive to bitter flavors. The taste buds on the
throat and palate are more receptive to sour
and bitter flavors.

Three different nerves are involved in
transmitting information on taste: the facial
nerve, the glossopharyngeal nerve, and the
vagus nerve.

Taste bud

The taste buds are the chemoreceptors for the
sense of taste. Food and drink reacts with saliva
in the mouth, with the information reaching the
cells through taste pores. Once the cells are
stimulated, they relay information on taste to
centers in the brain.

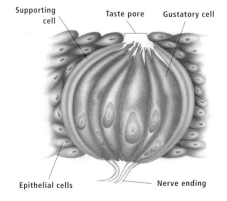

Supporting
cell

Taste pore

Gustatory cell

Epithelial cells

Nerve ending

The facial nerve relays information from taste buds located on the front two-thirds of the tongue. The glossopharyngeal nerve relays information from taste buds at the back of the tongue. Taste buds in the throat convey impulses via the vagus nerve.

Information sent along the three nerves is conveyed firstly to the medulla oblongata in the brain stem, and from there it is relayed to the thalamus and on to processing centers in the parietal lobe of the brain for identification.

Our sense of smell is closely allied to the sense of taste, with smell being attributable to determination of a large proportion of what we taste.

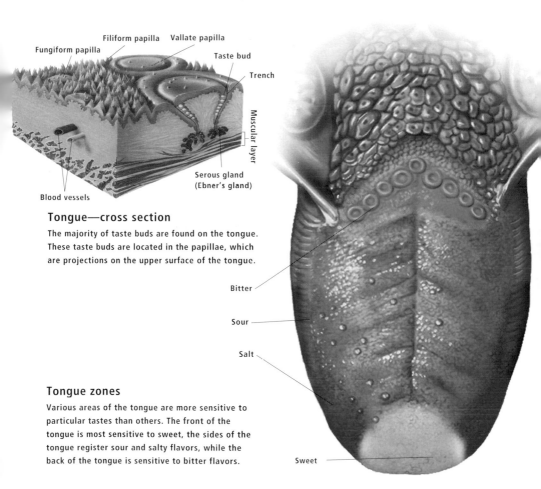

Fungiform papilla
Filiform papilla
Vallate papilla
Taste bud
Trench
Muscular layer
Serous gland (Ebner's gland)
Blood vessels

Tongue—cross section

The majority of taste buds are found on the tongue. These taste buds are located in the papillae, which are projections on the upper surface of the tongue.

Bitter

Sour

Salt

Tongue zones

Various areas of the tongue are more sensitive to particular tastes than others. The front of the tongue is most sensitive to sweet, the sides of the tongue register sour and salty flavors, while the back of the tongue is sensitive to bitter flavors.

Sweet

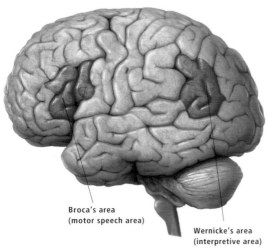

Broca's area
(motor speech area)

Wernicke's area
(interpretive area)

Speech

Primarily under the control of the nervous system, speech is created by the combined efforts of many parts of the body. The action of muscles in the throat, larynx, and mouth adapt raw sound into distinguishable words and sounds.

Sound is produced as air exits the lungs. This air passes through the vocal cords, and dependent on their tension and position, can produce sound. During breathing, the vocal cords are separated to allow optimum levels of oxygen into the body. When speaking the cords are brought together, and air is forced between them, causing them to vibrate. The length and

Speaking and understanding speech

Specialized centers in the brain coordinate our speech functions and our understanding of speech and language. Broca's area, located in the frontal lobe of the brain, is the control center for speech production. It instructs the breathing muscles, laryngeal muscles, pharynx, tongue, and lips, coordinating the efforts of all the structures involved in speech production. Comprehension and understanding of speech and language is dealt with by Wernicke's area, located in the temporal lobe of the brain.

Lip movement for speech

The movement of the lips modifies the sounds that come from the larynx and vocal cords into speech. For example, the lips need to join to make the sound "m," need to touch the teeth (with the bottom lip) to make "f," and round to make "o." The accurate movements that speech requires are made possible by a complex arrangement of muscles around the mouth and cheek areas.

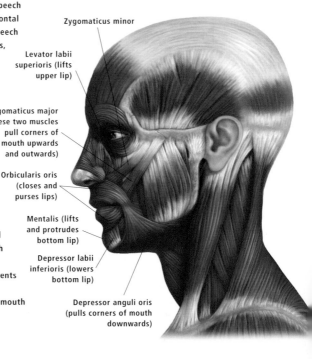

Zygomaticus minor

Levator labii superioris (lifts upper lip)

Zygomaticus major (these two muscles pull corners of mouth upwards and outwards)

Orbicularis oris (closes and purses lips)

Mentalis (lifts and protrudes bottom lip)

Depressor labii inferioris (lowers bottom lip)

Depressor anguli oris (pulls corners of mouth downwards)

tension of the vocal ligaments can be adjusted to produce sounds of different pitch, while volume is determined by the force used when breathing air through the vocal cords. The vocal cords stretch across the laryngeal cavity, connected to cartilage at either side. The cords attached to the front are fixed, but at the back the cords attach to cartilages which move freely, allowing the cords to vibrate and produce sound.

Vowels are usually produced by modifying the shape of the air column in the throat, mouth, and nose, while consonants are produced by introducing noise into the pure tone of the vibrating column (for example, by touching the tongue to the back of the teeth or the roof of the mouth).

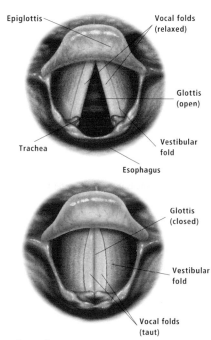

Producing sound

During breathing, the vocal cords are in the open position. For speech production, the cords come together, and air from the lungs is forced between the two cords, which causes them to vibrate. The raw sound produced by these vibrations is adjusted by the muscles of the mouth, cheeks, and lips to create the variety of sounds required for speech.

Vocal cords

The vocal cords, or vocal folds, stretch across the cavity of the larynx, connected to the cartilage structure. Firmly fixed to the thyroid cartilage at the front, they are attached to more mobile components of the larynx at the back, allowing them to be relaxed or stretched tight, and brought close together or held open. The two folds of mucous membrane are necessarily flexible, to allow the movements required for speech production.

Diseases and Disorders of the Nervous System and Sense Organs

There are a wide range of diseases and disorders associated with the nervous system, and affecting many parts of the body. Some of the more common conditions are discussed below.

Alzheimer's disease

More common in older people, though it can occur in middle age, Alzheimer's disease is caused by deterioration and death of brain cells. Generally this loss—which can account for up to 20 percent of the normal brain volume—occurs in the temporal, parietal, and frontal lobes of the brain. Its occurrence in these areas affects functions such as memory and intellectual processing. The cause of Alzheimer's disease is unknown, and at present there is no known cure. Medications are now available that slow the progress of the disease.

Paralysis

Paralysis affects the control of certain parts of the body. Varying degrees of paralysis can be experienced, determined by the location of injury. Injury or damage to the brain, spinal cord, nerves, or muscles can result in paralysis, with the degree of paralysis

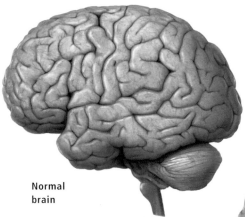

Normal brain

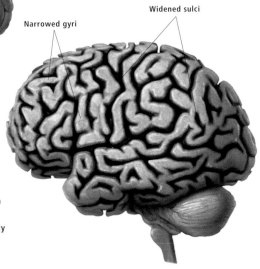

Narrowed gyri

Widened sulci

Alzheimer's disease

The capacity for recent memory is most often affected in Alzheimer's disease, although older memories generally remain unaffected during the early stages of the disease. The disease is characterized by deterioration and death of brain cells, which causes some brain shrinkage, while the gaps between brain matter increase, gradually reducing actual brain volume by as much as 20 percent.

dependent on whether the injury affects the peripheral nervous system or the central nervous system. Damage to peripheral nerves can result in paralysis in a specific area and subsequent muscle wasting. Damage to the central nervous system can result in hemiplegia, paraplegia, or quadriplegia. Hemiplegia, often occurring as the result of a stroke, indicates damage to the motor cortex of the brain, affecting the limbs on one side of the body. Paraplegia, most commonly the result of motor or sporting injuries,

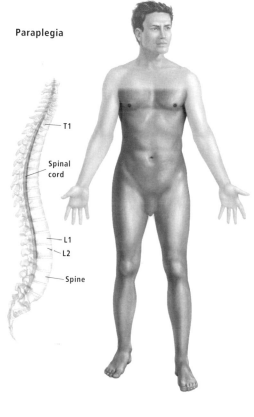

Paraplegia

- T1
- Spinal cord
- L1
- L2
- Spine

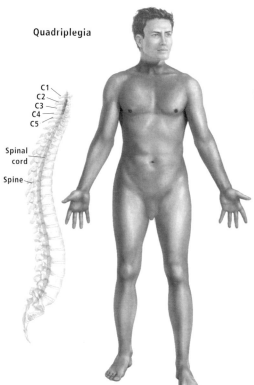

Quadriplegia

- C1
- C2
- C3
- C4
- C5
- Spinal cord
- Spine

Paralysis

Most often occurring as a result of brain or spinal injuries, paralysis is the loss of movement in an area of the body. The degree of loss of movement is dependent on the nerves affected or damaged. For instance, paraplegia is the result of spinal nerve injury between T1 and L2 vertebrae. Paraplegia results in loss of movement in parts of the body below the injury area. Quadriplegia is usually the result of injury to the spinal cord at the site of C4 or C5 vertebrae. Injury at this location generally results in loss of movement from the neck down.

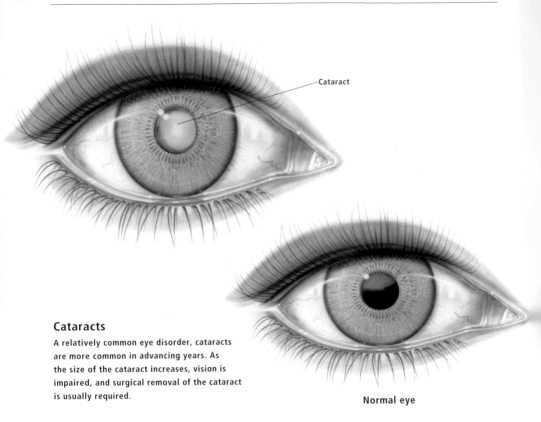

Cataract

Cataracts

A relatively common eye disorder, cataracts
are more common in advancing years. As
the size of the cataract increases, vision is
impaired, and surgical removal of the cataract
is usually required.

Normal eye

causes loss of movement from below the
point of damage in the spinal cord between
the T1 vertebra in the thoracic region and
the L2 vertebra in the lumbar region.

Quadriplegia, most often caused by injury to the cervical vertebrae, in particular
C4 and C5, results in loss of movement
from the neck down.

Cataracts

Occurring as a result of the ageing process,
or less commonly, as a result of injury,

cataracts are cloudy spots affecting the lens
of the eye. These cloudy spots gradually
increase in size, affecting vision. Cataract
removal is usually a simple routine procedure, although some cataracts respond to
medication that dilates the pupil.

Otitis media

Most commonly caused by bacteria entering via the eustachian tube, otitis media is
inflammation of the middle ear. The presence of bacteria can cause the build-up of

pus in the middle ear area, causing painful earache, fever, and deafness in the affected ear. In some cases, the build-up of pus can cause the eardrum to rupture. The symptoms are usually treated with antibiotics and painkillers. Children often suffer recurring bouts of otitis media, commonly known as glue ear, and in these cases, the symptoms are often alleviated by the surgical insertion of drainage tubes known as grommets.

Vertigo

Vertigo often produces symptoms such as dizziness, confusion, vomiting, and nausea. It is most often caused by disorders of the inner ear that can affect the delicate balance organs, the vestibular organs, or can be a result of disorders in the central nervous system.

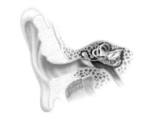

Otitis media

Bacterial infection, generally entering via the eustachian tube connecting the middle ear to the throat, is the most common cause of otitis media.

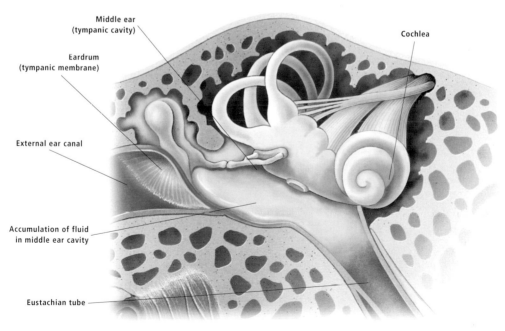

Middle ear
(tympanic cavity)

Cochlea

Eardrum
(tympanic membrane)

External ear canal

Accumulation of fluid
in middle ear cavity

Eustachian tube

THE AUTONOMIC NERVOUS SYSTEM

A division of the nervous system, the autonomic nervous system is primarily concerned with the operation of body functions and mechanisms outside our voluntary control. The autonomic nervous system is perhaps the most important system in homeostasis, with its responsibilities including the maintenance and control of those organs essential to life, keeping conditions in the body's interior at a constant rate. The autonomic nervous system is also responsible for providing our survival and self-preservation instincts and reactions, along with the ability to respond to stressful and dangerous situations.

Numerous mechanisms are activated by the autonomic nervous system, when we are nervous or under duress, fearful, or anxious. These mechanisms raise the body's defenses, and prepare it for response to crises.

The autonomic nervous system consists of nerve cells in the brain and spinal cord, their fibers which leave the central nervous system, collections of nerve cells in the various body cavities (ganglia), and nerve fibers which are distributed in the internal organs.

The autonomic nervous system is divided into two divisions: a sympathetic division and a parasympathetic division. The two divisions must work in harmony, complementing the activities of one another, in order to maintain the body's well-being. While these two divisions are quite separate entities, they do have areas of overlap. For instance, both divisions are involved in the control of the urinary system. The sympathetic division constricts the sphincter muscles of the bladder, thus controlling continence, while the parasympathetic division relaxes the muscles, allowing elimination.

Ganglion

In the case of the autonomic nervous system, a ganglion (plural, ganglia) refers to a group of nerve cell bodies. Sympathetic and parasympathetic ganglia are located along the spinal cord and in the body cavities.

Sympathetic Division

When we are presented with dangerous, frightening, or stressful situations, the sympathetic division of the autonomic nervous system, known as our "fight-or-flight" system, is activated.

When faced with such situations, we often experience rapid heartbeat, accelerated breathing, and enlarged pupils—the sympathetic division produces all of these reactions automatically. The division also often incorporates other measures, such as increased blood pressure and increased blood supply to muscles, in an effort to increase our ability to cope with crises.

Autonomic nervous system—sympathetic division

The sympathetic division of the autonomic nervous system is responsible for preparing the body for emergencies and crises. The sympathetic ganglion originate out of the thoracic and lumbar spinal regions.

Sympathetic division

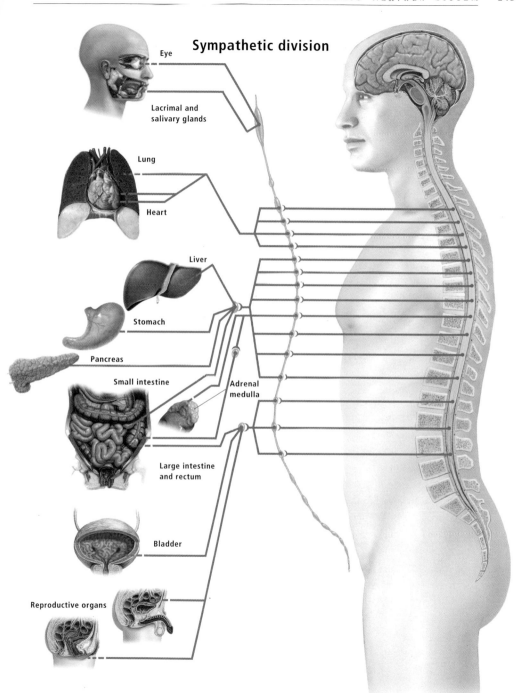

Eye

Lacrimal and salivary glands

Lung

Heart

Liver

Stomach

Pancreas

Small intestine

Adrenal medulla

Large intestine and rectum

Bladder

Reproductive organs

The sympathetic division also influences body temperature, either by stimulating sweat glands to lower body temperature, or by activating hair follicles in the skin to reduce heat loss.

The nerve cells of the sympathetic division lie in the thoracic and lumbar regions of the spinal cord, with a long chain of nerve cells lying alongside the backbone, called the sympathetic trunk, which supplies plexuses (networks of nerves), and internal organs.

Parasympathetic Division

The parasympathetic division is at its busiest when the body is at rest. The parasympathetic division seeks to restore the body to a normal operational state, promoting the activities which aid normal functioning. Saliva production, movement of the digestive tract muscles, promotion of digestion and peristalsis, and elimination of urine and feces are all under the control of the parasympathetic division. While the body is at rest, the parasympathetic division lowers heart rate and blood pressure, activates the glands of the respiratory airways, thus increasing secretions, and constricts the pupils of the eyes. All of these measures help the body to conserve energy and replenish and restore internal body functions to normal.

In the male reproductive system, the parasympathetic division is responsible for erection of the penis by increasing blood flow to the cavernous spaces of the penis, and it is also involved in ejaculation of semen.

Vagus nerve

The vagus nerve is responsible for the control of various upper body functions that are outside our voluntary control. These functions include breathing, swallowing, speaking, heart beat, the constriction of blood vessels and bronchial tubes, and digestion.

Interaction Between Sympathetic and Parasympathic Divisions

In the thoracic and the abdominopelvic cavities, sympathetic nerve fibers mix with parasympathetic nerve fibers in a series of plexuses.

Solar plexus

The solar plexus, or celiac plexus, is a network of nerve cells located on the abdominal aorta, behind the stomach. The nerve cells of the solar plexus extend out to all the abdominal organs. The coverage achieved by these nerves allows the autonomic nervous system to maintain constant body functions in the abdominal region. These nerves target the intestines, maintaining movement of the smooth muscle, and the adrenal secretions, as well as maintaining constancy in the kidneys, spleen, liver, and pancreas.

Autonomic nervous system— parasympathetic division

The parasympathetic division of the autonomic nervous system goes to work when the body is at rest, slowing down many of the body's functions, thus allowing energy levels to be refuelled and restored. The parasympathetic ganglia originate in the brain stem, and sacral region of the spine.

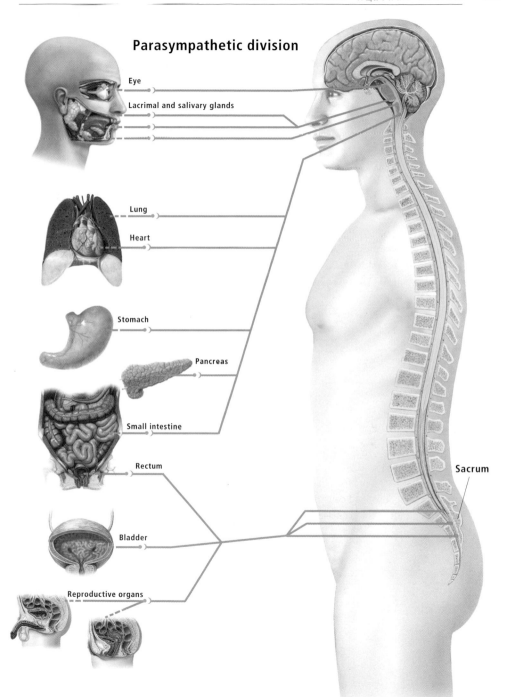

Parasympathetic division

Eye

Lacrimal and salivary glands

Lung

Heart

Stomach

Pancreas

Small intestine

Rectum

Sacrum

Bladder

Reproductive organs

Diseases and Disorders of the Autonomic Nervous System

There are a range of disease and disorders affecting the autonomic nervous system. Some of the more common ailments and conditions are discussed below.

Incontinence

While there are many causes of incontinence, the shared influence of both the sympathetic and parasympathetic divisions of the autonomic nervous system can affect the smooth coordinated operation of the bladder. The sympathetic division controls constriction of the sphincter muscle while the parasympathetic division controls relaxation of the muscle. If either division fails to adhere to its part of the operation, then bladder incontinence can result, causing the involuntary release of urine.

Raynaud's disease

Problems with the autonomic nervous system may be attributable to the symptoms of Raynaud's disease, also known as Raynaud's phenomenon.

Generally affecting the extremities of the fingers and toes, the symptoms are caused by constriction of the blood vessels, which manifest as whitening of the digits, coupled with loss of body temperature in the affected area. The affected area then turns blue, before regaining color when blood flow returns to normal.

Male incontinence

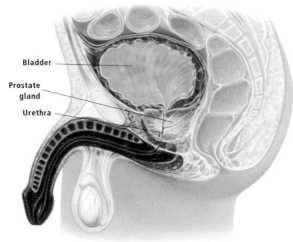

Bladder

Prostate gland

Urethra

Incontinence

The autonomic nervous controls the sphincter muscle of the bladder, with the sympathetic division controlling constriction and the parasympathetic division controlling relaxation of the muscle. Incontinence can result if the nerves are damaged or disrupted.

Female incontinence

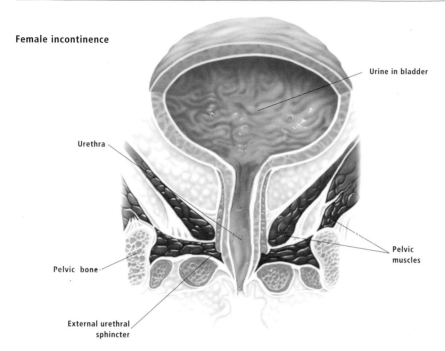

Urine in bladder

Urethra

Pelvic muscles

Pelvic bone

External urethral sphincter

Horner's syndrome

Caused by disruption or damage to the sympathetic nerves in the cervical region, the symptoms of Horner's syndrome manifest as a small pupil, a drooping eyelid, and inactive facial sweat glands.

Impotence

Impotence is a man's inability to achieve or sustain an erection—a condition more commonly affecting older men. Sometimes, the underlying cause is psychological rather than physical, and the duration of the condition varies from temporary to long term, depending on the cause.

Impotence does not affect sperm production levels, which remain constant.

If the underlying cause is physical, it can be as a result of some medical conditions, such as diabetes mellitus, stroke, or alcoholism; or due to malfunction of the governing glands of the penis, such as the pituitary gland or the testes. Impotence can even occur as a side effect of some prescription drugs, such as those taken for high blood pressure, or following relatively mild infections.

If the underlying cause is of a psychological nature, then various options may be considered, including counselling, or advice and treatment by a sex therapist.

Determination of the underlying cause must be addressed first, before a plan of action can be initiated.

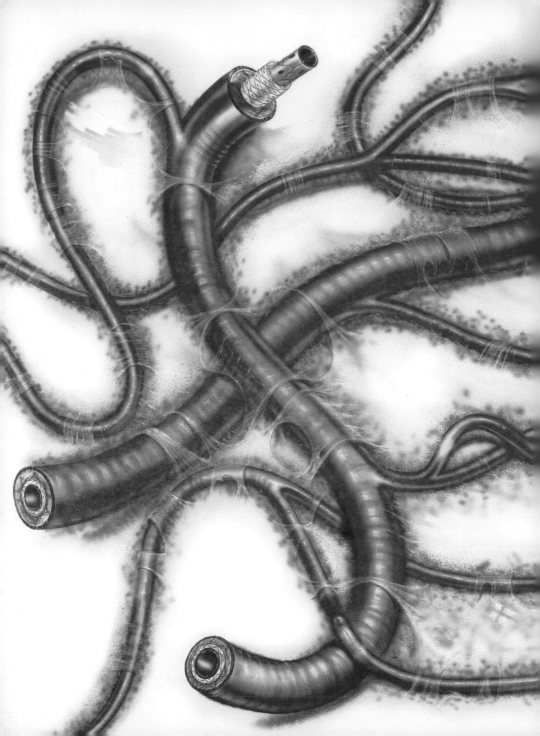

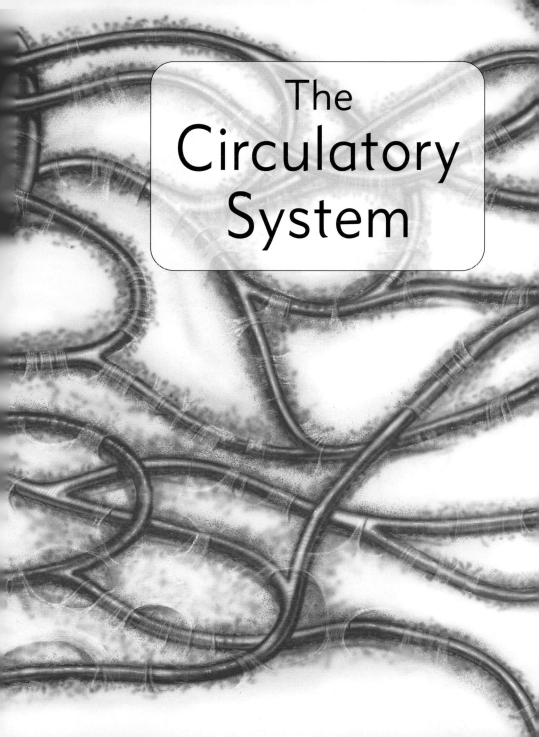

The
Circulatory
System

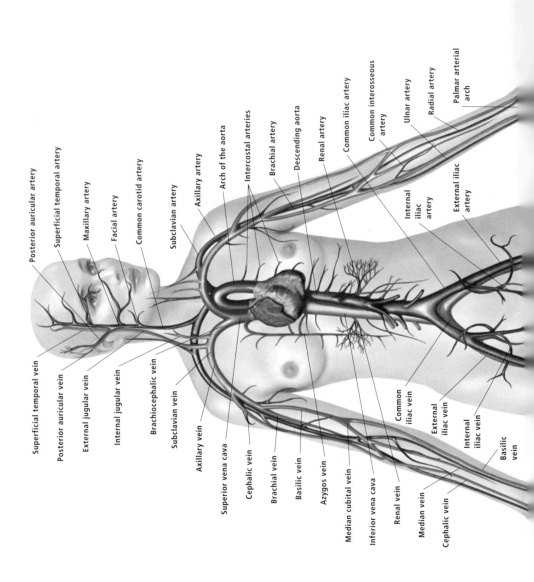

Posterior auricular artery
Superficial temporal artery
Maxillary artery
Facial artery
Common carotid artery
Subclavian artery
Axillary artery
Arch of the aorta
Intercostal arteries
Brachial artery
Descending aorta
Renal artery
Common iliac artery
Common interosseous artery
Ulnar artery
Radial artery
Palmar arterial arch
Internal iliac artery
External iliac artery

Superficial temporal vein
Posterior auricular vein
External jugular vein
Internal jugular vein
Brachiocephalic vein
Subclavian vein
Axillary vein
Superior vena cava
Cephalic vein
Brachial vein
Basilic vein
Azygos vein
Median cubital vein
Inferior vena cava
Renal vein
Median vein
Cephalic vein
Common iliac vein
External iliac vein
Internal iliac vein
Basilic vein

THE CIRCULATORY SYSTEM

The heart and blood vessels are the major components of the circulatory system. The heart is the pump for the system, maintaining blood flow at a constant rate throughout the body. Blood runs through a system of arteries, arterioles, veins, venules, and capillaries, which form a complete circuit.

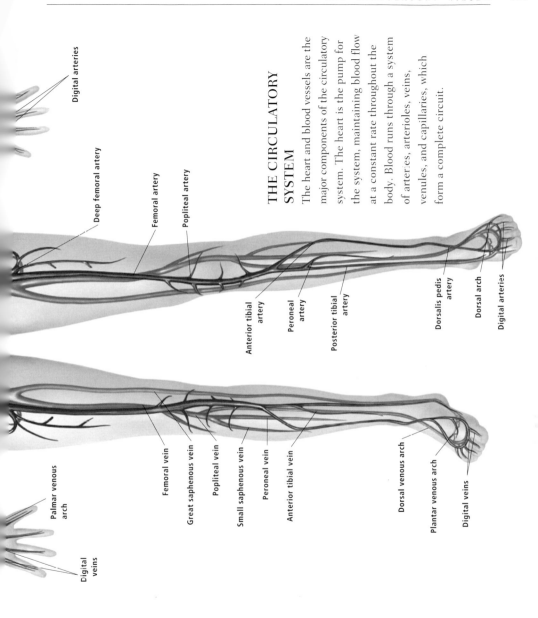

Digital arteries

Deep femoral artery

Femoral artery

Popliteal artery

Anterior tibial artery

Peroneal artery

Posterior tibial artery

Dorsalis pedis artery

Dorsal arch

Digital arteries

Palmar venous arch

Digital veins

Femoral vein

Great saphenous vein

Popliteal vein

Small saphenous vein

Peroneal vein

Anterior tibial vein

Dorsal venous arch

Plantar venous arch

Digital veins

There are two separate circulations in the circulatory system—the systemic circulation and the pulmonary circulation—with the heart and blood vessels playing a major role in both systems.

The systemic circulation carries nutrient-rich, oxygenated blood throughout the body. The oxygen and nutrients are deposited in body tissues, and any waste products and gases are transferred to the blood.

On the completion of the circuit of the body, the blood returns to the heart. By the time this blood has completed the full circuit of the body, its oxygen levels are depleted, and it is instead laden with carbon dioxide and waste products.

The pulmonary circulation takes the oxygen-depleted blood from the heart to the lungs. Gas exchange takes place in the lungs, and the blood is once more enriched with oxygen, and returned to the heart to enter the systemic circulation again.

The four chambers of the heart coordinate to maintain blood flow at a steady rate and to ensure that blood oxygen levels are optimized.

Arteries

Oxygenated blood from the heart is transported throughout the body by an extensive network of arteries.

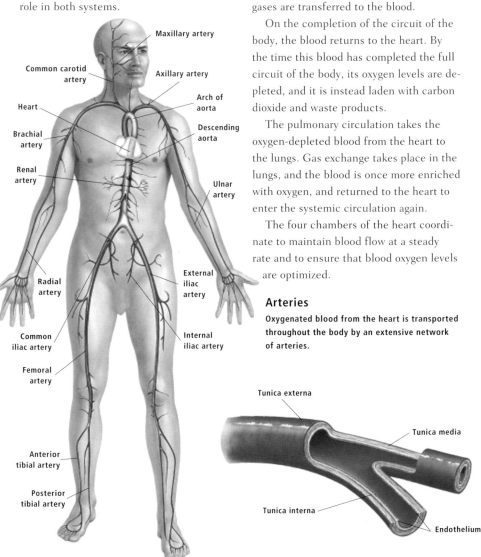

Maxillary artery

Common carotid artery

Axillary artery

Arch of aorta

Heart

Descending aorta

Brachial artery

Renal artery

Ulnar artery

Radial artery

External iliac artery

Common iliac artery

Internal iliac artery

Femoral artery

Anterior tibial artery

Posterior tibial artery

Tunica externa

Tunica media

Tunica interna

Endothelium

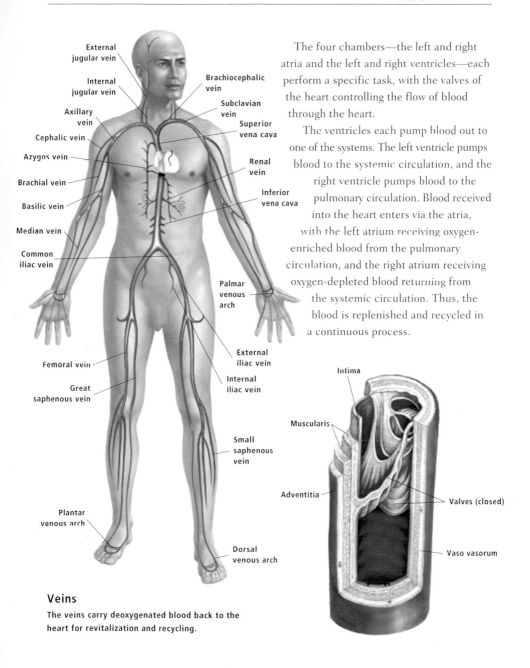

The four chambers—the left and right atria and the left and right ventricles—each perform a specific task, with the valves of the heart controlling the flow of blood through the heart.

The ventricles each pump blood out to one of the systems. The left ventricle pumps blood to the systemic circulation, and the right ventricle pumps blood to the pulmonary circulation. Blood received into the heart enters via the atria, with the left atrium receiving oxygen-enriched blood from the pulmonary circulation, and the right atrium receiving oxygen-depleted blood returning from the systemic circulation. Thus, the blood is replenished and recycled in a continuous process.

External jugular vein

Internal jugular vein

Axillary vein

Cephalic vein

Azygos vein

Brachial vein

Basilic vein

Median vein

Common iliac vein

Brachiocephalic vein

Subclavian vein

Superior vena cava

Renal vein

Inferior vena cava

Palmar venous arch

Femoral vein

Great saphenous vein

External iliac vein

Internal iliac vein

Small saphenous vein

Plantar venous arch

Dorsal venous arch

Intima

Muscularis

Adventitia

Valves (closed)

Vaso vasorum

Veins

The veins carry deoxygenated blood back to the heart for revitalization and recycling.

Artery

Arterioles

Capillaries

Venules

Vein

Capillary bed

The capillaries are the smallest blood vessels, and it is here that the arterial and venous systems meet. Nutrients permeate through the walls of the capillaries as they are deposited at target cells and tissues by arterial blood. Waste products from the tissues and cells enter through the walls of the capillaries. The capillaries are grouped together in a capillary bed which serves to join an artery and a vein. The deoxygenated blood, containing waste products, is drained through the veins of the venous system.

Blood Vessels

Blood leaving the heart for the systemic circulation is pumped through the aorta, the largest artery in the body, and on through the arterial system to supply the cells of body tissues and organs with oxygen and nutrients. Nutrient transfer takes place in tiny capillaries that connect the arteries and veins. This blood then travels through the veins and is returned to the heart.

Aorta

The aorta is the largest artery in the body. It is connected to the left ventricle of the heart, and forms a hairpin bend of thick elastic tube, which travels down through the back of the thorax and into the abdomen.

Arising from the thoracic section of the aorta are the coronary arteries that supply the heart, the carotid arteries that supply the head and neck, and the subclavian arteries that supply the upper limbs. In the abdominal region, the aorta supplies the abdominal organs. It then divides into two major branches, the left and right common iliac, which supply the pelvis and lower limbs.

Superior and inferior vena cavae

The superior and inferior vena cavae are the two largest veins in the body. Veins draining the head, neck, and upper limbs eventually join up with the superior vena cava, which drains into the heart from above. Veins draining the pelvic region, the lower limbs, and the abdominal organs eventually join up with the inferior vena cava, which drains into the heart from below.

Arteries

Arteries are comprised of three layers: an outer layer of tunica adventitia, a middle layer of tunica media, and an inner layer of tunica intima. These layers form the thick walls of the arteries that carry the blood supply.

Veins

Blood returning to the heart travels through the venous system. While veins are comprised of the same three layers as arteries—the tunica adventitia, tunica media, and tunica intima—the layers are much thinner.

Some veins, particularly those found in the lower limbs, contain valves that control the flow of blood. These valves operate with the assistance of muscle, with contracting muscles compressing the valves, forcing blood toward the heart.

Capillaries

Capillaries pass oxygen and nutrients to the body cells and tissues. As they approach their connection at the capillary bed, both arteries and veins become gradually smaller in diameter, with the arteries giving way to arterioles. These arterioles join with tiny venules of the venous system at the capillary bed.

Fenestrated capillary

Continuous capillary

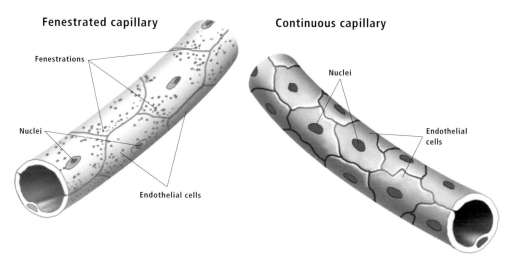

Fenestrations

Nuclei

Nuclei

Endothelial cells

Nuclei

Endothelial cells

Blood

The body of each adult contains around 10 pints (approximately 5 liters) of blood, which travels a complete circuit of the body (at rest) in around one minute. The components of blood include: red blood cells (erythrocytes), white blood cells (leukocytes), platelets, proteins, chemicals, and plasma.

The disk-shaped red blood cells (erythrocytes) are produced in the bone marrow. These blood cells contain hemoglobin, which carries oxygen, and is responsible for the characteristic red color of blood. On reaching the recipient cells and tissues, the oxygen in hemoglobin is exchanged for carbon dioxide, which is carried back to the lungs for expiration.

There are several types of white blood cells, each performing specific tasks generally associated with defense against viruses and bacteria.

Neutrophils are usually the first white blood cells to arrive to fight infection or foreign invaders. After releasing bacteria-killing enzymes, the neutrophils then ingest micro-organisms and foreign particles. Next to arrive are the monocytes.

Monocytes usually circulate in the system for 1–2 days, before transforming into macrophages when they enter the body

Blood vessel

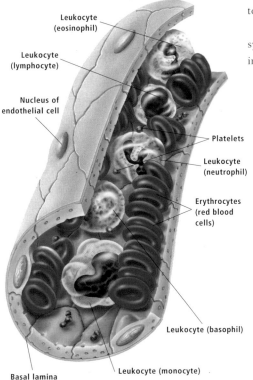

Leukocyte (eosinophil)

Leukocyte (lymphocyte)

Nucleus of endothelial cell

Platelets

Leukocyte (neutrophil)

Erythrocytes (red blood cells)

Leukocyte (basophil)

Leukocyte (monocyte)

Basal lamina

Blood

This illustration is representational only, showing the various components found in blood. On average, each cubic millimeter (0.006 cubic inches) of blood contains 4 to 6 million red blood cells and 4,000–10,000 white blood cells.

Red blood cell

tissues. Macrophages are designed to fight infection by engulfing foreign invaders. Eosinophils perform a similar duty to macrophages, cleaning up and disposing of foreign invaders, but they also release enzymes that cause allergic reactions and kill off some parasites.

Basophils also release substances, such as histamine, designed to combat invading allergens by increasing the body's response. Lymphocytes fall into three types: natural killer cells, T cells, and B cells. Natural killer cells and T cells attack invading viruses and bacteria, while B cells produce antibodies.

Platelets emit chemicals to attract more platelets to injury sites, thus triggering the formation of a clot.

The life span of a red blood cell is around 120 days, while most white blood cells have a life span of only a few days. The exceptions to this are T and B cells, which can last for years.

White blood cells

Monocyte

Macrophage

Neutrophil

Basophil

Eosinophil

Lymphocyte

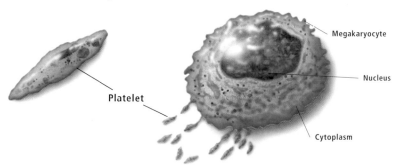

Platelet

Megakaryocyte

Nucleus

Cytoplasm

Heart

The heart lies in the center of the thorax, nestled between the lungs, and wrapped in a double layer of membrane called the pericardium. Consisting of four chambers, the heart is essentially divided through the middle, into two halves, by a septum, with each half then separated again by a septum. These divisions create the left atrium and left ventricle, and the right atrium and right ventricle. The septum creates a thick muscular wall between the chambers, ensuring each chamber is isolated from its neighbors. Strategically positioned between each of the atria and their respective ventricles are powerful valves,

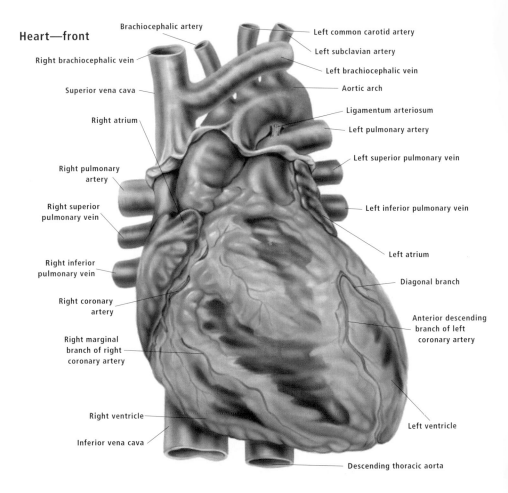

Heart—front

Brachiocephalic artery

Right brachiocephalic vein

Superior vena cava

Right atrium

Right pulmonary artery

Right superior pulmonary vein

Right inferior pulmonary vein

Right coronary artery

Right marginal branch of right coronary artery

Right ventricle

Inferior vena cava

Left common carotid artery

Left subclavian artery

Left brachiocephalic vein

Aortic arch

Ligamentum arteriosum

Left pulmonary artery

Left superior pulmonary vein

Left inferior pulmonary vein

Left atrium

Diagonal branch

Anterior descending branch of left coronary artery

Left ventricle

Descending thoracic aorta

the mitral and tricuspid valves. The aortic and pulmonary valves lie at the exit to each of the ventricles. These valves must operate together in a rhythmic, ordered routine to regulate blood flow.

In a precise, coordinated procedure, the powerful heart muscles pump fresh, oxygen-rich blood and deoxygenated blood to the systemic and pulmonary circulations respectively. The left atrium and ventricle are responsible for receiving and forwarding oxygen-rich blood to the systemic circulation. The right atrium and ventricle are responsible for receiving and forwarding oxygen-depleted, carbon dioxide-loaded blood to the pulmonary circulation for gas exchange.

Heart—cross section

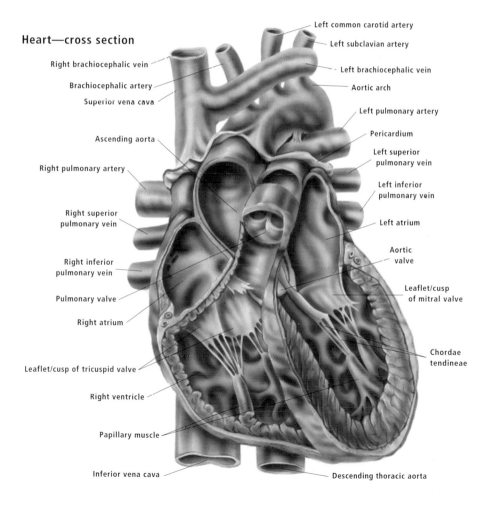

Left common carotid artery
Left subclavian artery
Right brachiocephalic vein
Left brachiocephalic vein
Brachiocephalic artery
Aortic arch
Superior vena cava
Left pulmonary artery
Ascending aorta
Pericardium
Left superior pulmonary vein
Right pulmonary artery
Left inferior pulmonary vein
Right superior pulmonary vein
Left atrium
Right inferior pulmonary vein
Aortic valve
Pulmonary valve
Leaflet/cusp of mitral valve
Right atrium
Chordae tendineae
Leaflet/cusp of tricuspid valve
Right ventricle
Papillary muscle
Inferior vena cava
Descending thoracic aorta

Ventricular systole

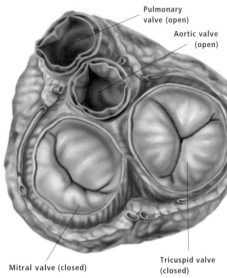

Pulmonary valve (open)

Aortic valve (open)

Mitral valve (closed)

Tricuspid valve (closed)

Ventricular diastole

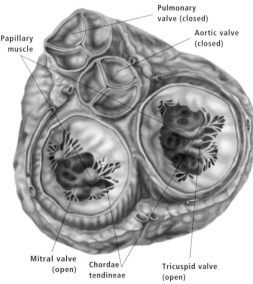

Papillary muscle

Pulmonary valve (closed)

Aortic valve (closed)

Mitral valve (open)

Chordae tendineae

Tricuspid valve (open)

Heart Valves

The four valves found in the heart are all designed to allow one-way flow only. Between each of the atria and ventricles are the atrioventricular valves. These valves operate to control outbound blood flow and prevent backflow of blood.

The mitral valve operates between the left atrium and left ventricle, while the tricuspid valve operates between the right atrium and right ventricle.

Deriving its name from the three leaflets or cusps, the tricuspid valve controls the flow of blood from the right atrium into the right ventricle. Blood draining from the veins of the systemic circulation enters the right atrium, and when the tricuspid valve opens, this blood then flows into the right ventricle.

The blood is then released from the right ventricle through the pulmonary valve,

Heart valves

When the ventricles contract (a process known as ventricular systole) the two valves between the atria and ventricles, the mitral and tricuspid valves, remain closed. At the same time, the pulmonary and aortic valves lying at the outlets of each of the ventricles are both open, allowing blood into the systemic and pulmonary circulation systems. When the ventricles dilate (a process known as ventricular diastole) the mitral and tricuspid valves open, allowing flow between the atria and ventricles, while the pulmonary and aortic valves close.

allowing the blood to enter the pulmonary circulation, where gas exchange takes place.

The mitral valve lies between the left atrium and left ventricle. Deriving its name from its resemblance to the miter worn by some members of the clergy, it is also referred to as the bicuspid valve.

The left atrium receives replenished blood from the pulmonary circulation—the flow of this blood into the left ventricle is controlled by the mitral valve.

The pulmonary and aortic valves, known as semilunar valves, are similar in structure, each having three half-moon shaped flaps, or pockets, which come together to form a seal when the valve is closed.

When there is outbound blood flow, these pockets are forced against the arterial walls, allowing free flow. However, should blood be forced back, these pockets fill up and seal the entrance to the ventricle, preventing blood from re-entering the ventricle.

Mitral valve

Lying between the left atrium and left ventricle, the two cusps of the mitral valve open and close in rhythm with the heart.

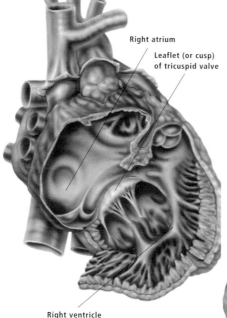

Right atrium

Leaflet (or cusp) of tricuspid valve

Right ventricle

Tricuspid valve

The tricuspid valve lies between the right atrium and right ventricle.

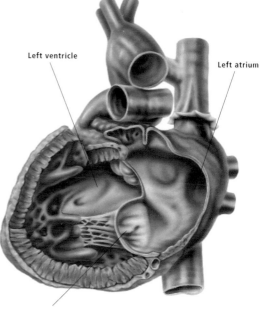

Left ventricle

Left atrium

Leaflet (or cusp) of mitral valve

Pumping Action of the Heart

The pumping of the heart is a precisely coordinated series of movements, with the left and right sides of the heart complementing each other's actions.

Located in the right atrium is the sinoatrial node, the heart's natural pacemaker, which sets the rate for the frequency of the heart's rhythmic contractions by the emission of electrical impulses.

As the heart chambers pass through a relaxation phase (diastole) and a contraction phase (systole), the components of the heart complete various tasks.

In atrial diastole (at the beginning of ventricular diastole), blood flows into the chambers of the left and right atria. Deoxygenated blood from the systemic circulation enters the right atrium via the superior vena cava, while at the same time, blood that has been freshly oxygenated in the lungs enters the left atrium via the pulmonary veins.

As ventricular diastole comes to an end, the atria contract (atrial systole) and force blood through the mitral and tricuspid valves into the left and right ventricles respectively.

When the ventricles have received their blood supply, the two atrioventricular valves close, preventing backflow into the atria. Then the right and left ventricles contract (ventricular systole), and pump blood through the aortic and pulmonary valves into the aorta and pulmonary artery, for circulation through the systemic and pulmonary circulations respectively. As the ventricles relax, blood once again enters the chambers of the left and right atria, and the cycle begins again.

This cyclic routine occurs approximately 70 times per minute when the body is at rest, and at a higher rate during physical exertion or stressful situations.

Heart cycle

The cardiac cycle is a synchronized routine involving the rhythmic contraction and relaxation of the chambers of the heart.

Heart cycle 1

In atrial diastole (at the beginning of ventricular diastole), deoxygenated blood from the systemic system enters the upper chamber of the right atria, while blood from the pulmonary circulation enters the left atria at the same time.

Heart cycle 2

As the end of the ventricular diastole phase nears, blood is pumped into the lower chambers of the left and right ventricle by the contracting atrial chambers.

Heart cycle 3

In ventricular systole, blood is pumped into the aorta and pulmonary arteries by the contracting right and left ventricles.

Heart cycle 4

The ventricles relax, and blood again enters the upper chambers of the left and right atria, as the heart cycle commences once more.

Heart cycle 1

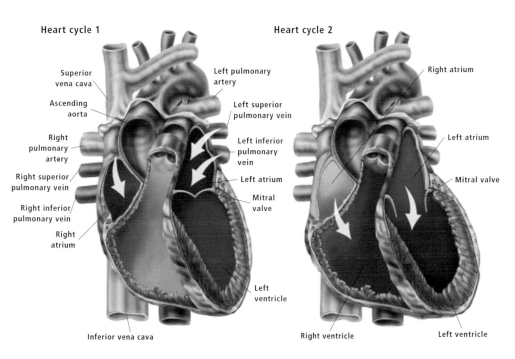

Superior vena cava

Ascending aorta

Right pulmonary artery

Right superior pulmonary vein

Right inferior pulmonary vein

Right atrium

Inferior vena cava

Left pulmonary artery

Left superior pulmonary vein

Left inferior pulmonary vein

Left atrium

Mitral valve

Left ventricle

Heart cycle 2

Right atrium

Left atrium

Mitral valve

Right ventricle

Left ventricle

Heart cycle 3

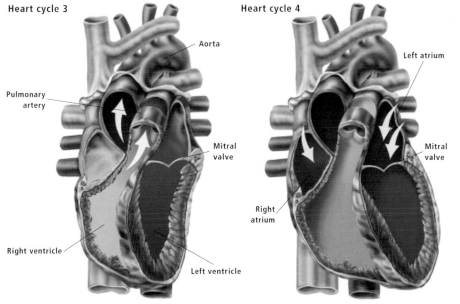

Aorta

Pulmonary artery

Mitral valve

Right ventricle

Left ventricle

Heart cycle 4

Left atrium

Mitral valve

Right atrium

Electrical
pathways
between
nerves in
the heart

Position of sinoatrial node

Heartbeat

The rhythmic contraction and relaxation of
the chambers of the heart is controlled by
the heart's natural pacemaker, the sinoatrial
node, which is located in the right atrium.
Emitting regular electrical impulses, the
sinoatrial node controls the heart rate.

The heartbeat can often be felt on the
lower left side of the chest, just below the
left nipple—this is known as the apex beat.
If a stethoscope is used to listen to
the heartbeat then two sounds are
discernible, a louder
sound followed by a
softer sound. The
first sound is created
by the closure of the
mitral and tricuspid valves,
while the second, softer sound
is created by the closure of
the aortic and pulmonary
valves.

The sinoatrial node main-
tains a regular, ordered
rhythm to the heart cycle
and is influenced by the autonomic nervous system. Both
the sympathetic and para-
sympathetic divisions cast an influence on the sinoatrial node, causing
either an increase or decrease in heart rate,
dependent on the situation.

Heartbeat

The rhythmic beating of the heart is controlled by
the sinoatrial node, which transmits electrical
impulses to initiate contraction of the heart muscle.

For instance, in stressful or frightening situations, the sympathetic division triggers an accelerated rate. When the body is at rest, the parasympathetic division often causes the heart rate to slow, to allow the body to revitalize and prepare for renewed activity.

Pulse

With each beat of the heart, blood is pumped into the arteries of the systemic circulation. The resultant pulse accompanying this surge can be felt at various points in the body.

The pulse can be monitored by placing a finger over the artery. The average adult pulse rate is around 60–70 beats per minute, although this rate is increased during physical activity and by excitement.

The pulse is generally detected by monitoring the radial artery at the wrist, or the carotid artery in the neck, although there are other pulse points that can be used to check the pulse rate.

Pulse points

The surge of blood entering the arteries from the heart creates a pulse that can be monitored. It is most easily detected at arteries that lie close to the skin, such as the radial artery in the wrist or the carotid artery in the neck.

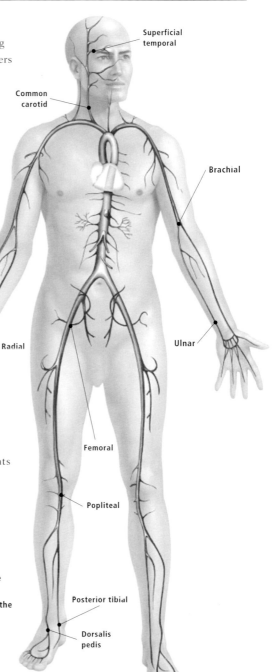

Superficial temporal

Common carotid

Brachial

Radial

Ulnar

Femoral

Popliteal

Posterior tibial

Dorsalis pedis

Coronary Arteries

The left and right coronary arteries, and their many branches, supply the muscle of the heart. Arising from the ascending aorta, these arteries are drained by a system of coronary veins.

The left coronary artery branches into the anterior interventricular artery and the circumflex artery. The anterior interventricular artery extends along a groove created by the interventricular septum at the front of the heart. The circumflex artery extends along the groove created between the left atrium and left ventricle.

The right coronary artery extends to the back of the heart along the groove formed by the junction of the right atrium and right ventricle, where it changes direction to follow the groove created by the interventricular septum.

Coronary arteries

The coronary arteries are so named for their crown-shaped formation around the heart. These arteries supply the muscle tissue of the heart.

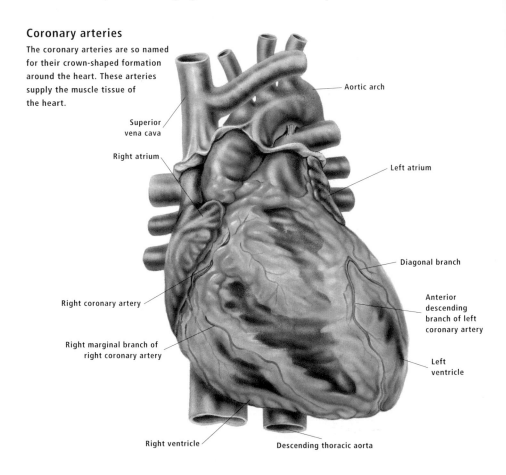

Aortic arch

Superior vena cava

Right atrium

Left atrium

Diagonal branch

Right coronary artery

Anterior descending branch of left coronary artery

Right marginal branch of right coronary artery

Left ventricle

Right ventricle

Descending thoracic aorta

Pericardium

Wrapped around the heart is the pericardium. Comprising a series of sacs, the pericardium serves to protect the heart and provide a friction-free environment for the pulsating movements of the heart muscle.

The inner set of sacs, called the serous pericardium, comprises an inner visceral layer, called the epicardium, and an outer parietal layer, with a thin film of fluid between the two layers. This combination provides a low-friction, fluid-filled space in which the muscular contractions of the heart can take place.

Lying outside the parietal layer of the serous pericardium is the fibrous pericardium. This strong layer of connective tissue joins to the top of the heart around the major vessels, holding the heart in position in the chest. The lower edge of the fibrous pericardium joins up to the diaphragm, merging with the central tendon of the diaphragm.

Pericardium

The pericardium holds the heart in place in the chest, with its layers of sacs providing a suitable environment to accommodate the muscular movements of the beating heart.

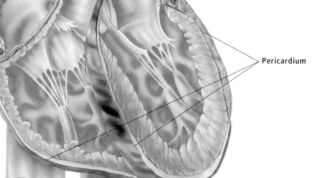

Pericardium

Oxygenated blood flows out of the lungs to the left side of the heart and is pumped out into the body for systemic circulation.

Oxygen-depleted blood enters the right ventricle of the heart and is pumped into the lungs to be oxygenated by the alveoli.

Pulmonary circulation

The pulmonary system carries deoxygenated blood to the lungs where gas exchange occurs. Carbon dioxide—picked up by the blood from cells and tissues in exchange for oxygen—is exchanged in the lungs for fresh supplies of oxygen. The carbon dioxide is then exhaled, while the now oxygen-rich blood is returned to the heart for another circuit of the body.

Pulmonary Circulation

The purpose of the pulmonary circulation is to re-oxygenate blood—this is essential to cell and tissue maintenance.

Once blood has completed a circuit of the systemic circulation, it is returned to the heart, entering the heart via the vena cavae. At this stage, the blood is relatively oxygen poor, having transferred oxygen to tissues and cells throughout the body, and in the process picking up any waste products in exchange.

The primary waste product collected is carbon dioxide. This deoxygenated blood enters the right atrium, and passes through the tricuspid valve into the right ventricle. The right ventricle then pumps this blood out through the pulmonary valve into the pulmonary trunk or artery, and along through the pulmonary arteries.

The pulmonary circulation is the only situation in the body where the roles of the arteries and veins are reversed, with arteries carrying deoxygenated blood, and the veins carrying oxygen-rich blood.

The pulmonary artery divides into the right and left pulmonary arteries, each entering their respective sides of the lungs. From here, these two main branches subdivide many times over, gradually becoming smaller, until they terminate at the tiny, thin-walled network of capillaries surrounding the alveoli, the air sacs of the lungs.

Here, carbon dioxide diffuses out of the blood into the lungs, from where it is exhaled. Simultaneously, oxygen contained in air inhaled into the lungs diffuses into the blood.

This freshly oxygenated blood then flows from the capillaries, through venules, and eventually into the four main pulmonary veins. These veins then deliver the revitalized blood to the left atrium, from where it passes through the mitral valve, and then into the left ventricle. The left ventricle then pumps fresh blood out through the aortic valve and into the systemic circulation.

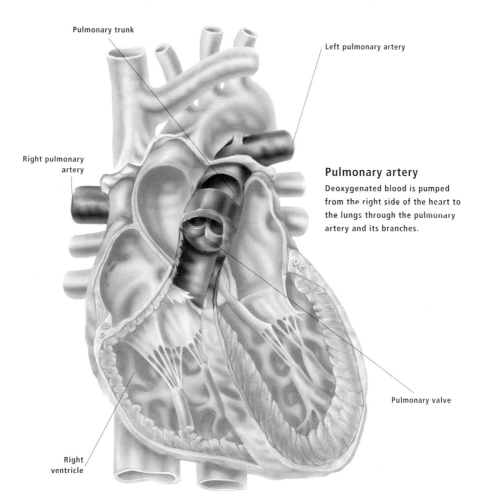

Pulmonary trunk

Left pulmonary artery

Right pulmonary artery

Pulmonary artery

Deoxygenated blood is pumped from the right side of the heart to the lungs through the pulmonary artery and its branches.

Pulmonary valve

Right ventricle

Cerebral Arteries

The brain demands the highest levels of oxygen of any organ in the body. As a consequence it has a network of blood vessels to provide this supply. The vessels originate from the aorta, which gives rise to the two major arteries extending into the brain: the carotid artery and the vertebral artery.

From the aorta, the carotid artery divides into two branches, the internal and external carotid arteries, with the external carotid responsible for blood supply to the face, neck, and internal structures of the neck and throat. The internal carotid ultimately reaches the brain, gaining entry through the skull

via each of the carotid foramen. The internal carotid gives rise to many branches, including the ophthalmic, the posterior communicating, the anterior cerebral, and the middle cerebral arteries. The front and middle of the brain are supplied by the branches of the internal carotid artery.

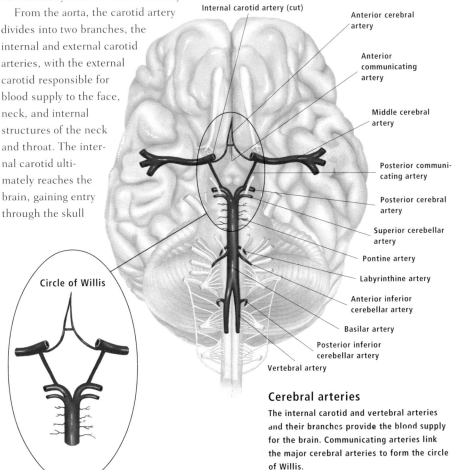

Internal carotid artery (cut)

Anterior cerebral artery

Anterior communicating artery

Middle cerebral artery

Posterior communicating artery

Posterior cerebral artery

Superior cerebellar artery

Pontine artery

Labyrinthine artery

Anterior inferior cerebellar artery

Basilar artery

Posterior inferior cerebellar artery

Vertebral artery

Circle of Willis

Cerebral arteries

The internal carotid and vertebral arteries and their branches provide the blood supply for the brain. Communicating arteries link the major cerebral arteries to form the circle of Willis.

The vertebral arteries arise from the subclavian artery. Entering through the transverse foramen of the skull, the vertebral arteries join at the base of the brain to become the basilar artery. These arteries supply the back of the brain, cerebellum, and brain stem.

The two middle cerebral arteries and the basilar artery are joined together by communicating arteries to form the circuit of blood vessels known as the circle of Willis.

Medial brain arteries

The illustration depicts the medial aspect of the cerebral hemisphere, showing the branches and distribution of the anterior cerebral artery.

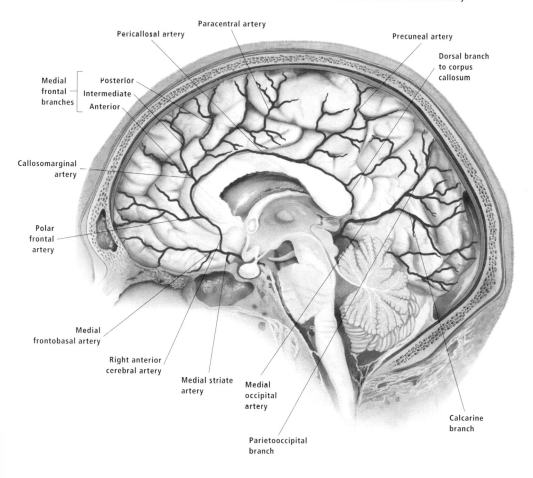

Paracentral artery

Pericallosal artery

Precuneal artery

Dorsal branch to corpus callosum

Medial frontal branches
- Posterior
- Intermediate
- Anterior

Callosomarginal artery

Polar frontal artery

Medial frontobasal artery

Right anterior cerebral artery

Medial striate artery

Medial occipital artery

Parietooccipital branch

Calcarine branch

Blood Vessels of the Head and Neck

The carotid arteries are the main arteries of the neck, receiving their supply from the aorta. Situated on either side of the neck, these major arteries supply blood not only to the neck and its internal structures, but the branches of the carotid artery supply the brain, face, and head.

The external carotid artery is the major supplier of blood to the head. Its branches include the superficial temporal artery, which extends to the top of the head; the facial artery, which extends across the cheek area, and also supplies the palate and submandibular gland;

Arteries of the head and neck

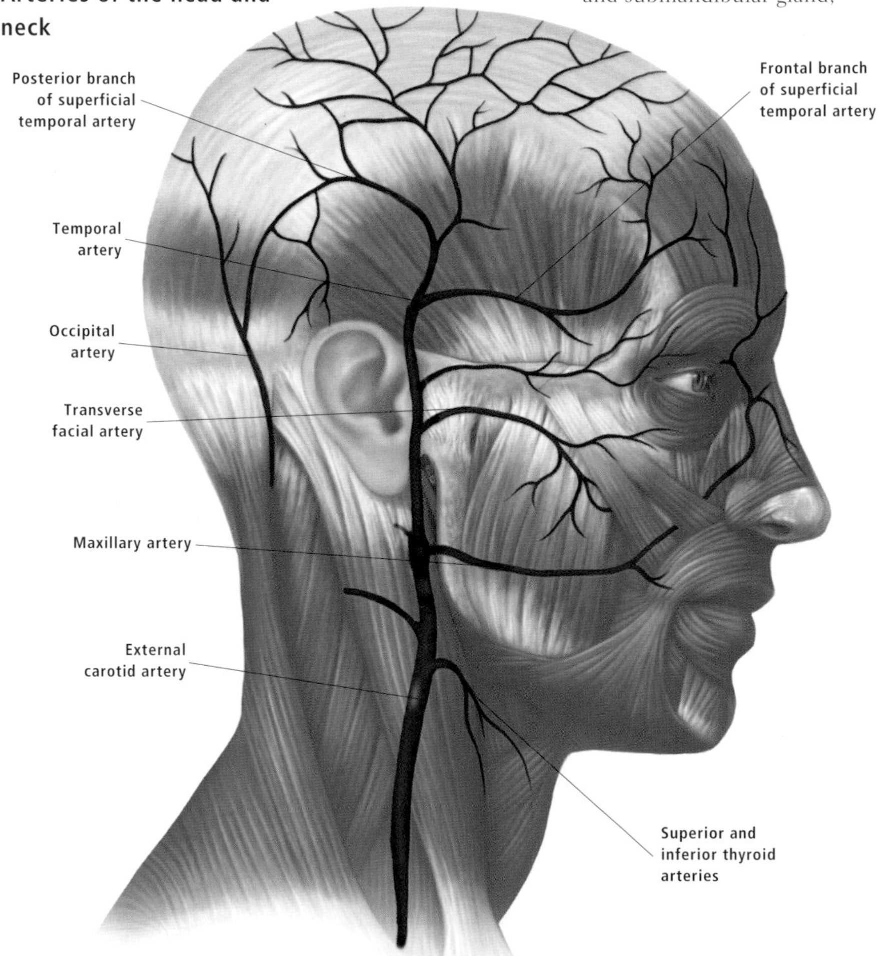

Posterior branch of superficial temporal artery

Frontal branch of superficial temporal artery

Temporal artery

Occipital artery

Transverse facial artery

Maxillary artery

External carotid artery

Superior and inferior thyroid arteries

the occipital artery, which extends over the temporal lobe, reaching to the occipital lobe; and the maxillary artery, which supplies the jaw, nose, and teeth.

The major superficial veins of the head are the temporal, facial, and maxillary veins. The temporal vein drains the superficial area around the frontal and parietal lobes; the facial vein drains the area around the cheeks and nose; and the maxillary vein drains the region around the nose, jaw, and teeth.

At each side of the neck are the internal and external jugular veins. The superficial veins of the head, face, and neck region drain into these two major veins to transport blood back to the heart.

Veins of the head and neck

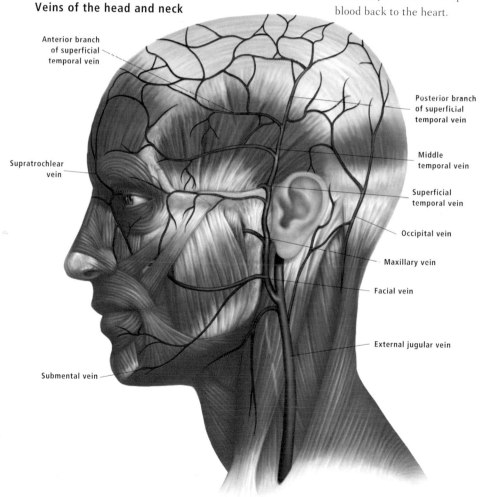

Anterior branch of superficial temporal vein

Posterior branch of superficial temporal vein

Supratrochlear vein

Middle temporal vein

Superficial temporal vein

Occipital vein

Maxillary vein

Facial vein

External jugular vein

Submental vein

Blood Vessels of the Eye

Running through the center of the optic nerve is the central artery of the retina. This artery enters the eyeball at the site of the optic disk, and divides into four main branches that cover the retinal surface, becoming gradually smaller until they form a capillary network.

This arterial network is drained by veins, which eventually drain into the central vein of the eye. The central vein of the eye exits at the optic nerve. Other veins that play an important role in drainage of the region include the vorticose and ciliary veins.

Other main arteries and veins supplying the eye include the ciliary arteries. The long posterior ciliary artery supplies the iris and the ciliary body and muscles, and the short posterior ciliary artery supplies the choroid layer of the eye. The choroid layer is highly vascular, supplying some of the nutrient requirements of the retinal layer. The choroid layer, ciliary body, and iris comprise the vascular tunic of the eye.

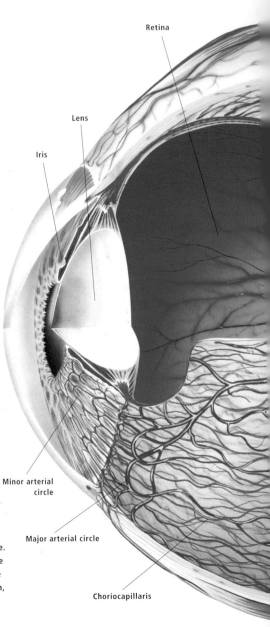

Retina

Lens

Iris

Minor arterial circle

Major arterial circle

Choriocapillaris

Blood vessels of the eye

A dense network of arteries and veins supplies the eye. The central artery of the eye enters the retina at the optic disk and its branches fan out across the entire surface. It is drained by branches of the central vein, which exits the region at the optic nerve.

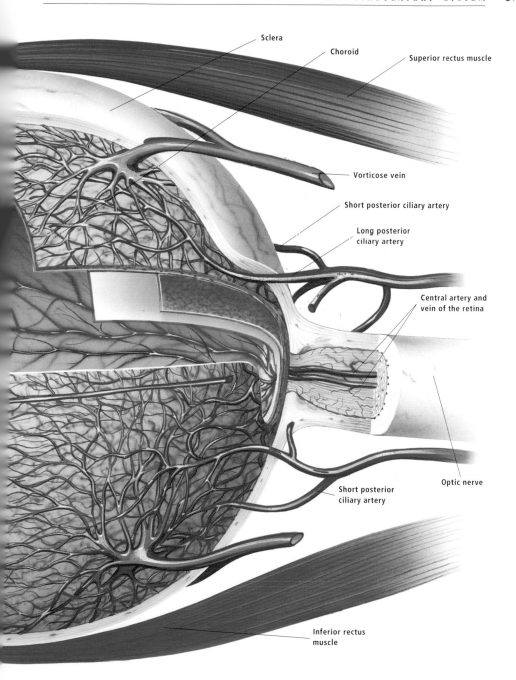

Sclera

Choroid

Superior rectus muscle

Vorticose vein

Short posterior ciliary artery

Long posterior
ciliary artery

Central artery and
vein of the retina

Optic nerve

Short posterior
ciliary artery

Inferior rectus
muscle

Hepatic portal system

Nutrients extracted by the digestive system are transported to the liver by the hepatic portal system. Entering the liver via the portal vein, this blood is processed, with nutrients extracted and stored for future use, toxins and ageing blood cells removed, and essential nutrients added for use by the body tissues.

The Hepatic Portal System

The hepatic portal system is the network of veins from the major parts of the digestive system, which lead back to the liver. Generally, the veins of the body transport deoxygenated blood directly back to the heart; however, in the case of the hepatic

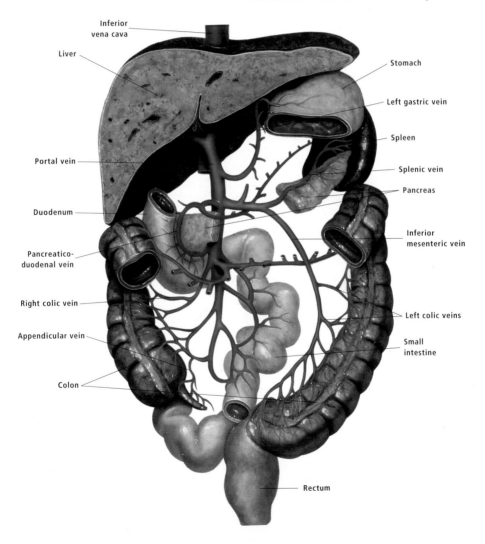

Inferior vena cava

Liver

Portal vein

Duodenum

Pancreatico-duodenal vein

Right colic vein

Appendicular vein

Colon

Stomach

Left gastric vein

Spleen

Splenic vein

Pancreas

Inferior mesenteric vein

Left colic veins

Small intestine

Rectum

portal system, the blood is first sent to the liver before being returned to the heart.

As processing of ingested food is carried out by the liver, small intestine, and large intestine, these organs extract nutrients needed for body fuel and upkeep. The extraction process results in not only nutrients being retrieved, but also a small amount of waste products and toxins. Blood carrying these products enters the venous system serving the intestinal area, the major components of which are the superior and inferior mesenteric veins.

The inferior mesenteric vein is joined by the splenic vein, which drains the accessory digestive organs of the spleen and pancreas, as well as part of the stomach. The superior and inferior mesenteric veins join to form the portal vein. The gastric vein draining the upper portion of the stomach, and the cystic veins draining the gallbladder, also drain into the portal vein. This short, wide vein transports the nutrient-rich blood into the liver, entering the liver through a slit-like opening called the porta hepatis—Latin for "door to the liver."

The blood flows through the liver, finally reaching the tiny capillaries called sinusoids, which permeate the liver. The sinusoids filter the incoming blood, extracting some nutrients for storage, removing toxins, waste products, and ageing blood cells for disposal, and adding nutrients required for body use.

The blood is then returned to the heart via the inferior vena cava.

Renal artery

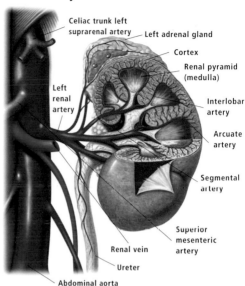

Celiac trunk left
suprarenal artery
Left adrenal gland
Cortex
Renal pyramid (medulla)
Left renal artery
Interlobar artery
Arcuate artery
Segmental artery
Superior mesenteric artery
Renal vein
Ureter
Abdominal aorta

Renal Arteries

Branching off from the abdominal aorta, the renal arteries supply blood to the kidneys. Branches of the renal arteries also supply the adrenal glands and the ureters.

Within the kidney, the renal arteries subdivide into segmental arteries, which branch into interlobar arteries. The interlobar arteries give rise to the arcuate arteries, which subdivide into interlobular arteries. The interlobular arteries supply the nephrons of the kidneys, which act as filtering units for the blood.

Veins corresponding to each of the arterial types drain the kidneys, ultimately emptying into the inferior vena cava.

Blood Vessels of the Arm

The major artery of the arm, the brachial artery, divides at the elbow into its two major branches, the radial artery and the ulnar artery, which supply the forearm and hand. Both the radial and ulnar arteries extend down the forearm, following the line of their companion bones, the radius and ulna respectively. Both extend over the wrist to supply the arteries of the hands and fingers. The radial artery runs particularly close to the surface of the skin, and is often used as a pulse point for monitoring the pulse rate.

The venous system of the arm includes the digital veins of the hands, the cephalic and median vein of the forearm, the basilic vein (which runs the length of the arm), and the brachial vein. These veins drain into the axillary vein and then into the superior vena cava, which returns deoxygenated blood to the heart.

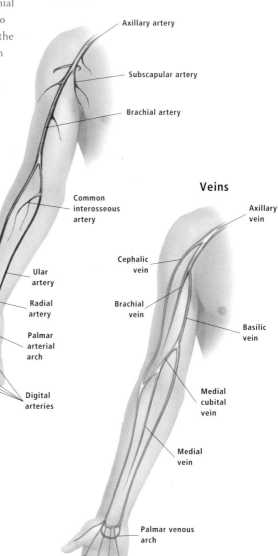

Arteries

- Axillary artery
- Subscapular artery
- Brachial artery
- Common interosseous artery
- Ular artery
- Radial artery
- Palmar arterial arch
- Digital arteries

Veins

- Axillary vein
- Cephalic vein
- Brachial vein
- Basilic vein
- Medial cubital vein
- Medial vein
- Palmar venous arch
- Digital veins

Blood vessels of the arm

Blood Vessels of the Leg

Blood supply to the leg is provided mainly by the external iliac artery, which becomes the femoral artery as it enters the leg. The major branches of the femoral artery include the profunda femoris artery in the thigh, and the anterior and posterior tibial arteries in the lower leg. The arteries of the lower leg extend past the ankle to supply the arteries of the feet and toes.

The venous system of the leg includes the great saphenous vein, the longest vein in the body. The saphenous veins are the major superficial veins of the leg. Much of the blood drained by the superficial veins is sent, via perforating veins, to the deep veins of the leg. The deep veins contain valves to assist the flow of venous blood back to the heart. The contractions of nearby muscles compress the valves, pushing blood in the direction of the heart. The veins of the leg drain into the external iliac vein.

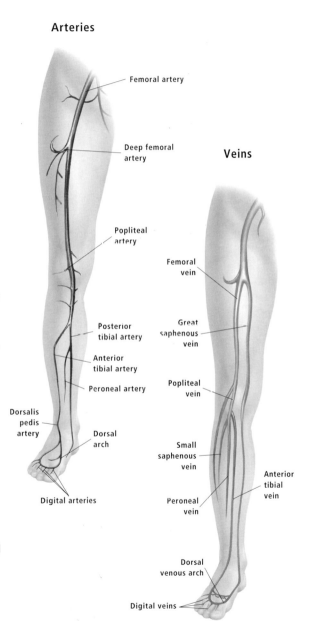

Arteries

Femoral artery

Deep femoral artery

Veins

Popliteal artery

Femoral vein

Posterior tibial artery

Great saphenous vein

Anterior tibial artery

Popliteal vein

Peroneal artery

Dorsalis pedis artery

Dorsal arch

Small saphenous vein

Anterior tibial vein

Digital arteries

Peroneal vein

Dorsal venous arch

Digital veins

Blood vessels of the leg

Stroke

Cerebral infarction

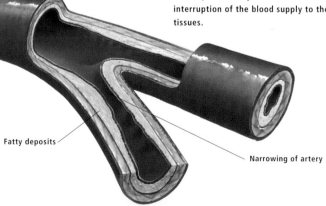

Blood and Circulatory Disorders

The free circulation of blood around the body is vital to sustain all body systems. Any disruption to blood flow, pressure, and content as a result of blood or circulatory disorders therefore has serious consequences.

Stroke

Stroke is one of the leading causes of death in the industrialized world. It is the loss of brain function due to cerebral infarction, which is a result of oxygen deprivation and death of tissue in part of the brain. This is caused by an interruption in the blood supply to the brain.

The interruption to the blood supply is usually the result of a clot forming in the arteries. These clots can form in the carotid

Cerebral infarction

When oxygen supply to the brain is interrupted, death of brain tissue occurs, resulting in cerebral infarction. The entire cerebral tissue can be affected, as can happen in the case of cardiac arrest. Localized areas of infarction, as shown in the illustration, are usually the result of blockage of an artery serving the brain.

Arteriosclerosis

Arteriosclerosis is often indicated as a cause of stroke. Arteriosclerosis is caused by a build-up of fat and calcium deposits in the wall of the artery. This build-up eventually blocks off the artery, causing interruption of the blood supply to the organ or tissues.

Fatty deposits

Narrowing of artery

Thrombosis

Thrombus sites

There are several sites in the body that are predisposed to the formation of blood clots in persons vulnerable to thrombosis. These sites include the deep veins of the leg, the renal veins, and the arteries of the heart.

arteries or their branches, blocking supply to the brain. Alternatively, a clot forming in these arteries can dislodge and travel to the brain, causing cerebral infarction.

Arteriosclerosis is also a major cause of stroke. The arteries become blocked by a build-up of fatty deposits, which eventually restrict blood flow to organs and tissues, or provide a likely site for the formation of a blood clot.

Symptoms of a stroke include loss of movement (paralysis) of a region of the body, weakness, decreased sensation, numbness, loss of coordination, vision problems, and speaking difficulty. These symptoms may have a sudden onset, or may develop over a period of days, fluctuating in severity.

Thrombosis

Thrombosis is the formation of a blood clot, or thrombus, in a blood vessel. It can occur

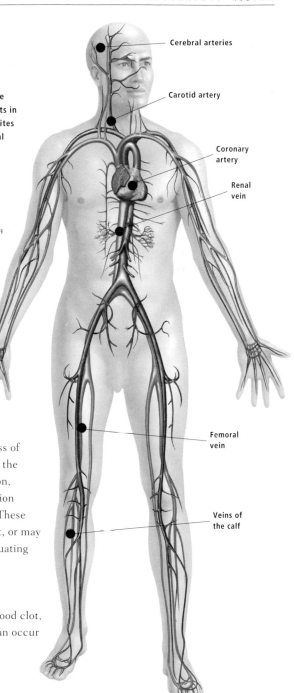

Cerebral arteries

Carotid artery

Coronary artery

Renal vein

Femoral vein

Veins of the calf

in arteries, veins, or capillaries, and has varying causes. Arterial thrombosis is most usually caused by disease of the arterial wall. Platelets in the blood attempt to heal the wall, attaching to its surface, but this build-up of platelets can create a suitable site for the formation of the blood clot, which results in loss of blood supply to the organ or tissue supplied by the artery, causing tissue death.

Venous thrombosis occurs most commonly in the deep veins of the legs, resulting in pain and swelling due to blockages in the blood leaving the legs. It can also be caused by disease of the vein wall, such as in varicose veins. Deep vein thrombosis can

occur after periods of immobility, such as following an operation. Treatment for venous thrombosis usually involves early post-operative mobilization, and the use of anticoagulants.

Anemia

Anemia is a condition where the oxygen-carrying capacity of the blood is reduced by

Anemia

There are several types of anemia, all affecting the red blood cells. In some cases insufficient red blood cells are produced; in others the size or shape of red blood cells is altered. In iron deficiency anemia the red blood cells are smaller and paler than normal. In sickle cell anemia, the red blood cells are sickle-shaped.

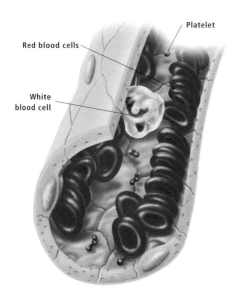

Red blood cells

White blood cell

Platelet

Normal blood

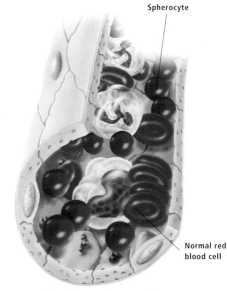

Spherocyte

Normal red blood cell

Hereditary spherocytosis

decreased amounts of hemoglobin in the red blood cells or a reduced number of red blood cells in the blood. There are many types of anemia, including iron deficiency anemia, pernicious anemia, aplastic anemia, sickle cell anemia, and hemolytic anemia.

Pernicious anemia is caused by inadequate absorption of vitamin B_{12}, while iron deficiency anemia is caused by inadequate absorption of iron, loss of iron, or excess use of iron supplies.

Aplastic anemia is due to reduced production of the red blood cells by the bone marrow, often brought about by the use of pharmaceutical drugs, such as immunosuppressant drugs or anticancer drugs.

In sickle cell anemia, the red blood cells become misshapen, forming their characteristic sickle shape. The misshapen blood cells cause blockages in the small blood vessels. These blockages inhibit blood supply to recipient tissues and organs, causing damage and pain.

In hemolytic anemia, red blood cell destruction occurs at a more rapid pace than new red blood cell production. One particular form of hemolytic anemia, hereditary spherocytosis, causes the formation of small spherocytes in place of red blood cells.

The approaches to anemia treatment are dependent on type, and often the underlying cause must be treated simultaneously.

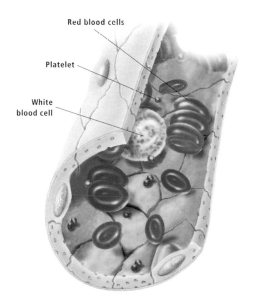

Iron deficiency anemia

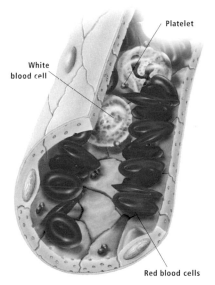

Sickle cell anemia

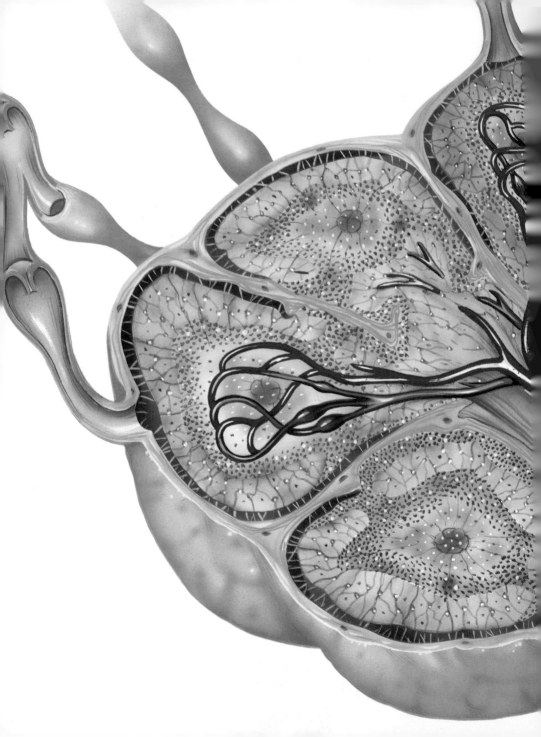

The Lymphatic/ Immune System

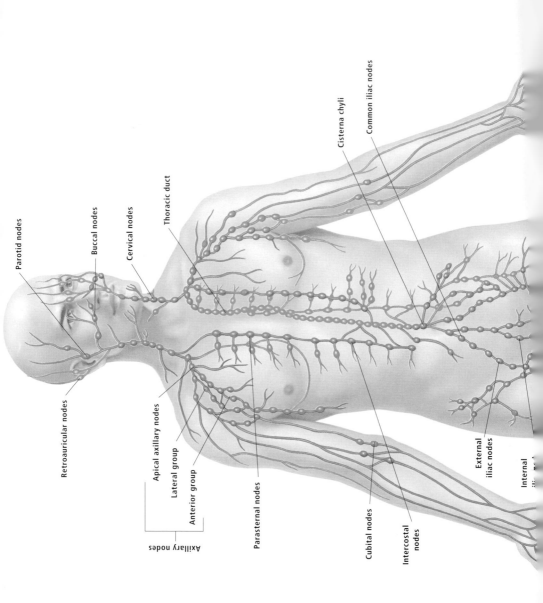

Parotid nodes

Buccal nodes

Cervical nodes

Thoracic duct

Cisterna chyli

Common iliac nodes

Retroauricular nodes

Apical axillary nodes

Lateral group

Anterior group

Axillary nodes

Parasternal nodes

Cubital nodes

Intercostal nodes

External iliac nodes

Internal iliac nodes

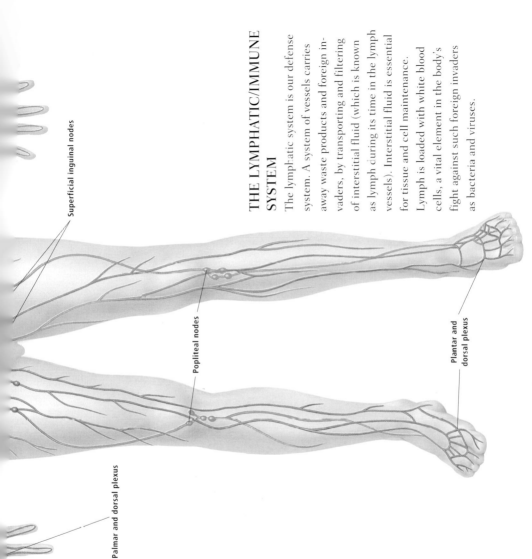

Superficial inguinal nodes

Popliteal nodes

Palmar and dorsal plexus

Plantar and dorsal plexus

THE LYMPHATIC/IMMUNE SYSTEM

The lymphatic system is our defense system. A system of vessels carries away waste products and foreign invaders, by transporting and filtering of interstitial fluid (which is known as lymph during its time in the lymph vessels). Interstitial fluid is essential for tissue and cell maintenance. Lymph is loaded with white blood cells, a vital element in the body's fight against such foreign invaders as bacteria and viruses.

Lymph

Lymph is an almost colorless fluid flowing through the vessels of the lymphatic system. Composed predominantly of water, it also contains protein molecules, salts, glucose, urea, and lymphocytes, which are specialized white blood cells with disease-fighting capabilities.

Lymph Vessels

Interstitial fluid from the cells and tissues of the body filters into the lymph system, which runs parallel to the arteries and veins

of the body. The endothelial cells of the lymph vessels are arranged in such a way that they often overlap, or are not closely joined. The resultant gaps between the

Lymph nodes

A fibrous capsule encloses each lymph node. Clustered along the route of the lymphatic vessels, the lymph nodes filter and clean incoming lymph supplied through afferent vessels. Once filtered, efferent vessels carry the lymph to the venous system.

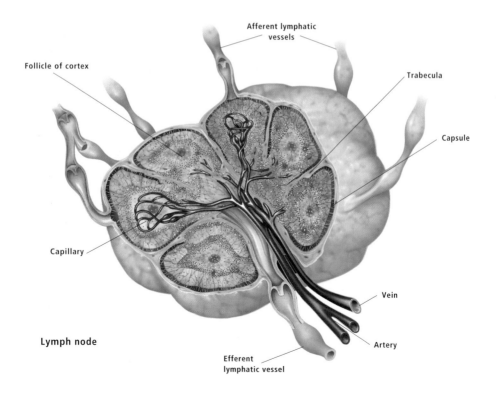

Afferent lymphatic vessels

Follicle of cortex

Trabecula

Capsule

Capillary

Lymph node

Vein

Artery

Efferent lymphatic vessel

Lymphatic vessel

The lymph vessels contain valves to prevent backflow of lymph through the system. These valves are controlled by the contractions of nearby muscles.

Lymph vessel

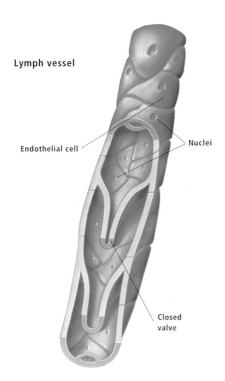

Endothelial cell

Nuclei

Closed valve

fluid in the system. As the lymphatic system does not have a pumping mechanism, the valves open and close in tune with the relaxation and contraction movements of neighboring muscles.

Lymph Nodes

Lymph nodes, often referred to as lymph glands, occur in clusters in the lymphatic system. Afferent vessels connect to the node, bringing lymph fluid to the node for filtering and cleaning. Lymphocytes and macrophages in the node engulf foreign bodies, bacteria, and debris. Once the unwanted materials have been removed, the protein-rich lymph is carried out in efferent vessels, which eventually drain to two large lymph vessels. These two large lymph vessels drain into the venous system at the neck.

While processing the contaminated lymph in the node, the macrophages perform another vital task. When engulfing invading bacteria, they extract identifiable elements of the bacteria (antigens), which they then present to the lymphocytes. Contact with these antigens stimulates the lymphocytes to produce antibodies to the particular bacteria, thus preparing a defense against a repeat invasion by the particular bacteria.

cells, along with pores in some of the cells, allow easy access for the interstitial fluid.

Once the fluid has entered the lymphatic system, it is known as lymph. The lymph vessels are a one-way system for the transport of lymph fluid. Beginning at blind-ended capillaries, the vessels become gradually larger, carrying lymph to the lymph nodes.

The larger lymph vessels are called lymphatics; these vessels have valve mechanisms in place to prevent the backflow of

Lymphatic tissue

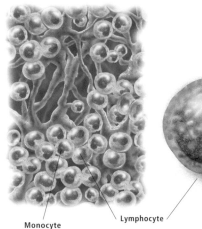

Monocyte

Lymphocyte

Lymph Circulation

The one-way system of vessels in the lymphatic system has a number of points along its route where lymph vessels converge.

The lymphatic vessels of the head and neck terminate at the internal jugular nodes. The lymphatic vessels of the upper limb converge at the axilla. It is here that lymph is also received from the chest wall, back, and breast.

The lymphatic vessels of the lower limbs drain to the inguinal nodes in the groin, and empty into the cisterna chyli, which in turn empties into the thoracic duct. Also draining into the thoracic duct is the lymph from the thoracic organs, the upper left limb, and the left half of the head and neck. The thoracic duct

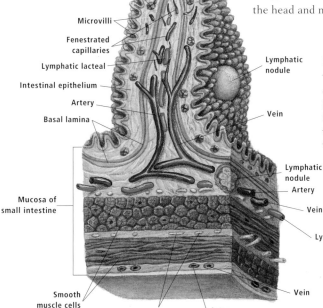

Lymphocytes

Microvilli

Fenestrated capillaries

Lymphatic lacteal

Intestinal epithelium

Artery

Basal lamina

Lymphatic nodule

Vein

Mucosa of small intestine

Lymphatic nodule

Artery

Vein

Lymphatic vessel

Smooth muscle cells

Nerves

Artery

Vein

Lacteals

The lymph vessels of the digestive system are known as lacteals. Lacteals transport lipids and nutrients in lymph fluid; these vessels drain into the cisterna chyli.

drains into the left internal jugular vein near the left subclavian vein. Lymph from the right half of the head, neck, and thorax, and the right upper limb drains into the right lymphatic duct. This duct empties into the junction of the right internal jugular and subclavian veins.

The lymph vessels in the walls of the digestive system are known as lacteals. These lacteals drain into the cisterna chyli.

Lymphoid Organs

The lymphoid organs include the thymus, spleen, and mucosa-associated lymphoid tissue.

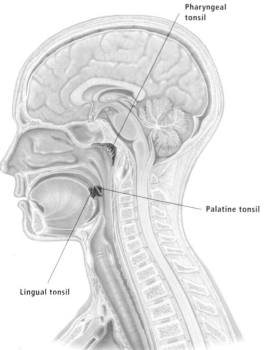

Pharyngeal tonsil

Palatine tonsil

Lingual tonsil

Organs of the lymphatic system

The organs of the lymph system include the thymus, spleen, and the concentrations of lymphatic tissue distributed in various parts of the body, including the gut and the tonsils.

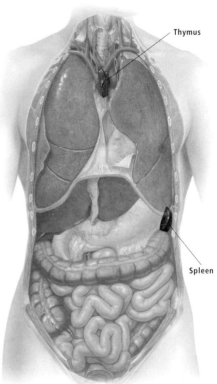

Thymus

Spleen

Tonsils

The three sets of tonsils—the palatine, pharyngeal, and lingual—are aggregations of lymphatic tissue. They offer the first line of defense against bacteria invading the respiratory and digestive systems via the mouth and nose.

Thymus

Located in the upper chest, between the heart and the sternum, the thymus is most active during childhood and adolescence, gradually shriveling to a small mass of fat and fibrous tissue by adulthood.

The thymus is responsible for the production of special white blood cells known as T lymphocytes. Lymphocytes mature into T lymphocytes as they travel through the central part of the thymus. This maturation process takes about 3 weeks, after which time, the T lymphocytes are ready to enter the bloodstream. As part of the body's defense system, T lymphocytes are programmed to recognize antigens of foreign invaders. Once the invader is identified, the T lymphocytes proliferate, attack, and destroy the foreign cells.

Although the thymus shrinks and becomes inactive following puberty, adequate numbers of T lymphocytes have been produced, and they continue to multiply and proliferate throughout life.

Thymus

T lymphocytes are produced in the thymus. These specialized white blood cells play a vital role in the body's immune system. After puberty, the thymus gradually shrivels and becomes inactive. However, it first lays the foundations for an adequate lifetime supply of T lymphocytes, which once produced continue to proliferate, providing the body with a defense against foreign invaders.

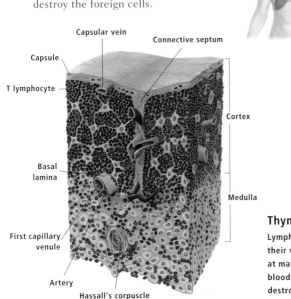

Capsular vein

Connective septum

Capsule

T lymphocyte

Basal lamina

First capillary venule

Artery

Hassall's corpuscle

Cortex

Medulla

Thymus

Left lobe

Right lobe

Thymus microstructure

Lymphocytes enter the thymus, gradually making their way through the cortex and medulla, emerging at maturity as T lymphocytes, specialized white blood cells designed to recognize and destroy foreign invaders.

Spleen

Located high in the upper left quadrant of the abdomen, the spleen has the highest concentration of lymphatic tissue in the body. Responsible for performing blood-filtering operations and for production and storage of lymphocytes, the spleen has two major tissue types. Blood filtering occurs in the red pulp of the spleen, where ageing blood cells are removed as they pass through the sinusoids in the red pulp. Lymphocyte production occurs in the white pulp of the spleen.

Mucosa-associated Lymphoid Tissue

The mucosa-associated tissue of the body includes lymphoid tissue found in the respiratory tract, urogenital tract, and digestive tract. These tissues contain B and T lymphocytes and form a line of defense around the entrances to the body. For instance, the tonsils guard the entrance to the digestive and respiratory systems.

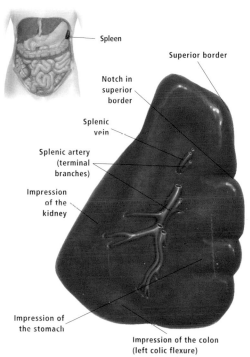

Spleen

Superior border

Notch in superior border

Splenic vein

Splenic artery (terminal branches)

Impression of the kidney

Impression of the stomach

Impression of the colon (left colic flexure)

Spleen

The large blood supply provided to the spleen gives it its characteristic red color and accounts for its soft, pulpy structure. The spleen's pliable consistency allows the surrounding organs to leave impressions on its surface. Two specific tissue types are found in the spleen: the red pulp and the white pulp. The red pulp area is responsible for blood-filtering activities, eliminating ageing blood cells from the circulation. The white pulp area is responsible for lymphocyte production, a vital element of the immune system.

Spleen microstructure

The channels in the red pulp are called sinusoids. It is in the sinusoids that ageing or abnormal red blood cells are broken down and destroyed. Concentrations of lymphocytes form the white pulp of the spleen.

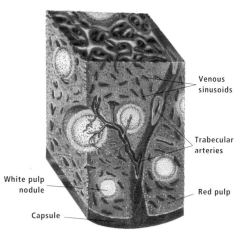

Venous sinusoids

Trabecular arteries

White pulp nodule

Red pulp

Capsule

Immunity

The body's response to invading organisms such as bacteria, viruses, fungi, and other types of pathogens is known as the immune response.

Several approaches to invasion by foreign substances are available, dependent on the nature and location of the invader. The two main options involve activation of either the humoral immune system, or the cell-mediated immune system.

The body gradually builds immunity to many common pathogens. This process begins at the initial exposure to the invading pathogen. The response to initial exposure to a new pathogen is often slow, while the immune system sets about determining the nature of the new invader. Once this is determined, an appropriate response is launched to overcome the invader.

Once the body has developed a response to a particular invader, subsequent exposure to the same invader activates a quicker response, because the body has special white blood cells called "memory" B and T cells, which recognize the invader from past exposure and respond immediately. At this point, the person is considered to have immunity to that particular pathogen.

Immunity can be acquired through vaccination against a particular pathogen. Vaccination has proven extremely effective in eliminating or severely reducing the incidence of many previously rampant and serious diseases. Vaccination introduces the body to a weakened or killed virus or bacteria, or occasionally a genetically engineered virus. Once introduced into the body, the body's immune system activates to produce antibodies to combat the introduced pathogen. Many vaccines offer lifetime protection from disease; others require scheduled boosters or additional "top-up" doses to maintain immunity.

Immunity is acquired through either active or passive means. Active immunity describes the body's own determination and response to disease or by vaccination. Passive immunity describes the introduction

Humoral immune response

The white blood cells produced in bone marrow, B lymphocytes, are responsible for production of antibodies to help recognize and destroy invading antigens (carried by bacteria or viruses). In conjunction with T lymphocytes and macrophages, they form the body's defenses against attack.

(a) Entering through cells at the surface, virus particles quickly proliferate and invade tissue.

(b) Macrophages (scavenging white blood cells) engulf the virus particles.

(c) Breaking down the virus, macrophages then present antigens to the circulating T lymphocytes, which in turn release proteins to attract additional T and B lymphocytes to help fight off attack.

(d) B lymphocytes split into memory B cells and plasma B cells. Plasma B cells make antibodies which target the particular virus. Memory B cells are programmed to remember the virus in the event of future attack.

(e) The virus particles are captured by the antibodies.

(f) Macrophages primed to recognize the antibody engulf the virus, breaking it down, and protecting the body from infection.

Humoral immune response

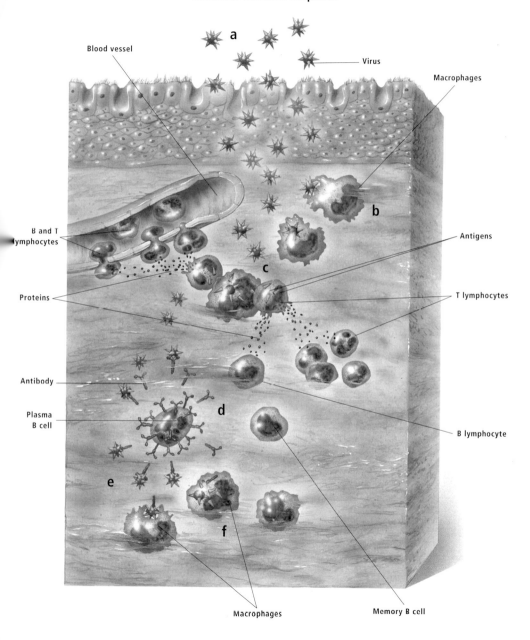

Blood vessel

Virus

Macrophages

a

b

B and T lymphocytes

Antigens

Proteins

T lymphocytes

Antibody

Plasma B cell

c

d

e

f

B lymphocyte

Macrophages

Memory B cell

of antibodies from elsewhere, either by injection of antibodies, or in the case of the fetus, from the mother. Unlike active immunity, which creates "memory" cells for immediate recognition of repeat invasions by a pathogen, passive immunity is usually only temporary.

The effective nature of the body's immune system can cause complications for organ transplant recipients. The body's immune system can cause rejection of the introduced organ, recognizing it as an invader, and mounting an immune response against it.

Humoral immune system

Humoral response is so named as it takes place in the body fluids (humors). When a foreign body, or antigen, enters the body, humoral response is activated. The antigen is identified by macrophages, scavenging white blood cells. Once identified, antibodies

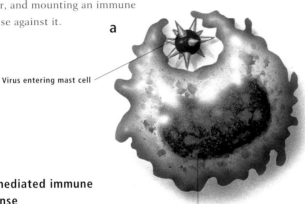

a

Virus entering mast cell

b

Mast cell

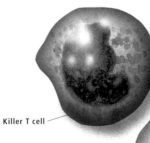

Killer T cell

Helper T cell

Cell-mediated immune response

Produced by the thymus, T lymphocytes are responsible for the delayed action of the cell-mediated response.

(a) The invading virus is engulfed by circulating mast cells.

(b) After processing the virus, mast cells then present antigens to T cells.

(c) The T cells produce clones, each playing a special role in the immune response: memory T cells remember the invading antigen for future attacks; helper T cells recruit B and T cells to the site of antigen attack; suppressor T cells inhibit the action of B and T cells; and killer T cells attach onto invading antigens and destroy them.

are produced by B lymphocytes, which then attach onto the foreign invader. The invader is then attacked and engulfed by macrophages, which recognize the antibody.

B lymphocytes are also the "memory" cells, which recognize subsequent attacks by a particular antigen.

Cell-mediated immune system

T lymphocytes, produced in the thymus, are responsible for cell-mediated immune response. In cell-mediated response, some T lymphocytes engulf and destroy invading pathogens, while other T lymphocytes destroy them directly.

On initial exposure to infection by invading pathogens, T lymphocytes proliferate, producing four different types of cells to combat the invader. These four cell types are: memory cells, suppressor cells, killer cells, and helper (T4) cells. In cell-mediated immune response, each of these cell types performs a specific task. Helper T cells and suppressor T cells are regulating cells—they control the activities of B cells produced by the bone marrow.

The body's initial exposure to invading viruses is slow, while the T cells determine their response. Firstly, the circulating mast cells alert the T cells to the presence of an invading virus. Then the T cells divide into their four types to combat the invader.

The helper T cells call up B cells to assist in the attack—the activity of both these cell types is regulated by the suppressor T cells. Memory T cells remember the invading antigen for quick response in future attacks, while the killer T cells attack and eliminate the invader.

Antibodies are not involved in cell-mediated immune response.

Antigens

C

T cell

Memory T cell

Suppressor T cell

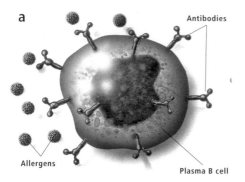

a

Antibodies

Allergens

Plasma B cell

Allergic Response

An allergic reaction is the body's response to a usually harmless substance, which the body mistakenly identifies as harmful, and mounts an often inappropriate and exaggerated response to the substance. The overproduction of antibodies to the substance can cause irritating or harmful, sometimes life-threatening, effects.

Any substance that will produce an allergic reaction is an allergen. Many common

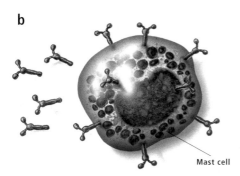

b

Mast cell

Allergic reaction

The body releases histamine in response to exposure to invading allergens. This release of histamine causes the symptoms associated with allergies.

(a) Plasma B cells produce antibodies in response to initial exposure to an allergen.

(b) Antibodies attach to circulating mast cells.

(c) These antibodies on the mast cells capture the allergens when they next enter the body.

(d) In response to the allergens, mast cells release histamine, which produces the symptoms of allergy.

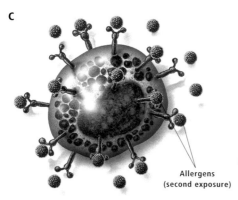

c

Allergens
(second exposure)

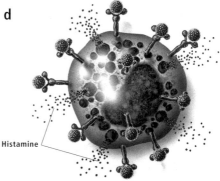

d

Histamine

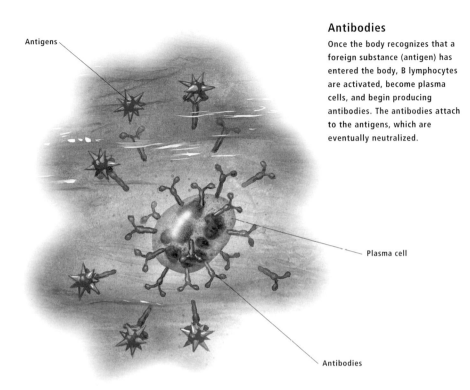

Antibodies

Antigens

Once the body recognizes that a foreign substance (antigen) has entered the body, B lymphocytes are activated, become plasma cells, and begin producing antibodies. The antibodies attach to the antigens, which are eventually neutralized.

Plasma cell

Antibodies

substances that are harmless to most people can produce an allergic reaction in others. Common causes of respiratory allergies include pollens, mold spores, some noxious weeds, dust mites and their droppings, dandruff, and animal hairs and fur.

Allergies to drugs can cause problems, even to people who have previously had no adverse reaction. Food, insect bites, chemicals, and even sunlight can all cause allergic reaction to those with a low allergic threshold. The physical effect of exposure

to an allergen is that the body will produce antibodies, normally manufactured in response to invading germs.

In an allergic reaction, allergens and antibodies act together to cause the release of histamine that then acts on capillaries, mucous glands, and muscles of the stomach and internal organs, stimulating them into excessive, often harmful, activity such as sneezing and rash. Repeated exposure to the allergen can produce increasingly severe reactions.

Diseases and Disorders of the Immune System

Immune system problems can cover a range of disorders, from allergic reactions to seemingly harmless substances to autoimmune disease, in which the body's immune system fails to recognize normal cells and mounts an attack against them. Immunodeficiency diseases, such as AIDS, reduce the body's immunity to even mild infections, causing serious complications for AIDS sufferers.

HIV

HIV is the human immunodeficiency virus —a retrovirus in which RNA is encased within a protein envelope. The virus attacks the white blood cells known as helper T lymphocytes, or T4 helper cells. These particular white blood cells are vital to the immune response of the body, and their destruction by HIV can compromise the immune response, rendering it vulnerable to opportunistic attack by bacteria and virus. The body's inability to respond to even relatively harmless invaders can lead to serious complications.

AIDS is acquired immune deficiency syndrome, and is the last stage of HIV disease. A person can be HIV-positive for many years without developing AIDS—the illness can lie dormant for as long as ten years. As yet, there is no cure for AIDS, and the complications caused by inadequate response of the immune system to opportunistic infections, rather than AIDS itself, can cause serious illness and death.

HIV transmission is usually through sexual contact or needle sharing, or it is passed from a pregnant woman to the fetus; a nursing mother can pass the virus to her baby through breast milk.

Now a major world health problem, AIDS has taken on epidemic proportions in many developing countries of the world. AIDS awareness in developed countries, advocating safe sex, and measures such as needle exchange programs for intravenous drug users, have contributed to public awareness and a resultant drop in newly reported cases of HIV.

Human immunodeficiency virus (HIV)

The human immunodeficiency virus (HIV) belongs to a specific group of viruses, known as retroviruses. While the genetic material of most viruses consists of DNA, the genetic material of retroviruses consists of RNA.

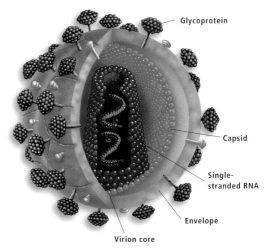

Glycoprotein

Capsid

Single-stranded RNA

Envelope

Virion core

AIDS

AIDS (acquired immune deficiency syndrome) is the final stage of HIV disease. As the body becomes increasingly unable to fight off disease, its vulnerability allows opportunistic infections and cancers to take hold.

Fungal disease

Fungal infections are often experienced by AIDS sufferers. Oral thrush, caused by the fungus Candida albicans, is particularly common.

Lymphomas

Reduced immunity, due to decreased T4 cell production, can result in lymphomas (tumors of the lymph system), such as non-Hodgkins lymphoma.

Kaposi's sarcoma

Kaposi's sarcoma, the most frequently occurring cancer in AIDS sufferers, produces raised skin lesions that are purple to brown in color. As the disease progresses, the lungs and other organs may also become affected.

AIDS dementia

About 50 percent of AIDS patients develop brain disorders. Dementia is a condition in which the brain atrophies, and concentration and memory fail.

Retinopathy

Retinopathy is a non-inflammatory disease of the retina, which can result in loss of vision.

Herpes simplex

With reduced immunity, the body is vulnerable to many virus strains, particularly herpes simplex.

Pneumonia

Lung infections may develop into pneumonia, which may be untreatable and is often the cause of death in late-stage AIDS.

Hodgkin's disease

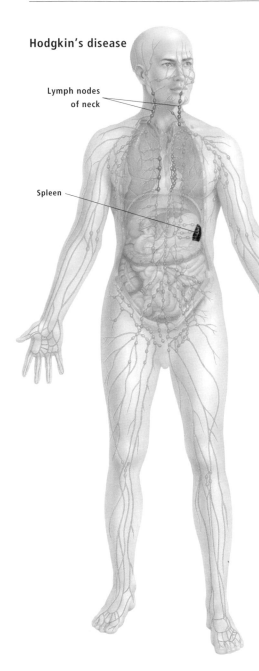

Lymph nodes
of neck

Spleen

Hodgkin's disease

Hodgkin's disease is characterized by a lymphoma (a cancer that originates in the lymph nodes). Most commonly developing during young- to middle-adult life, this type of cancer is usually slow spreading. It moves progressively through the lymph nodes, eventually involving the spleen and other organs.

Its slow spread and specific target area make Hodgkin's disease more responsive to cancer treatments than many other types of cancer, although control of the disease is dependent on the extent of spread at the time of diagnosis.

Symptoms of Hodgkin's disease include fever, enlarged lymph nodes, night sweats, and weight loss.

Lupus erythematosus

There are two types of lupus erythematosus—discoid lupus erythematosus (DLE) and systemic lupus erythematosus (SLE).

Both forms are caused by the body's immune system turning on itself to attack its own connective tissue. DLE affects the skin, and most commonly affects middle-aged women. In DLE, the skin acquires thickened, reddish patches on the face, cheeks, and forehead.

The skin is similarly affected in SLE, with sufferers often developing a characteristic butterfly-shaped rash on the face. The joints and internal organs are also affected, and arthritis, particularly in the hands, often develops as a result of SLE.

Tonsillitis

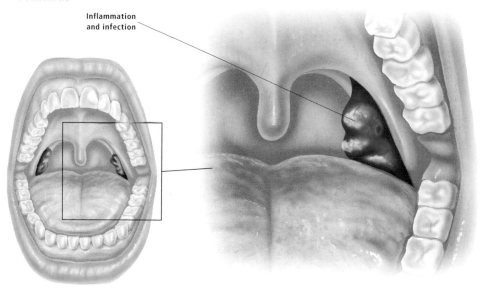

Inflammation
and infection

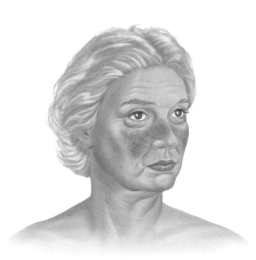

Lupus erythematosus

Corticosteroid creams are usually recommended for the skin, while the use of anti-inflammatory drugs can alleviate joint pain and arthritis. In more severe cases, immunosuppressive drugs may be used.

Tonsillitis

The purpose of the tonsils is to guard the entrances to the respiratory and digestive tracts against infection. However, when bacteria take over the tonsils, they can cause tonsillitis.

The inflamed tonsils develop areas of pus-filled spots, and can cause sore throat and fever. Tonsillectomy, the surgical removal of the tonsils, is often considered for patients with recurring tonsillitis.

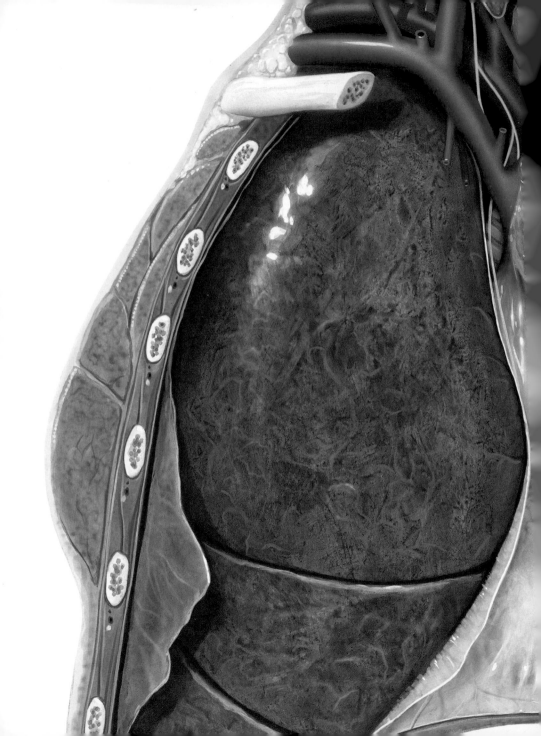

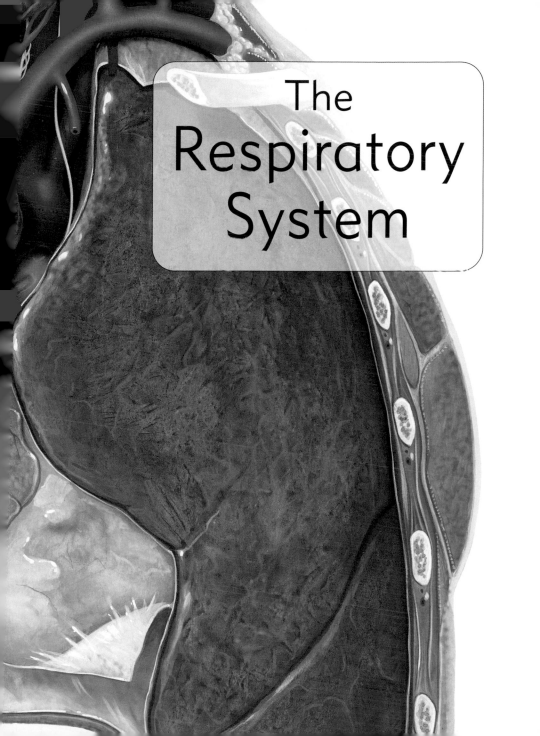

The
Respiratory
System

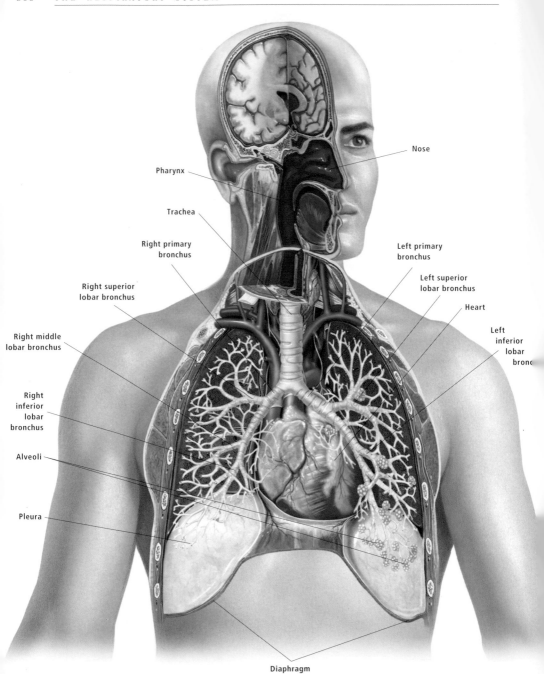

Nose

Pharynx

Trachea

Right primary
bronchus

Left primary
bronchus

Right superior
lobar bronchus

Left superior
lobar bronchus

Heart

Right middle
lobar bronchus

Left
inferior
lobar
bronc

Right
inferior
lobar
bronchus

Alveoli

Pleura

Diaphragm

THE RESPIRATORY SYSTEM

The act of breathing is known as respiration or ventilation. The respiratory system provides the body's means of acquiring the oxygen necessary for cell and tissue maintenance and disposing of unwanted carbon dioxide, and for the energy needed to support life. Respiration also includes the transfer of oxygen from the lungs to the tissues of the body, as well as the transfer of carbon dioxide from the tissues to the lungs.

The upper part of the respiratory system (also known as the upper respiratory tract) comprises the nose, mouth, larynx, pharynx, and trachea.

Air is taken in (inhaled) through the nose. It then moves down the trachea and passes into the lungs. Oxygen is absorbed by the tiny air sacs (alveoli) in the lungs, and passes into the bloodstream. The alveoli in the bronchial tree dispose of unwanted carbon dioxide. In the bronchi there are small hair-like structures known as cilia that make sure that the bronchial airways remain clear.

Breathing is caused by the actions of the muscles between the ribs (the intercostal muscles) and the diaphragm—a dome of muscle that divides the chest and abdomen. When air is breathed in, the intercostal muscles move the ribs upward and outward, and the diaphragm is pushed downward, thus taking air into the expanded lungs.

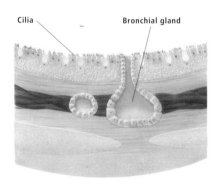

Cilia   Bronchial gland

Cross section of bronchi

NB: In the illustration at left, the front two-thirds of the lungs have been removed to show the heart and bronchial tree.

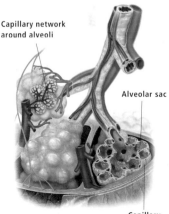

Alveoli

Capillary network around alveoli

Alveolar sac

Capillary

Nose

The role of the nose in the respiratory system is to provide an entry and exit point, and a passageway for air. The bony framework of the nose consists of the nasal bone and the maxilla. The remaining framework of the nose consists of cartilage and the nasal septum, which is part cartilage and part bone, and separates the two nostrils.

The nostrils lead to the nasal cavity. Inhaled air is warmed and any particles inhaled with the air are trapped by tiny hairs (vibrissae) that line the nostrils.

In the nasal cavity, a large surface area is created by the curved structure of the three nasal conchae, located on the side wall of the nose. The blood vessels supplying this large surface area provide the capacity for

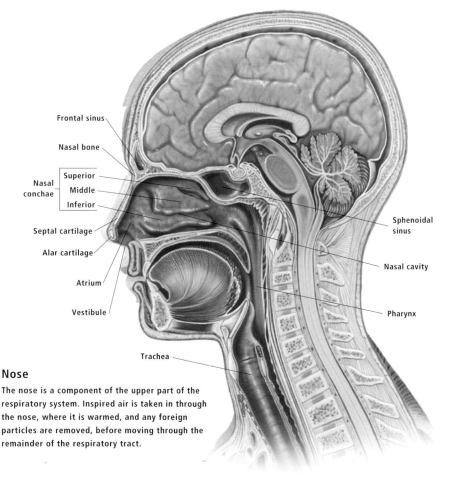

Frontal sinus

Nasal bone

Nasal conchae
— Superior
— Middle
— Inferior

Septal cartilage

Alar cartilage

Atrium

Vestibule

Trachea

Sphenoidal sinus

Nasal cavity

Pharynx

Nose

The nose is a component of the upper part of the respiratory system. Inspired air is taken in through the nose, where it is warmed, and any foreign particles are removed, before moving through the remainder of the respiratory tract.

warming of inhaled air, while the mucous lining, or respiratory mucosa, of the conchae picks up any remaining inhaled particles. The tiny hairs (cilia) lining the respiratory mucosa are responsible for this task, while tiny glands in the mucosa produce a watery secretion that serves to protect the walls of the nasal cavity, and evaporates to humidify the inspired air.

Paranasal Sinuses

Cavities in the frontal, ethmoid, maxillary, and sphenoid bones of the skull form the paranasal sinuses. These cavities lighten the weight of the bones, and add resonance to the voice. Connected to the nose by passageways, these sinuses are each lined with mucous membrane that secretes mucus over the surfaces of the nasal cavity.

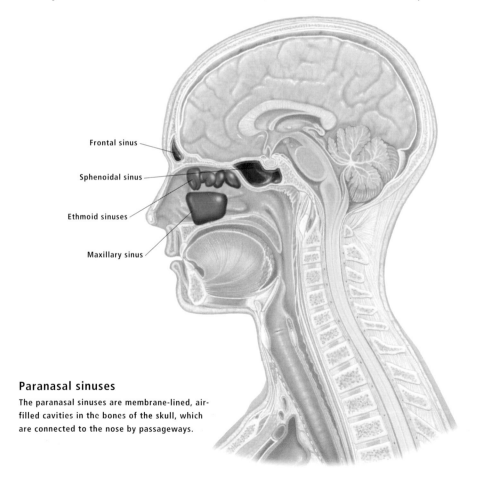

Frontal sinus

Sphenoidal sinus

Ethmoid sinuses

Maxillary sinus

Paranasal sinuses

The paranasal sinuses are membrane-lined, air-filled cavities in the bones of the skull, which are connected to the nose by passageways.

Throat

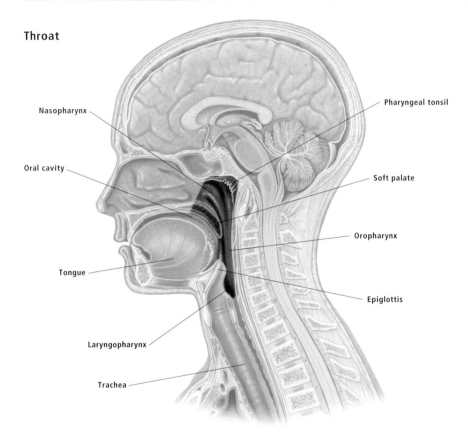

Nasopharynx

Oral cavity

Tongue

Laryngopharynx

Trachea

Pharyngeal tonsil

Soft palate

Oropharynx

Epiglottis

Throat

Located in the front section of the neck, the throat comprises the fauces and the pharynx. The fauces is the opening at the back of the mouth leading to the pharynx, and the pharynx is the major structure of the throat, comprising three regions: the nasopharynx, the oropharynx, and the laryngopharynx.

While each of these regions is distinct, together they combine to create the dual passageway leading to both the respiratory and digestive tracts, and assist in actions such as breathing, speaking, and swallowing.

Pharynx

The pharynx is an elongated tube, surrounded by muscle, which connects to the mouth, nose, and larynx, giving rise to three separate regions: the nasopharynx, oropharynx, and laryngopharynx.

Directly beneath the base of the skull is the region of the nasopharynx. Lying at the back of the nasopharynx are the pharyngeal tonsils (adenoids), aggregations of lymphatic tissue designed to protect the body against bacteria and viruses gaining entry via the nasal passageways.

The eustachian tubes from the middle ear have their opening in the side walls of the nasopharynx. The eustachian tubes equalize pressure in the middle ear to match the external environment. This balance allows the eardrum to function effectively.

The oropharynx lies behind the mouth and provides a passage for air, water, and food, and also receives air and mucus from the nasopharynx lying above. As with the nasopharynx, the entrances to the respiratory and digestive tracts located in the oropharynx are guarded by tonsils.

The laryngopharynx includes the area from the tip of the epiglottic cartilage to the lower edge of the larynx, and contains the entrance to the airways of the trachea and also leads to the esophagus. The epiglottis operates like a trapdoor, sealing off the airways when food passes through the laryngopharynx, and opening up to allow the passage of air in and out of the lungs.

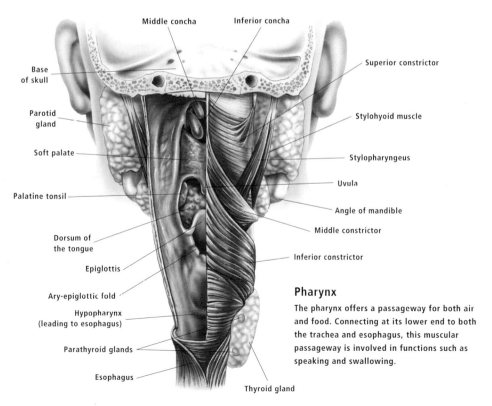

Pharynx

The pharynx offers a passageway for both air and food. Connecting at its lower end to both the trachea and esophagus, this muscular passageway is involved in functions such as speaking and swallowing.

Larynx—anterior view

Hyoid bone greater horn

Epiglottis

Thyrohyoid membrane

Thyroid cartilage of larynx

Superior tubercle

Laryngeal prominence

Inferior tubercle

Cricothyroid muscle

Thyroid gland

Trachea

Larynx

Nine cartilages, held together by ligaments, make up the majority of the structure of the larynx, or voice box, which serves to carry air from the pharynx to the trachea, and to produce a source of air vibration for the voice.

The elastic cartilage of the epiglottis lies at the top of the larynx. Muscles control this flap of cartilage, folding it down to close off the entrance to the trachea during swallowing, and lifting it up to allow the passage of air through the respiratory tract.

Larynx—posterior view

Hyoid bone greater horn

Epiglottis

Thyrohyoid membrane

Opening for internal laryngeal nerve

Corniculate cartilage

Superior horn of thyroid cartilage

Stem of epiglottis

Lamina of thyroid cartilage

Quadrangular membrane

Arytenoid cartilage

Inferior horn of thyroid cartilage

Capsule of cricoarytenoid joint

Cricothyroid joint

Cricoid cartilage

Tracheal muscle

Tracheal cartilage

Larynx

The larynx, or voice box, is composed of cartilages, which are joined by ligaments, and held in place and controlled by skeletal muscles. Ligaments, which form the vocal folds, stretch from front to back between the cartilages.

The largest of the laryngeal cartilages is the thyroid cartilage. The two plates of this cartilage meet at the front of the throat, creating a distinct ridge that can be felt at the front of the throat. In males, it is the rapid growth of this cartilage and the subsequent elongation of the attached vocal ligament, which results in the deepening of the voice, and creates an even more distinct ridge and notch in the thyroid cartilage, which we know as the Adam's apple.

At the base of the larynx lies the cricoid cartilage. This cartilage forms a ring around the top of the trachea, with the two structures connected by ligaments. The vocal cords are attached to the sides of the cricoid cartilage, and extend up to attach to the thyroid and arytenoid cartilages. The paired arytenoid cartilages are situated above, and articulate with, the cricoid cartilage. Lying above, and articulating with, the arytenoid cartilages are the tiny corniculate cartilages. Vocal and vestibular ligaments stretch across the larynx, connecting the thyroid and arytenoid cartilages, and the thyroid and corniculate cartilages respectively. The corniculate cartilages, along with the tiny pair of cuneiform cartilages, serve to strengthen the folds of membrane around the laryngeal entrance.

Closed epiglottis

The cartilage structure of the epiglottis is controlled by muscles. The muscles close the epiglottis over the glottis during swallowing, to prevent food and liquids from entering the airways.

Position of the larynx

The larynx lies in the front section of the neck, below the hyoid bone and in front of the laryngopharynx.

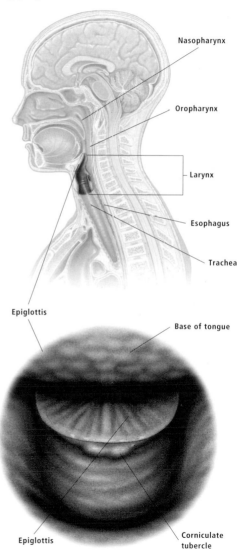

Nasopharynx

Oropharynx

Larynx

Esophagus

Trachea

Epiglottis

Base of tongue

Epiglottis

Corniculate tubercle

Trachea

Forming part of the respiratory tract, the trachea, or windpipe, joins to the base of the larynx, and terminates at the right and left bronchi. This short tube measures about $3\frac{1}{2}$– 5 inches (9–12 centimeters) long and $\frac{3}{5}$ inch (1.5 centimeters) wide, and conveys air from the upper respiratory area to the lungs. Designed for movement, its structure forms a tube of fibro-elastic ligaments and muscle, reinforced with in-complete rings of cartilage. The ends of the cartilage are bridged by trachealis muscle. This combination of elements creates a flexible structure capable of moving in con-junction with the movements of the larynx (in swallowing and breathing), and the dia-phragm (in breathing). The trachea is lined with tiny hairs (cilia) and mucous glands. The mucus attracts any stray particles that may have entered the passageway, while the tiny hairs move the mucus upward, away from the lungs.

Trachea

The muscular fibro-elastic tube of the trachea is reinforced with cartilage. This flexible tube forms a passageway between the larynx and the bronchi of the lungs.

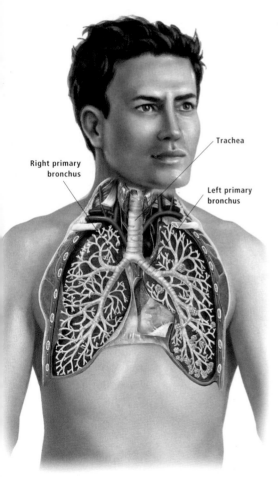

Trachea

Right primary bronchus

Left primary bronchus

Trachea cross section

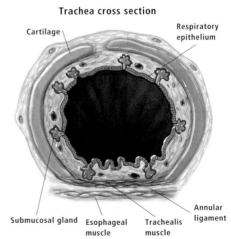

Cartilage

Respiratory epithelium

Submucosal gland

Esophageal muscle

Trachealis muscle

Annular ligament

Bronchus

The bronchi carry air from the trachea to the lungs. The two main bronchi divide into lobar bronchi, with the lobar bronchi subdividing into gradually smaller bronchi and bronchioles, which finally terminate at the tiny air sacs of the lungs.

Bronchus cross section

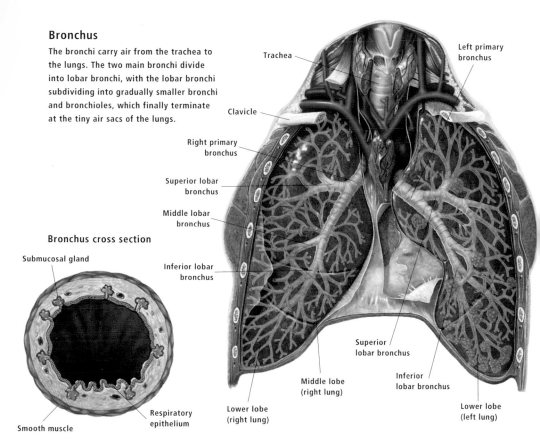

Trachea

Clavicle

Right primary bronchus

Superior lobar bronchus

Middle lobar bronchus

Inferior lobar bronchus

Left primary bronchus

Submucosal gland

Smooth muscle

Respiratory epithelium

Lower lobe (right lung)

Middle lobe (right lung)

Superior lobar bronchus

Inferior lobar bronchus

Lower lobe (left lung)

Bronchus

The trachea terminates at the right and left bronchi. These two major passageways for air serve the right and left lung respectively. The structure of the two primary bronchi is the same as that of the trachea, with the two ends of the horseshoe-shaped cartilage joined by trachealis muscle to form a ring. The inner respiratory mucosa is the same composition as is found in the trachea, with tiny hairs and mucous glands lining the inner surface of the bronchi, though the bronchial lining also contains lymphoid tissue. The two primary bronchi enter the lungs and divide first into lobar bronchi, which serve each lobe of the lung. These lobar bronchi then subdivide into smaller (tertiary) bronchi, gradually becoming the tiny bronchioles. The bronchioles become terminal bronchioles, which culminate at the alveoli, the tiny air sacs of the lungs.

Bronchial tree

The bronchial tree is composed of bronchi, which branch out into gradually smaller and smaller airways, ending in the tiny bronchioles. This massive network of airways enable the lungs to supply the body with the essential oxygen required for maintenance of body tissues and cells.

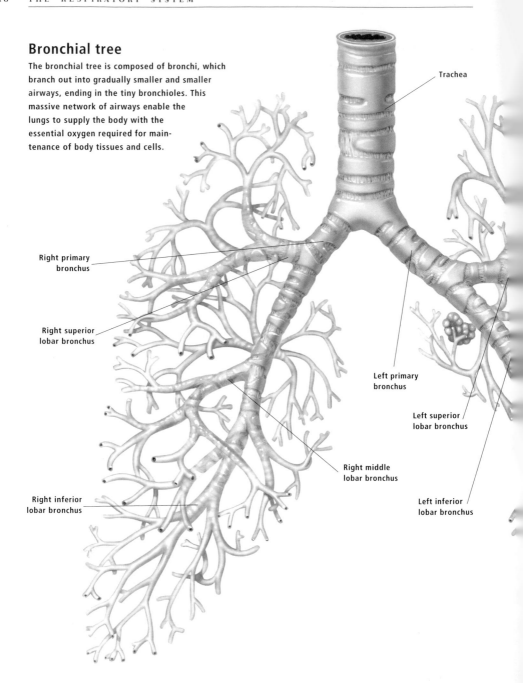

Trachea

Right primary bronchus

Right superior lobar bronchus

Left primary bronchus

Left superior lobar bronchus

Right middle lobar bronchus

Right inferior lobar bronchus

Left inferior lobar bronchus

Bronchial Tree

The collective bronchi, bronchioles, and alveoli make up the bronchial tree. Each primary bronchus divides into lobar bronchi, which are the main bronchi of each of the lobes of the lung. The right lung, which has three lobes, contains the superior lobar bronchus, the middle lobar bronchus, and the inferior lobar bronchus. The left lung, with two lobes, contains the superior lobar bronchus and the inferior lobar bronchus.

Each of the lobar bronchi then divides to form smaller (tertiary) bronchi. While the bronchi have encircling rings of cartilage surrounding them, the amount of cartilage in the bronchi structure diminishes as the branching bronchi become smaller. The tertiary bronchi subdivide many times, which eventually separate into the tiny, thread-like, terminal bronchioles. The lungs contain about 64,000 of these minute bronchioles. The bronchioles are composed of smooth muscle, and contain no cartilage component. Clustered at the end of each of the terminal bronchioles are the alveoli, the tiny air sacs of the lungs, each separated by interalveolar septa.

Each lung contains around 300 million alveoli, which provide the lungs with a massive surface area of around 1,000 square feet (93 square meters). It is here, in the alveoli, that carbon dioxide and oxygen are exchanged, replenishing the oxygen supply vital to the body and transferring unwanted carbon dioxide for expiration. The two gases are exchanged via the thin-walled blood vessels of the alveoli.

Terminal bronchioles

Alveoli

Alveoli

Each terminal bronchiole terminates with the alveoli, a cluster of air sacs resembling bunches of grapes. The alveoli are separated from each other by the interalveolar septa, with the combined area of all the alveoli providing a vast surface area for gas exchange.

Lungs

Protected by the encircling ribs, the paired
organs of the lungs are encased by the pleura,
a thin, two-layered membrane, which lines
the lung and chest cavity and enables the
lungs to move smoothly against the chest
wall during breathing.

The outer parietal layer of the pleura is
attached to the rib cage, while the inner
visceral layer is in contact with the lungs.
A thin film of fluid separates the two layers,
with this fluid aiding the smooth movement
of the lungs.

Nestled between the pleural sacs of
each lung are the heart, esophagus, trachea,
and major vessels and nerves; this region
is known as the mediastinum. The location
of these organs, with the apex of the heart
pointing to the left, result in the left lung
being smaller than the right lung.

Each lung is composed of elastic tissue,
divided into lobes. The left lung usually
has two lobes, separated by an oblique
fissure. The larger right lung has three
lobes, created by an upper horizontal
fissure and a lower oblique fissure.

Branching from the base of the trachea,
the left and right bronchi enter the lungs
at the hilum, located on the mediastinal
surface of the lungs. The major blood vessels serving the lungs also enter and leave
at this location.

The base of the lungs rests on the
diaphragm, the principal muscle involved
in inspiration.

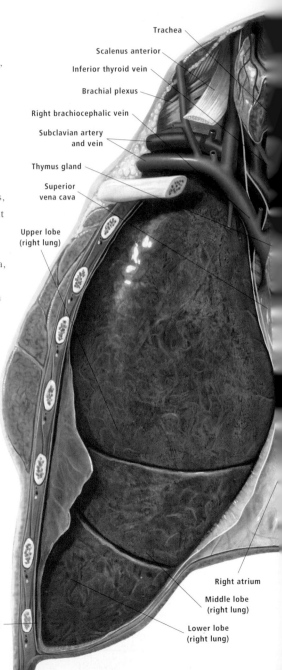

Trachea

Scalenus anterior

Inferior thyroid vein

Brachial plexus

Right brachiocephalic vein

Subclavian artery
and vein

Thymus gland

Superior
vena cava

Upper lobe
(right lung)

Right atrium

Middle lobe
(right lung)

Costodiaphragmatic sinus

Lower lobe
(right lung)

Lungs

Sitting within the thoracic cage are the lungs. Segmented into lobes by fissures, the two lungs are positioned around the major organ of the heart, with the positioning of the heart resulting in the left lung being slightly smaller than the right lung. Air is drawn into the lungs by the muscular action of the diaphragm and the intercostal muscles of the ribs. These muscles work in unison to increase the capacity of the thoracic cage to allow the lungs to expand on inspiration.

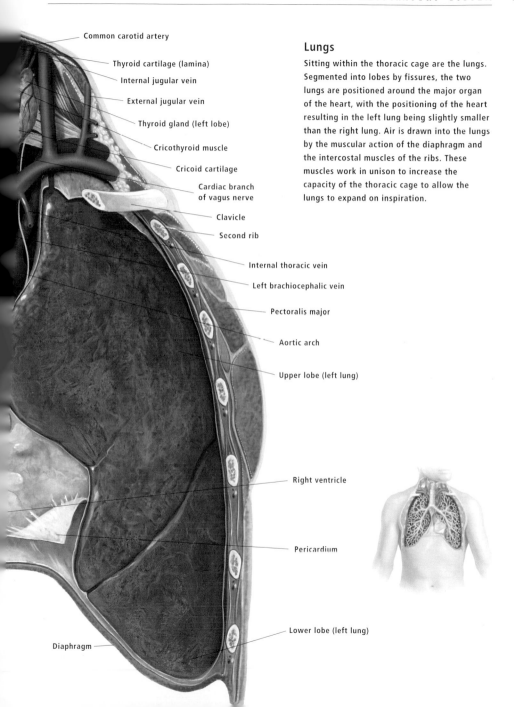

Common carotid artery

Thyroid cartilage (lamina)

Internal jugular vein

External jugular vein

Thyroid gland (left lobe)

Cricothyroid muscle

Cricoid cartilage

Cardiac branch of vagus nerve

Clavicle

Second rib

Internal thoracic vein

Left brachiocephalic vein

Pectoralis major

Aortic arch

Upper lobe (left lung)

Right ventricle

Pericardium

Lower lobe (left lung)

Diaphragm

Breathing

Respiration is the name given to the act of breathing. It also refers to the transfer of oxygen to the body tissues, and of carbon dioxide from the tissues to the lungs.

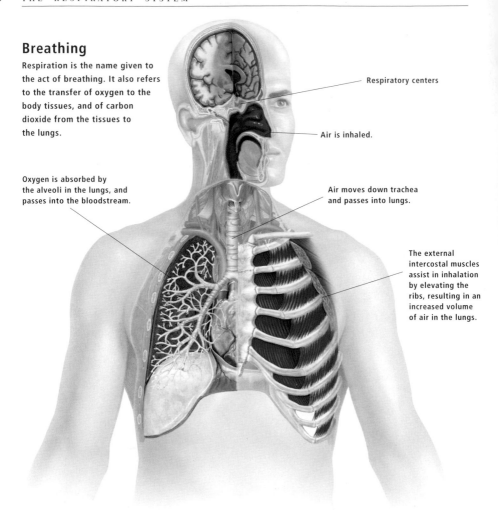

Respiratory centers

Air is inhaled.

Air moves down trachea and passes into lungs.

Oxygen is absorbed by the alveoli in the lungs, and passes into the bloodstream.

The external intercostal muscles assist in inhalation by elevating the ribs, resulting in an increased volume of air in the lungs.

Breathing

Breathing is also referred to as respiration or ventilation. It involves the process of taking in of air (inspiration), and expelling of air (expiration).

While the diaphragm is the prime mover in inspiration, it is the coordinated move-ments of the diaphragm, the intercostal muscles, and the abdominal wall combined that increase lung capacity, allowing the intake of air in inspiration.

The purpose of breathing is to replenish the oxygen levels in the blood by inspiration of oxygen-loaded air, and the elimination of carbon dioxide accumulated as waste product from the tissues and cells of the body. Carbon dioxide is expelled during the breathing out process (expiration). Therefore, the effective function of the lungs in inspiration and expiration is crucial to this oxygen replenishment requirement.

On average, the body at rest breathes in about 10 to 15 times per minute. This rate increases during physical activity, or stressful situations, as the rapid consumption of oxygen by tissues and cells demands an immediate increase in supply of oxygen. Any increase in oxygen demands by the body results in a faster breathing rate.

Breathing sequence

Air is drawn into the lungs as a result of the contraction of the diaphragm and the elevation of the rib cage by the intercostal muscles. Relaxation of the tension in these muscles results in expiration, or breathing out.

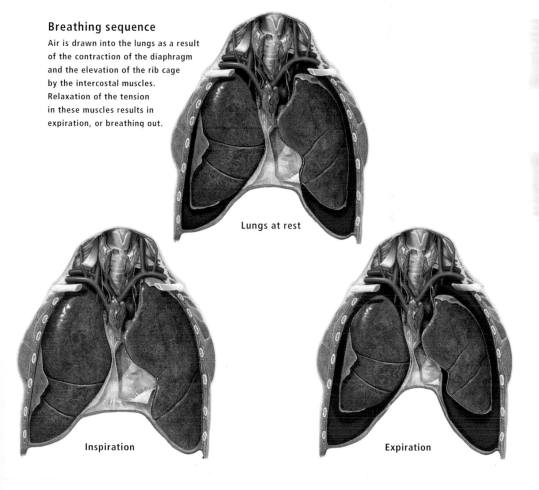

Lungs at rest

Inspiration

Expiration

Gas Exchange

Gas exchange takes place at the alveoli. The tiny alveoli resemble clusters of grapes, with each lung containing about 300 million of these minute air sacs.

Blood that has circulated around the body has passed on much of its oxygen and nutrients to the body's cells and tissues for cellular metabolism and, in turn, has absorbed unwanted carbon dioxide from the cells and tissues.

This oxygen-poor blood is directed from the heart to the lungs for gas exchange to take place. The blood is then transported through the lungs by the pulmonary arteries, eventually making its way to the alveoli.

The tiny chambers of the alveoli are separated from one another by extremely thin interalveolar septa and are supplied by an extensive network of thin-walled capillaries.

Alveoli

The millions of alveoli create a huge surface area for gas exchange. The alveoli are separated from each other by thin septa, which have a fine coating of pulmonary surfactant. This substance reduces surface tension and thus prevents collapse of the alveoli during expiration.

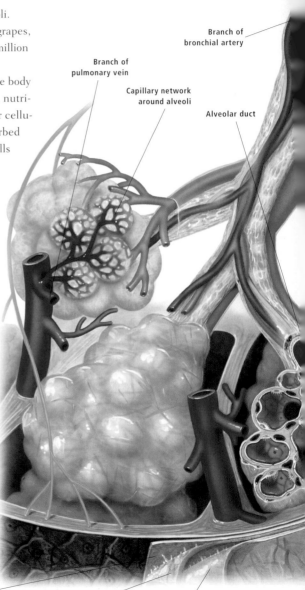

Branch of bronchial artery

Branch of pulmonary vein

Capillary network around alveoli

Alveolar duct

Visceral pleura

Endothoracic fascia

Parietal pleura

Bronchiole

Branch of
pulmonary
artery

It is through these capillaries that oxygen is absorbed from inhaled air, and carbon dioxide in the blood diffuses into the alveolar sac and is then expelled when breathing out.

Red blood cells contain hemoglobin, which is responsible for the transport of oxygen. When hemoglobin and oxygen combine, this is known as oxyhemoglobin, and bears the characteristic bright red color of blood. When depleted of oxygen, the deoxygenated hemoglobin takes on a much darker color. Normal levels of hemoglobin in the blood are approximately 14–17 grams per 100 milliliters for males and 12–15 grams per 100 milliliters for females. Since the hemoglobin is the crucial element in red blood cells responsible for transporting oxygen from the lungs to the body tissues, and carbon dioxide back to the lungs, it is therefore a vital element in the overall metabolism of the body.

Gas exchange is most effective when at sea level, and is driven by the pressure difference for that particular gas between the alveolar air and the blood.

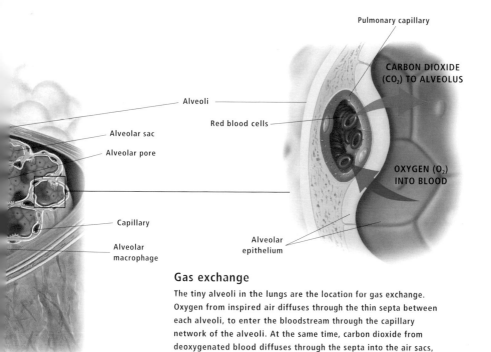

Pulmonary capillary

CARBON DIOXIDE (CO_2) TO ALVEOLUS

Alveoli

Red blood cells

OXYGEN (O_2) INTO BLOOD

Alveolar sac

Alveolar pore

Capillary

Alveolar macrophage

Alveolar epithelium

Gas exchange

The tiny alveoli in the lungs are the location for gas exchange. Oxygen from inspired air diffuses through the thin septa between each alveoli, to enter the bloodstream through the capillary network of the alveoli. At the same time, carbon dioxide from deoxygenated blood diffuses through the septa into the air sacs, ultimately to be expelled from the lungs during expiration.

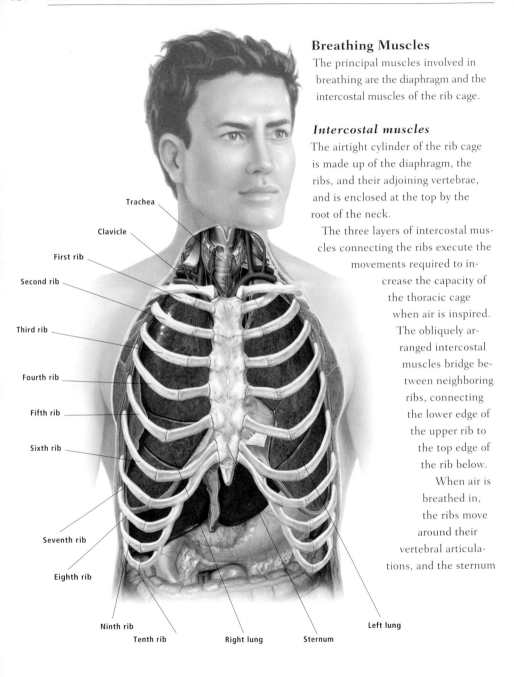

Breathing Muscles

The principal muscles involved in breathing are the diaphragm and the intercostal muscles of the rib cage.

Intercostal muscles

The airtight cylinder of the rib cage is made up of the diaphragm, the ribs, and their adjoining vertebrae, and is enclosed at the top by the root of the neck.

The three layers of intercostal muscles connecting the ribs execute the movements required to increase the capacity of the thoracic cage when air is inspired. The obliquely arranged intercostal muscles bridge between neighboring ribs, connecting the lower edge of the upper rib to the top edge of the rib below. When air is breathed in, the ribs move around their vertebral articulations, and the sternum

Trachea

Clavicle

First rib

Second rib

Third rib

Fourth rib

Fifth rib

Sixth rib

Seventh rib

Eighth rib

Ninth rib

Tenth rib

Right lung

Sternum

Left lung

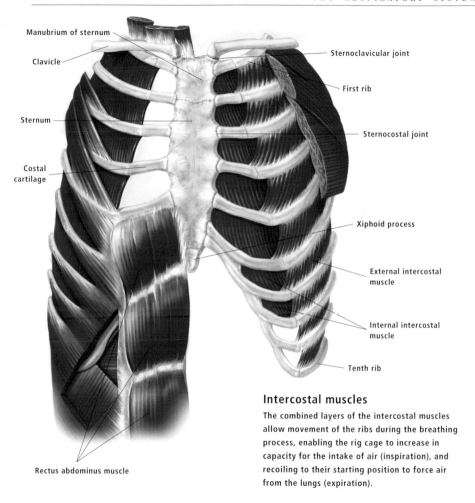

Manubrium of sternum

Clavicle

Sternum

Costal
cartilage

Sternoclavicular joint

First rib

Sternocostal joint

Xiphoid process

External intercostal
muscle

Internal intercostal
muscle

Tenth rib

Rectus abdominus muscle

Intercostal muscles

**The combined layers of the intercostal muscles
allow movement of the ribs during the breathing
process, enabling the rig cage to increase in
capacity for the intake of air (inspiration), and
recoiling to their starting position to force air
from the lungs (expiration).**

connected to the true ribs is elevated, thus
raising the ribs. This movement increases
the front-to-back and side-to-side dimensions of the rib cage.

In conjunction with these movements, the
diaphragm moves downward, resulting in
expansion of the vertical size of the thoracic
cage. The increased capacity of the rib cage

results in a drop in intrathoracic pressure,
which falls below atmospheric pressure,
allowing air to be sucked into the airways.

Breathing out is a passive process. The
muscles of the thoracic cage recoil to their
resting position, returning the thoracic cage
to its original size, and pushing air from the
lungs as a result.

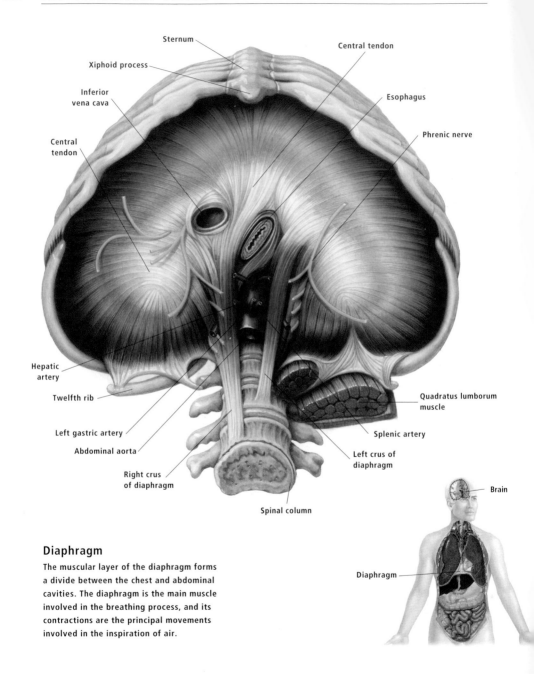

Sternum

Xiphoid process

Inferior
vena cava

Central
tendon

Central tendon

Esophagus

Phrenic nerve

Hepatic
artery

Twelfth rib

Left gastric artery

Abdominal aorta

Right crus
of diaphragm

Spinal column

Quadratus lumborum
muscle

Splenic artery

Left crus of
diaphragm

Brain

Diaphragm

Diaphragm

The muscular layer of the diaphragm forms
a divide between the chest and abdominal
cavities. The diaphragm is the main muscle
involved in the breathing process, and its
contractions are the principal movements
involved in the inspiration of air.

Diaphragm

The diaphragm is the principal muscle involved in breathing. It lies at the base of the thoracic cage, effectively separating the chest cavity from the abdominal cavity.

The diaphragm is attached to the vertebral column at the back of the rib cage, to the ribs along the side of the chest, and to the sternum at the front of the chest, with the phrenic nerves leading from the upper spinal cord controlling the movements of the diaphragm.

When the diaphragm contracts, it moves downward, increasing the capacity of the chest cavity, thus forcing the lungs to expand and drawing in air to fill the void. This expansion is assisted by movement of the ribs, which work in conjunction with the diaphragm during breathing. The contraction of the diaphragm causes the intrathoracic pressure to drop, and the resultant intake of air then equalizes the pressure.

Air is expelled from the lungs as the muscles involved in breathing—the diaphragm and the intercostal muscles—relax the tension in the soft and hard tissues. This is a passive process, and the diaphragm does not play an active role in expiration.

Respiratory centers located in the pons and medulla areas of the brain stem control the breathing rate. Information on blood oxygen levels is transmitted to these areas from monitoring areas within the body. The respiratory centers then adjust the rate and depth of breathing in response to information received.

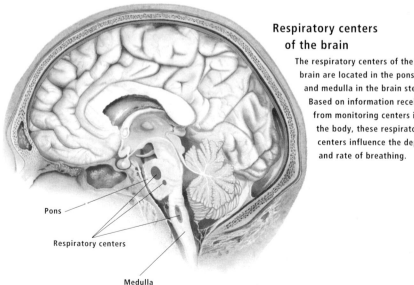

Respiratory centers of the brain

The respiratory centers of the brain are located in the pons and medulla in the brain stem. Based on information received from monitoring centers in the body, these respiratory centers influence the depth and rate of breathing.

Pons

Respiratory centers

Medulla

Heart–Lung Relationship

The heart lies between the two lungs, in an area known as the mediastinum. With the apex of the heart pointing to the left, it results in the left lung being smaller than the right, to accommodate the heart. Each organ has a protective outer layer, with the double layered pleura surrounding the lungs and the pericardium surrounding the heart. The flexible surrounds of each of these organs ensure that the movements required for both breathing and pumping of the heart can be achieved.

The teamwork of the heart and lungs ensures that the body has a constant supply of oxygen for cellular and tissue activity, and that the major waste product of the cells and tissues—carbon dioxide—is constantly removed. This all occurs through the pulmonary circulation, with the heart supplying blood that has completed a circuit of the body to the lungs.

As blood travels through the body, its oxygen supply is delivered to the cells and tissues. Carbon dioxide is offloaded by the cells and tissues, so the blood returning to the heart is oxygen poor and loaded with carbon dioxide.

The now dark red blood is returned to the heart from the head, limbs, and internal organs. It is sent from the heart via the pulmonary trunk and arteries to the lungs, where gas exchange takes place at alveolar level through the extensive network of capillaries associated with the alveoli. Oxygen is replenished and carbon dioxide diffuses into the alveolar sac for exhalation. The blood is then returned to the heart by the pulmonary venous system, ready for another circuit of the systemic circulation.

Cross section through heart and lungs

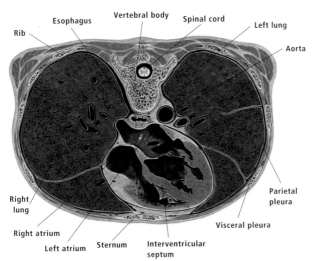

Esophagus · Vertebral body · Spinal cord · Left lung · Rib · Aorta · Right lung · Right atrium · Left atrium · Sternum · Interventricular septum · Visceral pleura · Parietal pleura

Heart and lungs

The heart lies in the mediastinum, the region between the two lungs. Each organ has a protective outer layer, with the double layered pleura surrounding the lungs, and the pericardium surrounding the heart. The heart and lungs work together in the pulmonary circulation, with the two organs connected by the pulmonary vessels.

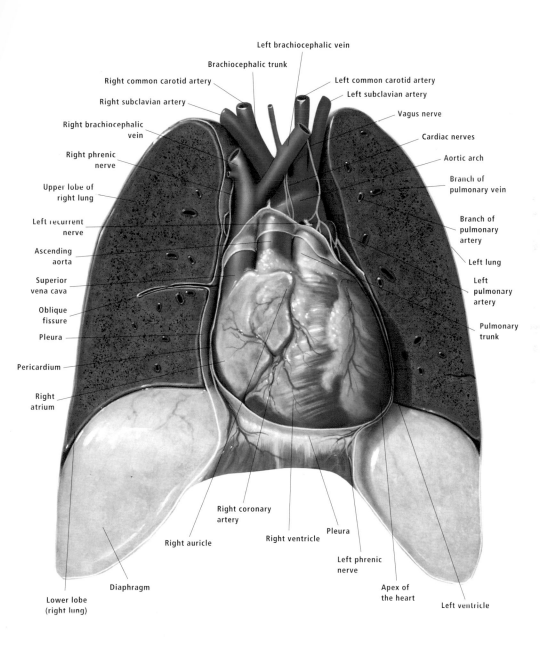

Left brachiocephalic vein

Brachiocephalic trunk

Right common carotid artery

Left common carotid artery

Left subclavian artery

Right subclavian artery

Vagus nerve

Right brachiocephalic vein

Cardiac nerves

Right phrenic nerve

Aortic arch

Upper lobe of right lung

Branch of pulmonary vein

Left recurrent nerve

Branch of pulmonary artery

Ascending aorta

Left lung

Superior vena cava

Left pulmonary artery

Oblique fissure

Pulmonary trunk

Pleura

Pericardium

Right atrium

Right coronary artery

Right auricle

Right ventricle

Pleura

Left phrenic nerve

Diaphragm

Apex of the heart

Lower lobe (right lung)

Left ventricle

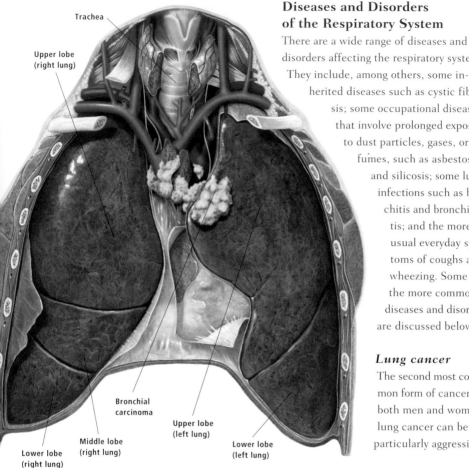

Trachea

Upper lobe
(right lung)

Bronchial
carcinoma

Upper lobe
(left lung)

Middle lobe
(right lung)

Lower lobe
(right lung)

Lower lobe
(left lung)

Diseases and Disorders of the Respiratory System

There are a wide range of diseases and disorders affecting the respiratory system. They include, among others, some inherited diseases such as cystic fibrosis; some occupational diseases that involve prolonged exposure to dust particles, gases, or fumes, such as asbestosis and silicosis; some lung infections such as bronchitis and bronchiolitis; and the more usual everyday symptoms of coughs and wheezing. Some of the more common diseases and disorders are discussed below.

Lung cancer

The second most common form of cancer for both men and women, lung cancer can be a particularly aggressive

Lung cancer

Lung cancer is a particularly insidious form of cancer, often spreading through the walls of the airways, entering the bloodstream and lymphatic channels, and generally avoiding detection until secondary growths or other symptoms present themselves. This illustration shows the presence of bronchial carcinoma in the airways.

form of cancer, and is the most common cause of cancer death in most countries.

Tobacco smoking, especially in the form of cigarettes, is particularly indicated in lung cancer, though pipe and cigar smoking can also be culprits.

Lung cancers generally develop in the epithelium of the airways, and several different types of cancer cell are indicated. Often the first indication of lung cancer stems from a secondary growth, often occurring in the bones, liver, or brain.

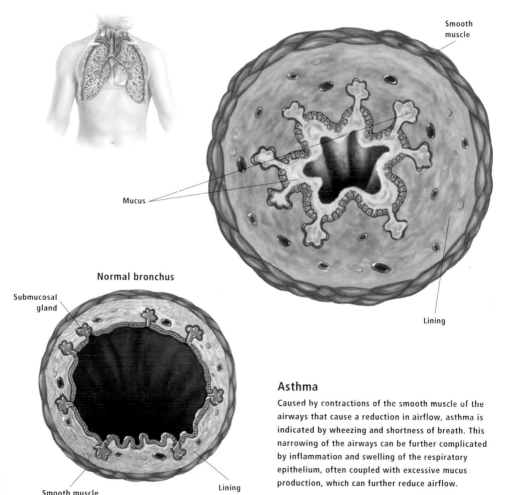

Smooth muscle

Mucus

Normal bronchus

Submucosal gland

Lining

Lining

Smooth muscle

Asthma

Caused by contractions of the smooth muscle of the airways that cause a reduction in airflow, asthma is indicated by wheezing and shortness of breath. This narrowing of the airways can be further complicated by inflammation and swelling of the respiratory epithelium, often coupled with excessive mucus production, which can further reduce airflow.

Early detection is essential in the fight against lung cancer, as often this insidious invader is not easily detectable until it has become firmly established and spread significantly, often invading neighboring structures, or entering the bloodstream or lymphatics, or both.

Lung cancer can be treated by surgery, radiation therapy, or chemotherapy, depending on the exact nature and location of the cancer.

Asthma

In industrialized countries, asthma is becoming increasingly prevalent, with reported new cases on the rise. Generally, asthma is more common in children than adults, although childhood sufferers often outgrow the disorder by adulthood.

Asthma involves muscular contractions of the airways, which reduce airflow and cause labored breathing, shortness of breath, coughing, and wheezing.

Asthma is often triggered by airborne irritants, such as dusts, pollens, cigarette smoke, and animal hair. Certain pharmaceutical drugs, exercise, and respiratory infections are among the many other triggers of asthma attack.

Symptoms can range from mild to serious, and for milder symptoms the disorder can often be kept in check with the use of inhaled medication, which acts to relax the muscles, reinstating airflow. Severe asthma attacks require urgent medical attention and hospitalization.

Cough

A cough can be instigated by a variety of triggers, and can sometimes be an indication of more serious lung conditions.

Often triggered by airborne irritants, infection, inflammation, inhaled particles, or a growth in the airways, cough is the body's reflex action to eliminate the cause of the irritation. Muscular contractions of the diaphragm and chest cause a rapid expiration of air in an attempt to clear the airways.

A cough is often associated with the common cold or respiratory tract infections. A cough can also indicate a more serious condition, such as emphysema or lung cancer, and therefore the underlying cause of the cough must be determined to ensure the most appropriate method of treatment.

Cough

A Irritants are inhaled and stimulate nerve receptors in larynx, trachea, and bronchi.

B Nerve receptors in larynx, trachea, and bronchi send signals to brain stem via vagus nerve.

C Brain stem triggers coughing reflex via phrenic nerve.

D Diaphragm rises and chest muscles contract, forcing air out of lungs as a cough.

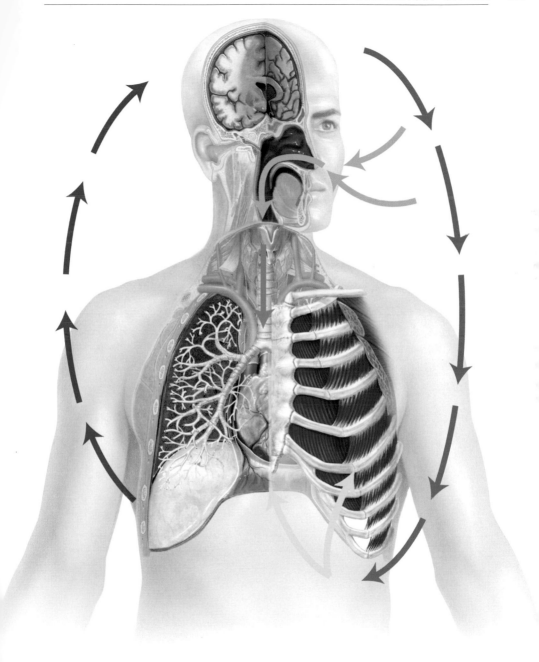

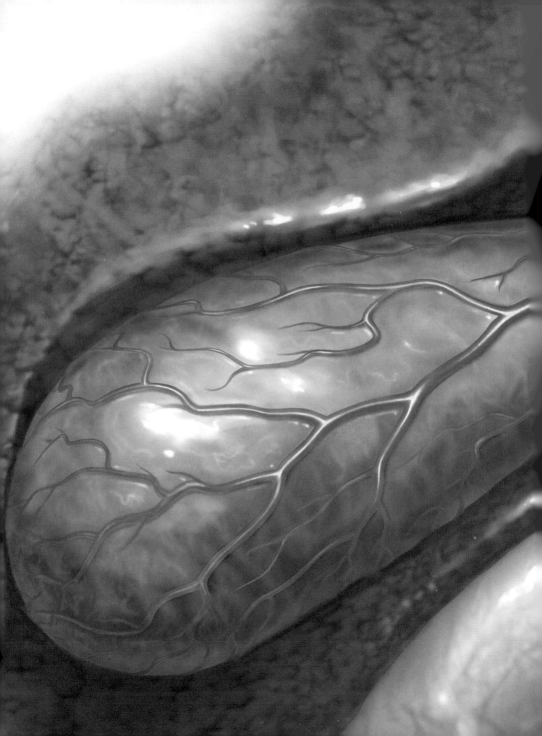

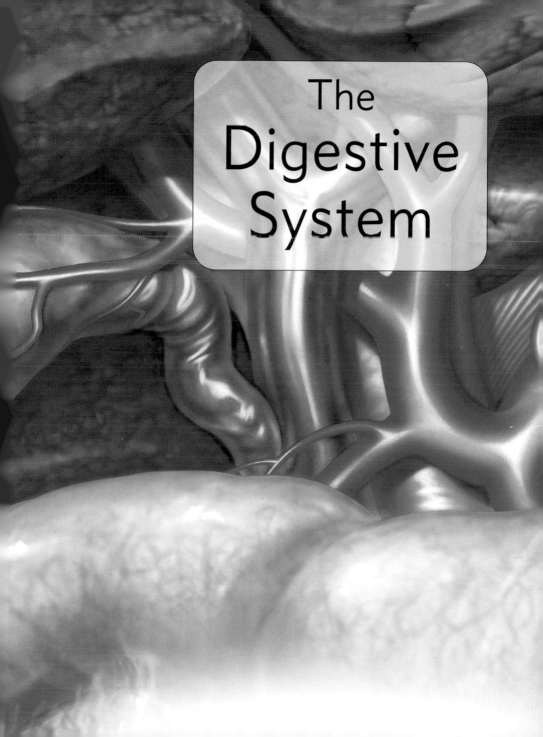

The
Digestive
System

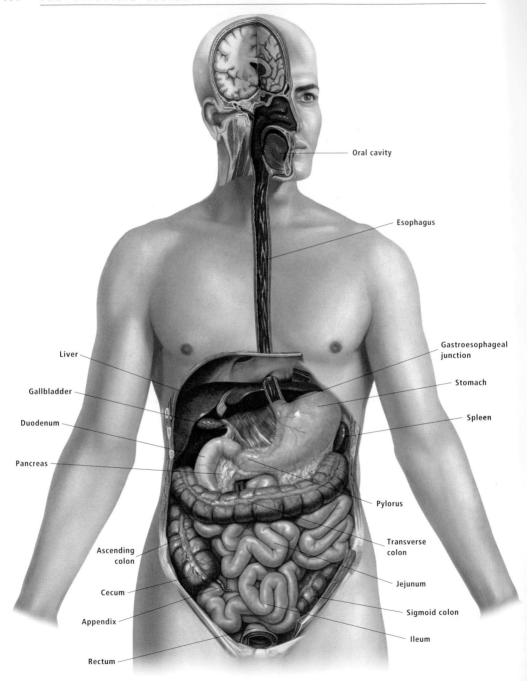

Oral cavity

Esophagus

Gastroesophageal junction

Liver

Stomach

Gallbladder

Spleen

Duodenum

Pancreas

Pylorus

Ascending colon

Transverse colon

Cecum

Jejunum

Appendix

Sigmoid colon

Ileum

Rectum

THE DIGESTIVE SYSTEM

The combined structures belonging to the digestive system are responsible for processing the food we eat, extracting the essential elements necessary for body upkeep, and eliminating the remainder as waste matter. Eventually, food is broken down into molecules that are absorbed by the body and used in the construction of human cells.

Mechanically, the digestive tract is composed of smooth muscle fibers that run up and down, as well as in a circular motion. In certain areas, the circular fibers are packed down into a ring, known as a sphincter. This contracts, shutting down the opening within the tract—the lumen. During the process known as peristalsis, the circular fibers contract in sequence along the tract, kneading or compressing the contents with the digestive enzymes.

From the moment food enters the mouth, and throughout its journey through the digestive system, it is systemically broken down into simpler components. As they pass through the various structures of the digestive system, these components are extracted and absorbed, entering the bloodstream to be transported to target cells, tissues, and organs. The remainder—which is of no use to the body—is excreted as feces.

The digestive system is synergistically designed so that its mechanical actions

NB: In this illustrationn the liver has been lifted up to show the gallbladder.

ensure the optimal efficiency of the chemical actions of the enzymes. These digestive enzymes break down the food groups into their chemical components: carbohydrates are converted to glucose; lipids are converted to fatty acids and glycerol; and proteins are converted to amino acids. There are about 25 amino acids that occur in most animal species. Human proteins are composed of the amino acids that come from animal protein.

Vitamins and minerals are essential to the efficient functioning of the body, and its ability to maintain a balanced state. As the body is unable to produce most of the vitamins and minerals required, these essential elements must be obtained from dietary intake.

The minerals essential to well-being include: calcium, chloride, iron, magnesium, phosphorus, potassium, and sodium. The body's requirements for each of these minerals varies, and many are required only in minute amounts. In some cases, too much of a certain mineral can be toxic to the body.

Vitamins are required for many functions in the body, with some aiding in cell regeneration, such as Vitamin A, which promotes pigment formation necessary for good vision, and Vitamin B_2, which aids in repair and maintenance of skin and body tissues. Healthy bones are achieved with the help of Vitamins D and K, while Vitamins B_{12} and E are important to red blood cell production and red blood cell maintenance respectively.

Digestion and Absorption of Nutrients

Nutrition is the intake of food and its use by the body. The nutrients in food must supply the body's requirements for energy, cell renewal and replacement, tissue renewal and replacement, maintain chemical levels in the body, and supply all the chemicals and organic compounds that the body cannot manufacture for itself.

A healthy, balanced, varied diet is the basis of good nutrition, and ensures that the body acquires all the essential nutrients for well-being.

Chewed food is pushed from the mouth to the esophagus and down to the stomach, where it mixes with stomach acids. The stomach contents are churned and mixed for some time, then are gradually released into the duodenum. The majority of processing, absorption, and extraction takes place in the duodenum and jejunum of the small intestine. The contents then pass to the ileum where more nutrients are absorbed, and the remaining material is transferred to the large intestine, where it is passed to the rectum to be excreted as feces.

The nutrients from food are categorized broadly as carbohydrates, fats and oils, cholesterol, proteins, water, minerals, and vitamins.

Carbohydrates are the main sources of fuel energy for the body. The body's requirement for fats and oils is small; fats and oils contribute vitamin E and provide an energy resource for extremely active persons, though they are often stored as fat in others. Proteins aid in cell building and repair; minerals supply iron and calcium for the body requirements; vitamins bring about the chemical reactions that make and repair cells, as well as provide energy; and water is crucial to hydration levels in the body.

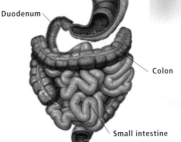

Esophagus

Duodenum

Stomach

Duodenum

Colon

Small intestine

Stomach

Food goes from the mouth to the stomach via the esophagus. The gastric juices are then stimulated into action, mixing with the food by muscular contractions of the stomach wall. Small amounts of the resultant mixture (chyme), are released into the duodenum, where absorption of nutrients begins.

Digestion and absorption of nutrients

Carbohydrates, proteins, fats, minerals, vitamins, and water are all essential to well-being, and these nutrients are all found in the dietary intake. The body has digestive processes in place to extract these nutrients from the food we eat, thereby ensuring dead cells are replaced with new ones, existing cells are maintained at peak performance, and that chemical and water balances in the body are kept stable.

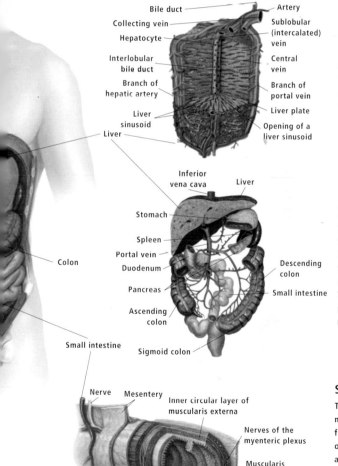

Liver

The liver contains thousands of lobules, each a tiny processing center for the blood carried back from the gastrointestinal tract. The nutrients in this blood are extracted, with many stored in situ in the cells of the liver.

Hepatic portal system

Nutrients are absorbed through the lymphatic vessels and capillaries in the intestines. The lymphatic vessels, known as lacteals, carry nutrients through the lymph system. The capillaries drain to the hepatic portal system, transporting absorbed nutrients to the liver for processing.

Small intestine

The small intestine is responsible for most of the absorption of nutrients from food, with the absorption occurring mostly in the duodenum and jejunum. The tiny villi that line the small intestine create a huge surface area for absorption. Any remaining matter will pass along the colon to be excreted as feces.

Structure of the mouth (with tongue down)

Mouth

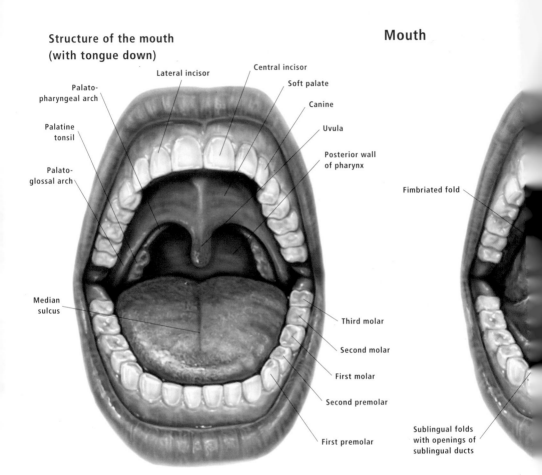

- Lateral incisor
- Central incisor
- Palato-pharyngeal arch
- Soft palate
- Palatine tonsil
- Canine
- Uvula
- Palato-glossal arch
- Posterior wall of pharynx
- Fimbriated fold
- Median sulcus
- Third molar
- Second molar
- First molar
- Second premolar
- Sublingual folds with openings of sublingual ducts
- First premolar

Mouth

The entrance to the body for the digestive system, the mouth contains the elements to commence the first part of the digestive process. The teeth, tongue, and salivary glands are all located in the mouth, which is surrounded at the sides by the muscles of the cheeks, primarily the buccinator muscles, and encircled at the front by the versatile circular muscle of the lips, the orbicularis oris. The roof is formed by the soft and hard palates, while the floor is formed by the tongue and underlying tissue spanning from the teeth on the left side to those on the right. The hard palate is mostly formed by the maxilla, while the soft palate is comprised of muscle with an outer layer of

Salivary glands (with tongue up)

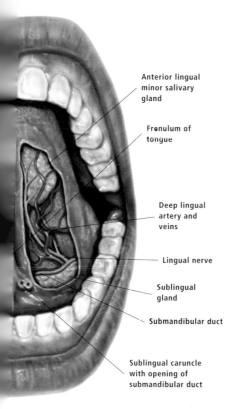

Anterior lingual minor salivary gland

Frenulum of tongue

Deep lingual artery and veins

Lingual nerve

Sublingual gland

Submandibular duct

Sublingual caruncle with opening of submandibular duct

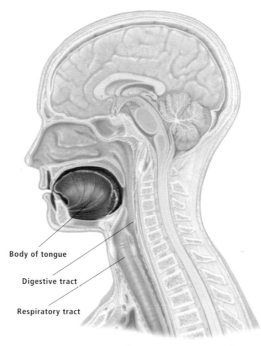

Body of tongue

Digestive tract

Respiratory tract

Mouth—entrance to body

Providing the entrance to the digestive system, the mouth commences the initial breakdown of food for transfer to the digestive tract. The mouth also connects the respiratory system, and plays a role in speaking and breathing.

mucous membrane. These two palates separate the oral and nasal cavities. At the back of the mouth is the entrance to the oropharynx, and this entrance is guarded by the dangling structure of the uvula.

Food is ground down by the teeth into smaller components, with the tongue and the buccinator muscles in the cheeks enabling the food to be moved around in the mouth. Saliva secreted by the salivary glands moistens the food, and the reaction between the food and saliva activates the taste buds. The saliva facilitates swallowing, and the chewed food is eventually formed into a ball, which is then pushed to the back of the mouth and into the pharynx.

Teeth

Several layers of substances cover the inner roots of the teeth. Although the teeth have a bone-like appearance, the outer visible part, known as the crown is, in fact, composed of enamel, the hardest substance in the body. Below the gum line, the outer surface of the teeth is composed of cementum, which is attached to the bone of the gum by the periodontal ligaments, thus holding the teeth firmly in place in their sockets.

The inner layer of the teeth, encased by the enamel and cementum, is made up of dentine, a softer, more sensitive tissue layer, which surrounds the pulp cavity.

The pulp cavity contains the nerve fibers and blood vessels supplying the tooth.

During childhood, the first set of teeth appears. These are known as the primary, milk, baby, or deciduous teeth. Numbering 20 teeth in total, these primary teeth are generally whiter in appearance than the second set of permanent teeth. Later in childhood, the primary teeth are gradually replaced by the second, permanent set of teeth.

There are 32 permanent teeth, 16 in each jaw, and comprise the incisors, canines, premolars, and molars; the final molars, often called the wisdom teeth, appear in late

Teeth—structure

At the center of each tooth is the central pulp containing the nerves and blood vessels of the tooth. The central pulp is encased by protective layers of dentine, cementum, and enamel.

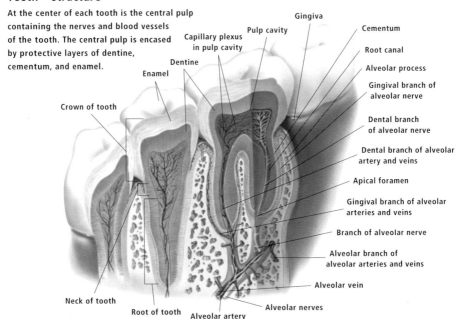

Crown of tooth

Capillary plexus in pulp cavity

Dentine

Enamel

Pulp cavity

Gingiva

Cementum

Root canal

Alveolar process

Gingival branch of alveolar nerve

Dental branch of alveolar nerve

Dental branch of alveolar artery and veins

Apical foramen

Gingival branch of alveolar arteries and veins

Branch of alveolar nerve

Alveolar branch of alveolar arteries and veins

Alveolar vein

Neck of tooth

Root of tooth

Alveolar artery

Alveolar nerves

adolescence. The different teeth perform different tasks in the process of chewing food: the incisors are designed for biting, the canines are designed for tearing, and the premolars and molars are designed for grinding food. The sum of the efforts of the different types of teeth result in food being converted into a bolus, or ball of food, which is able to easily pass into the pharynx, and on through the digestive tract.

Apart from their involvement in the digestive process, the teeth also play a role in speaking, and contribute to the shape of the face.

Teeth

The 20 primary teeth are gradually replaced by 32 permanent teeth, which include canines, incisors, and molars.

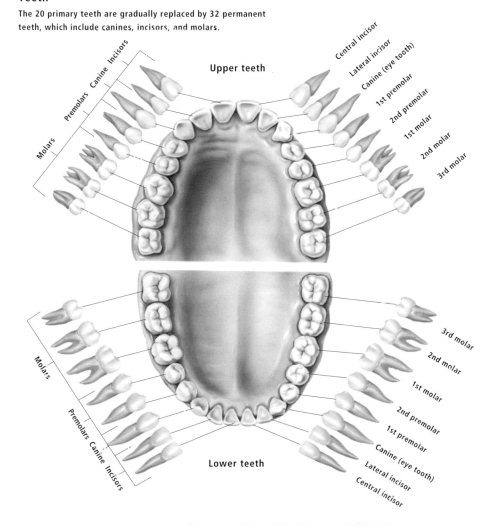

Incisors
Canine
Premolars
Molars

Upper teeth

Central incisor
Lateral incisor
Canine (eye tooth)
1st premolar
2nd premolar
1st molar
2nd molar
3rd molar

Molars

3rd molar
2nd molar
1st molar
2nd premolar
1st premolar
Canine (eye tooth)
Lateral incisor
Central incisor

Premolars Canine Incisors

Lower teeth

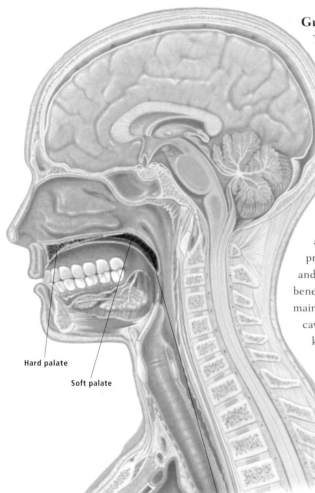

Hard palate

Soft palate

Uvula

Palate

Comprising two sections, the hard palate and the soft palate, the palate separates the oral and nasal cavities. The bony hard palate extends back from the top teeth, and when the mouth is closed, the tongue rests on the hard palate. The soft palate lies behind the hard palate and is composed of muscle fibers.

Gums

The teeth sit in sockets in the upper and lower jaw bones (maxillae and mandible). They are held firmly in place by the oral mucosa of the gums, or gingivae. The soft tissue of the gums extends from inside the lips, around and between the teeth, to the floor of the mouth (lower jaw), and the palate (upper jaw). The gums hold tightly around the neck of each tooth, to prevent the entry of food particles and bacteria into the sensitive area beneath the gums. As with the remainder of the structures in the oral cavity, the surfaces of the gums are kept moist by the secretions of the salivary glands.

Palate

Separating the oral and nasal cavities, and forming the roof of the mouth, is the palate, which is itself composed of two regions, the soft palate and the hard palate, which join together to form the complete unit.

The front section of the palate, the hard palate, extends from behind the front of the top teeth, and is formed primarily by the upper jaw bones of the maxilla, in conjunction with the palatine bone of the skull. Overlying the bony structure is a layer of mucous membrane.

The soft palate, which lies behind the hard palate and extends to the back of the mouth, is composed of muscle fibers, and, like the hard palate, is overlain by a layer of mucous membrane. Hanging like a pendulum from the middle of the back of the soft palate is the uvula.

The soft palate and the uvula prevent food and liquid from entering the nasal cavity, by moving upward during the swallowing action.

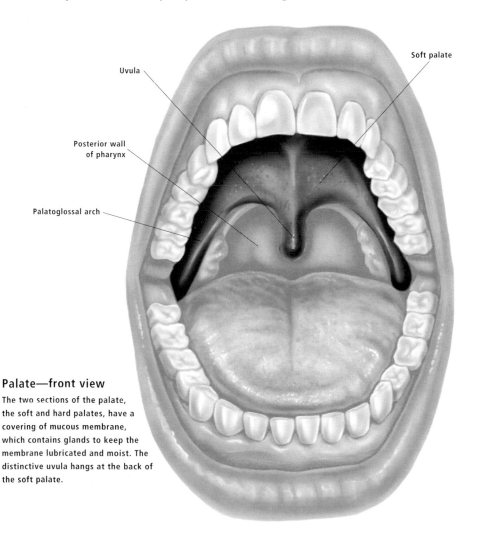

Uvula

Soft palate

Posterior wall of pharynx

Palatoglossal arch

Palate—front view

The two sections of the palate, the soft and hard palates, have a covering of mucous membrane, which contains glands to keep the membrane lubricated and moist. The distinctive uvula hangs at the back of the soft palate.

Tongue

Responsible for our sense of taste and for aiding in the chewing action needed to prepare food for processing by the digestive organs, the tongue is a highly versatile structure composed mainly of muscles.

Two groups of muscles move the tongue: the intrinsic tongue muscles and the extrinsic tongue muscles. The intrinsic muscles lie within the tongue and are arranged in different directions to enable maximum range of movement. The fibers of these muscles run vertically, horizontally, and longitudinally in the tongue. The contractions of each separate type of muscle allow the shape of the tongue to be altered.

The extrinsic muscles join the tongue to the jaw, skull, palate, and hyoid bones. These muscles allow the tongue to move forward, backward, upward, and downward.

The muscles of the tongue allow food to be moved around the mouth, and once reduced to a manageable bolus, suitable for forwarding to the pharynx, the tongue muscles move the bolus to the back of the mouth for swallowing.

The muscles of the tongue are also involved in speech, joining with other parts

Tongue

The muscular organ of the tongue coordinates movement to enable it to move food around the mouth in readiness for forwarding to the digestive organs. Because of its muscular dexterity, it is involved in the formation of sounds in speech.

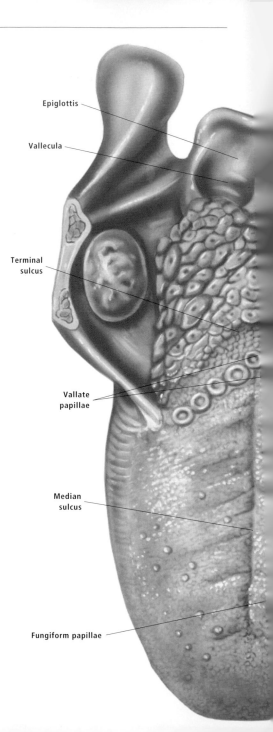

Epiglottis

Vallecula

Terminal sulcus

Vallate papillae

Median sulcus

Fungiform papillae

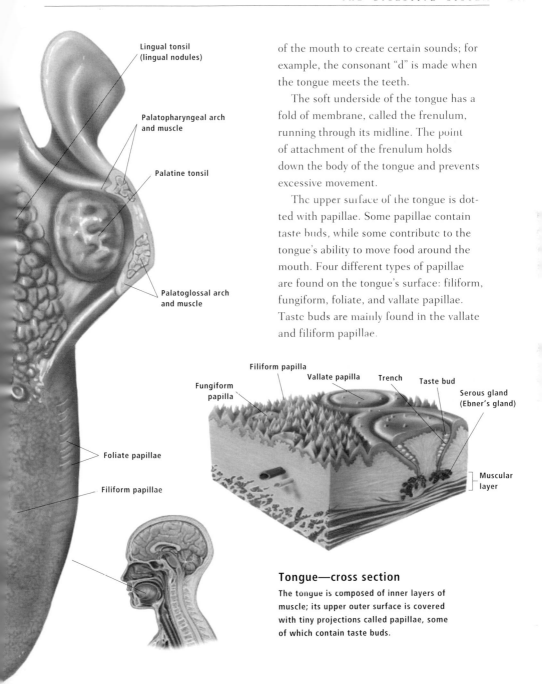

Lingual tonsil
(lingual nodules)

Palatopharyngeal arch
and muscle

Palatine tonsil

Palatoglossal arch
and muscle

Foliate papillae

Filiform papillae

Fungiform
papilla

Filiform papilla

Vallate papilla

Trench

Taste bud

Serous gland
(Ebner's gland)

Muscular
layer

of the mouth to create certain sounds; for example, the consonant "d" is made when the tongue meets the teeth.

The soft underside of the tongue has a fold of membrane, called the frenulum, running through its midline. The point of attachment of the frenulum holds down the body of the tongue and prevents excessive movement.

The upper surface of the tongue is dotted with papillae. Some papillae contain taste buds, while some contribute to the tongue's ability to move food around the mouth. Four different types of papillae are found on the tongue's surface: filiform, fungiform, foliate, and vallate papillae. Taste buds are mainly found in the vallate and filiform papillae.

Tongue—cross section

The tongue is composed of inner layers of muscle; its upper outer surface is covered with tiny projections called papillae, some of which contain taste buds.

Salivary Glands

Saliva produced by the salivary glands aids in digestion, activates our sense of taste, keeps the mouth moist, and assists in the fight against bacteria entering the body.

Three main pairs of salivary glands: the parotid gland, the submandibular gland, and the sublingual gland, supply the mouth, assisted by other microscopic glands dotted around the oral cavity.

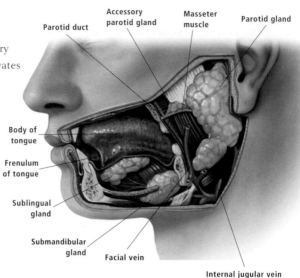

Parotid duct · Accessory parotid gland · Masseter muscle · Parotid gland · Body of tongue · Frenulum of tongue · Sublingual gland · Submandibular gland · Facial vein · Internal jugular vein

Salivary glands

Three main pairs of salivary glands—the parotid glands, the submandibular glands, and the sublingual glands—provide the saliva necessary for the commencement of digestion of food, and to enable the partially processed food to travel to the next stage of the digestive process.

While the parotid gland is the largest of the salivary glands, the majority of the saliva produced in the mouth is provided by the submandibular gland.

The parotid glands lie in front of the ears, and supply saliva via a duct that releases opposite the second molar of the upper teeth. The submandibular and sublingual glands secrete saliva through ducts that open out into the floor of the mouth. The submandibular glands and the small sublingual glands lie in the floor of the mouth, beneath the tongue.

Primarily controlled by the nervous system, the saliva produced by the salivary glands performs many functions. Its primary purpose is to commence the digestive process, by initiating the breakdown of food into simpler, smaller components.

The composition of saliva includes water, mucus, mineral salts, proteins, and enzymes, of which amylase is a particularly important component.

Amylase reacts with food, initiating the breakdown of starches, a process that continues until the partially processed food reaches the stomach. Once the partially processed food is turned into a bolus by the chewing action and movement of the tongue, the mucus in saliva enables easy movement of the bolus to the next stage of the digestive system.

The properties of saliva trigger the taste buds into action—when saliva and food combine, the resultant release of chemicals stimulates the taste buds, which then relay information to the brain. Saliva also contains antibodies, to keep harmful bacteria in check, and its ability to keep the mouth moist and clean aids in the prevention of tooth decay. Saliva also plays a role in maintaining the correct water balance in the body.

The parasympathetic and sympathetic divisions of the autonomic nervous system also exert an influence over saliva production, with the parasympathetic nerves stimulating saliva production, in response to the sight, smell, and thought of food, as well as to the actual presence of food in the mouth.

Stimulation by sympathetic nerves, which initiate the body's defensive mode in response to stressful or frightening situations, generally results in a dry mouth.

Salivary glands microstructure

Each of the salivary glands has a unique cellular structure, resulting in each of the glands producing saliva with slightly different properties.

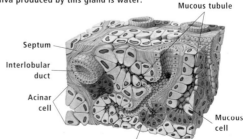

Parotid microstructure

The parotid glands secrete an enzyme-rich saliva, thin in consistency, which targets starches for breakdown.

Acinar cell
Intercalated ducts
Serous cell
Artery
Septum
Striated duct
Vein
Interlobular duct

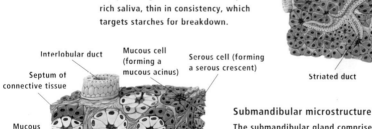

Interlobular duct
Mucous cell (forming a mucous acinus)
Serous cell (forming a serous crescent)
Septum of connective tissue
Mucous tubule
Serous crescent (serous demilune)

Submandibular microstructure

The submandibular gland comprises both enzyme-producing serous cells and mucus-producing cells. The main constituent of saliva produced by this gland is water.

Mucous tubule
Septum
Interlobular duct
Acinar cell
Mucous cell
Intercalated duct

Sublingual microstructure

The sublingual gland produces thicker, watery mucus, particularly in response to milk or cream, which helps to lubricate the mouth.

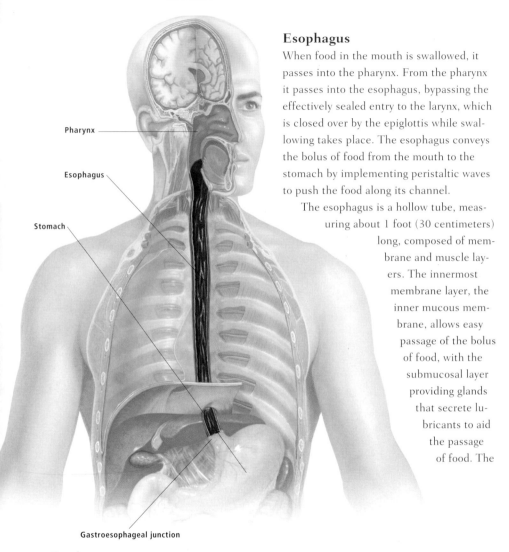

Pharynx

Esophagus

Stomach

Gastroesophageal junction

Esophagus

When food in the mouth is swallowed, it passes into the pharynx. From the pharynx it passes into the esophagus, bypassing the effectively sealed entry to the larynx, which is closed over by the epiglottis while swallowing takes place. The esophagus conveys the bolus of food from the mouth to the stomach by implementing peristaltic waves to push the food along its channel.

The esophagus is a hollow tube, measuring about 1 foot (30 centimeters) long, composed of membrane and muscle layers. The innermost membrane layer, the inner mucous membrane, allows easy passage of the bolus of food, with the submucosal layer providing glands that secrete lubricants to aid the passage of food. The

Esophagus

The strong peristaltic movements of the muscles of the esophagus convey food from the pharynx to the stomach. Effective sphincter muscles at the points of entry and exit regulate the volume of food passing into and out of the esophagus.

outer muscular layers are formed by a longitudinal muscle layer surrounding a circular muscle layer. The smooth muscles of the esophagus create peristaltic wave-like movements, which push the food down the esophagus toward the stomach.

Sphincter muscles regulate the flow of food in and out of the esophagus—the upper sphincter muscle relaxes to accept food from the pharynx, while the lower sphincter muscle relaxes to allow food to pass into the stomach.

The esophagus sits behind the trachea, and the arrangement of muscle and incomplete rings of cartilage in the trachea are such that when swallowing, the esophagus is allowed to expand into the flexible gap in the cartilage.

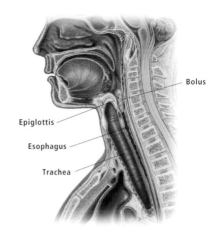

In a precisely coordinated action, the entrance to the airways is sealed off, allowing food to pass into the esophagus. Breathing is halted during the short time it takes for this part of the swallowing process to take place.

Smooth muscle

The outer wall of the esophagus, muscularis externa, has two layers—an inner circular layer and an outer longitudinal layer. The opposing movements of these two layers create the wave-like movements required to transport food along the length of the esophagus to the stomach.

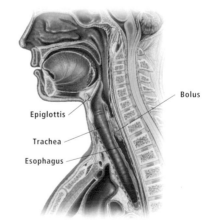

The bolus of food passes through the upper esophageal sphincter, from where the muscular movements of the smooth muscle walls of the esophagus push the bolus towards the stomach in rhythmic, wave-like movements.

Stomach

Connected to the esophagus by the cardio-esophageal or gastroesophageal junction, and lying beneath the diaphragm, the stomach is a sac-like structure that acts as a reservoir in the digestive tract.

The structure of the stomach facilitates the further processing of food. The stomach is composed of several layers, each performing a specific task. Since the stomach

Stomach

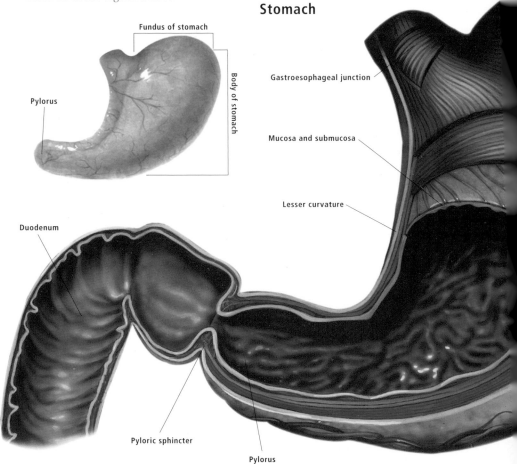

Fundus of stomach

Body of stomach

Pylorus

Gastroesophageal junction

Mucosa and submucosa

Lesser curvature

Duodenum

Pyloric sphincter

Pylorus

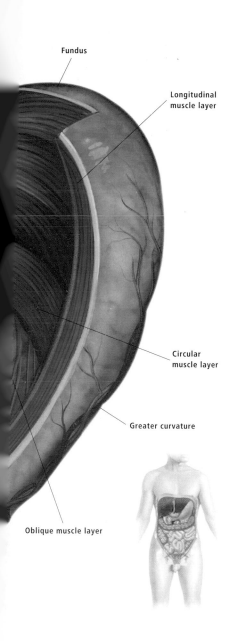

Fundus

Longitudinal
muscle layer

Circular
muscle layer

Greater curvature

Oblique muscle layer

expands in size to accommodate incoming food, and also moves in muscular waves when digesting—the outer serous layer is smooth and slippery to allow these movements. When empty, the stomach contains only a tiny amount of liquid, but when full has the capacity to hold up to 2 pints (1.2 liters) of contents.

Lying beneath the outer serous layer are the muscular layers of the stomach. Three layers, arranged in different directions, work together to provide the churning movements of the stomach, and also allow the stomach to expand when food enters. The three layers are the longitudinal muscle layer, the circular muscle layer, and the innermost muscle layer—the oblique muscle layer. These muscles slide over one another, creating the churning movements required for the processing of food.

The inner layers of the stomach are the submucosa and mucosa. These layers contain the cells that release the gastric juices needed to break down the food into smaller, more easily digestible amounts. Mucous cells in the inner lining produce mucus that lines and protects the stomach, which would otherwise be compromised and attacked by its own acidic juices. Other cells in the lining produce hydrochloric acid, while others produce enzymes—both crucial elements in the effective breakdown of digested food. Digested food, known as chyme, then passes through the pylorus, and is released in small amounts by the pyloric sphincter, into the duodenum.

Stomach Function

Food arriving in the stomach stimulates the cells in the stomach lining to produce gastric juices including pepsin, and enzymes, hormones, and acids, all essential to the stomach's contribution to the breakdown of food in the digestive process.

Parietal cells in the lining are responsible for the secretion of hydrochloric acid; zymogen cells produce enzymes to break down protein and fat; while other cells release the hormone gastrin, which controls gastric juice secretion.

Once the food begins to be broken down by the digestive juices, the muscular walls of the stomach begin to contract. The churning action provided by the muscular layers of the stomach mix the contents, until the food and digestive juices are converted into a thick, semiliquid substance known as chyme.

The mixing and churning process can take several hours, before the correct level of food breakdown has been achieved. Once the stomach has completed

its part of the digestive process, the partially digested food, now known as chyme, is sent toward the duodenum.

It must pass through the final part of the stomach—the pylorus and pyloric canal—and finally be released by the pyloric sphincter, in small, regulated amounts, into the duodenum. The pyloric sphincter effectively separates the stomach and duodenum, preventing backflow of chyme into the stomach.

Stomach function

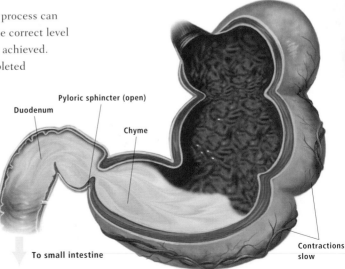

Exiting the stomach

After several hours of processing, the contractions slow, and the chyme moves to the pylorus, stimulating the pyloric sphincter to open. Small amounts of chyme are released into the duodenum.

Pyloric sphincter (open)

Duodenum

Chyme

To small intestine

Contractions slow

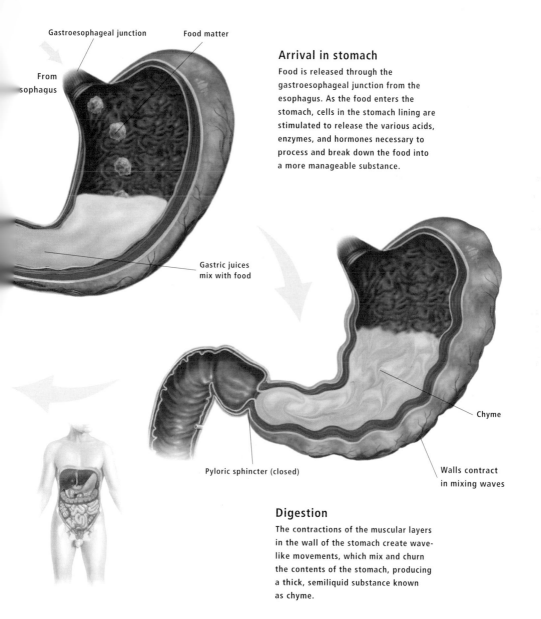

Gastroesophageal junction

Food matter

From esophagus

Gastric juices mix with food

Arrival in stomach

Food is released through the gastroesophageal junction from the esophagus. As the food enters the stomach, cells in the stomach lining are stimulated to release the various acids, enzymes, and hormones necessary to process and break down the food into a more manageable substance.

Chyme

Pyloric sphincter (closed)

Walls contract in mixing waves

Digestion

The contractions of the muscular layers in the wall of the stomach create wave-like movements, which mix and churn the contents of the stomach, producing a thick, semiliquid substance known as chyme.

Abdominal organs

The digestive organs dominate the abdominal region, with the stomach, small and large intestine, and their associated organs accounting for a large percent of the organs filling the abdominal cavity.

NB: In this illustration the liver has been pulled back to reveal the other abdominal organs.

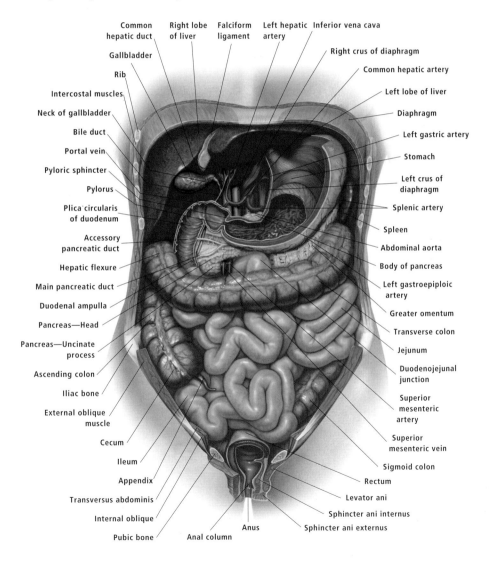

Common hepatic duct
Right lobe of liver
Falciform ligament
Left hepatic artery
Inferior vena cava
Gallbladder
Rib
Intercostal muscles
Neck of gallbladder
Bile duct
Portal vein
Pyloric sphincter
Pylorus
Plica circularis of duodenum
Accessory pancreatic duct
Hepatic flexure
Main pancreatic duct
Duodenal ampulla
Pancreas—Head
Pancreas—Uncinate process
Ascending colon
Iliac bone
External oblique muscle
Cecum
Ileum
Appendix
Transversus abdominis
Internal oblique
Pubic bone
Anal column

Right crus of diaphragm
Common hepatic artery
Left lobe of liver
Diaphragm
Left gastric artery
Stomach
Left crus of diaphragm
Splenic artery
Spleen
Abdominal aorta
Body of pancreas
Left gastroepiploic artery
Greater omentum
Transverse colon
Jejunum
Duodenojejunal junction
Superior mesenteric artery
Superior mesenteric vein
Sigmoid colon
Rectum
Levator ani
Sphincter ani internus
Sphincter ani externus
Anus

Intestines

Filling most of the lower region of the abdominal cavity, the intestines consist of two parts, the small intestine and the large intestine. The small intestine includes the duodenum, jejunum, and ileum, while the large intestine comprises the colon, rectum, and anus. It is in the small intestines that much of the nutrient requirements of the body are derived.

The large intestine serves to extract water and salts from digested food. The colon itself is composed of several regions: the ascending colon, the transverse colon, the descending colon, the sigmoid colon, and the rectum. At the beginning of the colon is the region known as the cecum, and it is from here that the tail-like appendage of the appendix hangs.

The large intestine lies around the margins of the lower abdomen, with the small intestine coiled within the boundaries of the large intestine.

Abdominal organs and peritoneum

The thin membrane of the peritoneum lines the abdominal cavity, protecting both the cavity and the organs, and providing a lubricated surface to allow the movement of the abdominal organs, such as those occurring during peristalsis. Mesenteries and omenta, folds in the peritoneum, supply blood vessels and nerves to the organs. In this illustration, the greater omentum is shown overlying the lower abdominal organs.

Peritoneum

The peritoneum lines the abdominal cavity. This thin lubricating membrane forms a protective layer around the cavity and around the organs within the cavity. Folds of the peritoneum attach the organs to the back of the abdominal cavity, while allowing the intestines to move relatively freely, in order to aid movement of food down the alimentary canal. Other folds of the peritoneum, called mesenteries and omenta, supply the abdominal organs with nerves, blood vessels, and lymph channels.

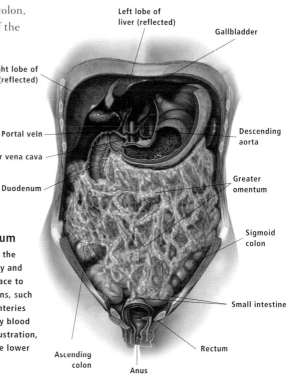

Left lobe of liver (reflected)

Gallbladder

Right lobe of liver (reflected)

Portal vein

Inferior vena cava

Duodenum

Descending aorta

Greater omentum

Sigmoid colon

Small intestine

Ascending colon

Anus

Rectum

The Small Intestine

The small intestine consists of the duodenum, the jejunum, and the ileum, and its primary function is the absorption of nutrients. Nutrients are absorbed both through lymphatic vessels (lacteals) in the intestines or transported through the lymph system or through capillaries and carried back to the liver via the hepatic portal system for processing, before returning via the systemic circulation, to the heart.

Forming the first section of the small intestine, the duodenum is joined to the stomach at the pyloric sphincter. The natural curve of the duodenum is nestled around the head of the pancreas, beyond which point, the duodenum joins with the middle part of the small intestine, the jejunum. The jejunum measures about 8 feet (2.5 meters) in length, and joins to the final section of the small intestine, the ileum. The ileum is the longest section of the small intestine, measuring about 12 feet (3.5 meters) in length.

The composition of all three components of the small intestines is similar, comprising outer muscular layers surrounding inner layers of submusosa and mucosa. The inner surface has many folds; these folds are covered with tiny projections known as intestinal villi (singular, villus). These tiny villi create a massive surface area for nutrient absorption. Also within the submucosal lining are the cells that release intestinal juice to protect the inner surface from attack by the acidic contents.

Much of the breakdown and absorption of nutrients occurs in the duodenum. It is here that bile from the liver and gallbladder, and digestive enzymes from

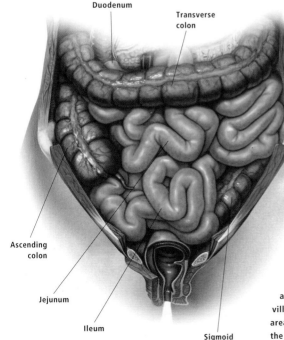

Duodenum

Transverse colon

Ascending colon

Jejunum

Ileum

Sigmoid colon

Small intestine

The duodenum, jejunum, and ileum comprise the small intestine. Each has a similar structure composed of outer muscular layers, and inner mucosal layers covered in tiny villi. The villi are responsible for creating a huge surface area, enabling maximum absorption of nutrients by the sections of the small intestine.

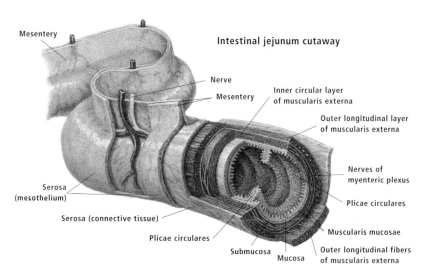

Intestinal jejunum cutaway

Mesentery

Nerve

Mesentery

Inner circular layer of muscularis externa

Outer longitudinal layer of muscularis externa

Nerves of myenteric plexus

Plicae circulares

Muscularis mucosae

Outer longitudinal fibers of muscularis externa

Mucosa

Submucosa

Plicae circulares

Serosa (connective tissue)

Serosa (mesothelium)

the pancreas, are received. As the contents of the stomach are gradually released into the duodenum, these digestive aids from the liver, gallbladder, and pancreas assist in the breakdown of food into the nutrients required by the body for the upkeep of cells and tissues.

Further absorption of nutrients occurs in the jejunum, and by the time the contents reach the ileum, much of the extraction has already been done. However, the ileum is responsible for the extraction of several key elements required by the body, including vitamin B_{12}, and bile salts.

An outer layer of longitudinal muscle surrounding an inner layer of circular muscle provides the muscular contractions to pulse the digested material along the small intestine to the large intestine.

Intestinal jejunum villus cross section

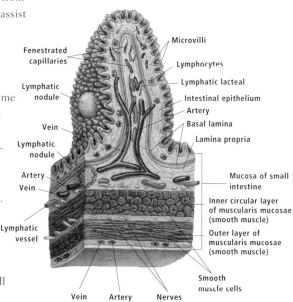

Fenestrated capillaries

Lymphatic nodule

Vein

Lymphatic nodule

Artery

Vein

Lymphatic vessel

Microvilli

Lymphocytes

Lymphatic lacteal

Intestinal epithelium

Artery

Basal lamina

Lamina propria

Mucosa of small intestine

Inner circular layer of muscularis mucosae (smooth muscle)

Outer layer of muscularis mucosae (smooth muscle)

Smooth muscle cells

Vein   Artery   Nerves

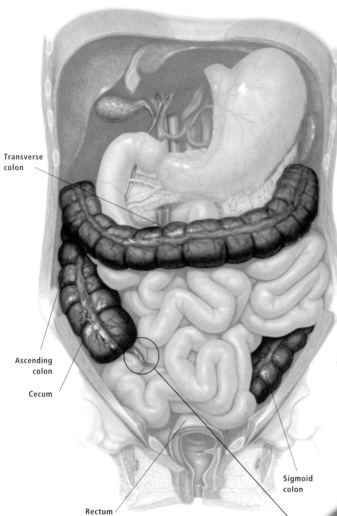

Transverse colon

Ascending colon

Cecum

Rectum

Sigmoid colon

The Large Intestine

The large intestine consists of the colon and rectum. While the small intestine's primary function is the absorption of nutrients, the large intestine's primary function is the reabsorption of water and the movement of waste material toward the anus.

The ileum of the small intestine joins the colon at the cecum. Attached to the base of the cecum is the thin, tail-like pouch of the appendix.

Forming the major part of the large intestine, the colon is divided into several regions, with the cecum being the first part, then the ascending colon, transverse colon, descending

Colon

The major component of the large intestine, the colon extends from the small intestine to the rectum. The function of the colon is to absorb water and bile salts remaining in digested material, and pass the remainder to the rectum, as feces.

colon, and finally the sigmoid colon. The colon serves to remove any remaining salt and water in the material sent from the small intestine, until all that now remains is waste material, which is passed to the rectum. The lining of the latter parts of the colon contains mucus-secreting glands to aid the movement of feces.

The final parts of the large intestine are the rectum and anus. The rectum receives fecal material from the sigmoid colon, storing the material for a short time until it is convenient to expel the stool. Feces pass through the anal canal, and feces are expelled via the anus, which opens and closes by an involuntary and a voluntary sphincter.

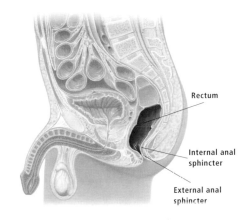

Rectum

Internal anal sphincter

External anal sphincter

Rectum

The rectum is the final part of the large intestine, receiving and storing waste material, until it is convenient to expel the fecal matter.

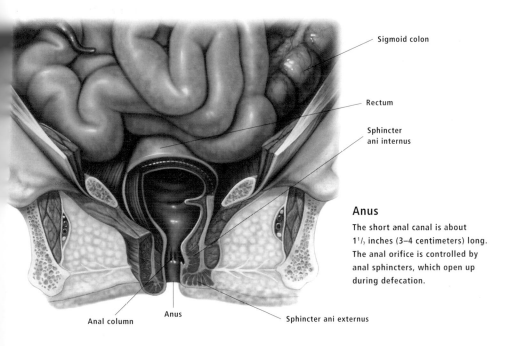

Sigmoid colon

Rectum

Sphincter ani internus

Anus

The short anal canal is about 1¹/₂ inches (3–4 centimeters) long. The anal orifice is controlled by anal sphincters, which open up during defecation.

Anal column

Anus

Sphincter ani externus

Liver

Lying in the upper right side of the abdomen, protected by the lower ribs, the liver is the heaviest organ in the body. The upper surface of the liver is in contact with the diaphragm, to which it is held by folds of membrane called the falciform, triangular, and coronary ligaments. Its visceral surfaces are in contact with many of the abdominal organs, including part of the stomach, the right kidney, the gallbladder, the right large bowel, and the upper section of the duodenum. Folds of membrane hold the liver to the stomach and duodenum, while the gallbladder is attached to the liver by connective tissue.

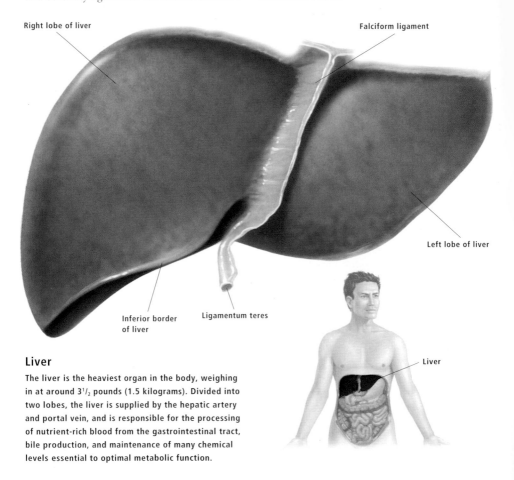

Right lobe of liver

Falciform ligament

Left lobe of liver

Inferior border of liver

Ligamentum teres

Liver

Liver

The liver is the heaviest organ in the body, weighing in at around 3¹/₂ pounds (1.5 kilograms). Divided into two lobes, the liver is supplied by the hepatic artery and portal vein, and is responsible for the processing of nutrient-rich blood from the gastrointestinal tract, bile production, and maintenance of many chemical levels essential to optimal metabolic function.

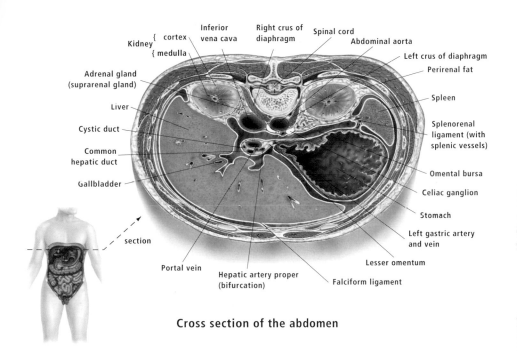

Kidney { cortex { medulla

Inferior vena cava

Right crus of diaphragm

Spinal cord

Abdominal aorta

Left crus of diaphragm

Perirenal fat

Adrenal gland (suprarenal gland)

Liver

Cystic duct

Common hepatic duct

Gallbladder

Spleen

Splenorenal ligament (with splenic vessels)

Omental bursa

Celiac ganglion

Stomach

Left gastric artery and vein

Lesser omentum

section

Portal vein

Hepatic artery proper (bifurcation)

Falciform ligament

Cross section of the abdomen

The large reddish-brown organ of the liver is divided into two lobes by a fold of peritoneum called the falciform ligament. At the lower end of the falciform ligament is the ligamentum teres.

Defunct after birth, this serves as the left umbilical vein during fetal development, supplying blood from the placenta to the fetus. On the visceral surface of the liver is the porta hepatis, the "doorway to the liver."

This slit-like opening is the point of entry for the major blood vessels serving the liver, principally the hepatic artery, which supplies oxygenated blood to the liver, and

the portal vein, which carries the nutrient-loaded blood from the digestive organs to the liver for processing.

Bile, a key element in the digestive process and essential to the breakdown of food, is produced by the liver. A series of ducts carry bile to the gallbladder for storage, and ultimately to the duodenum, where it acts on partially digested food, in particular targeting fats for emulsification and breakdown.

While the liver is heavily involved in the digestive process, it serves many other functions, and has a particular influence and involvement in metabolic functions.

Liver microstructure

The cells in the liver lobule—hepatocytes—are responsible for filtering blood returned to the liver from the gastrointestinal tract via the hepatic portal system. They extract nutrients required for cell and tissue maintenance and metabolic functions. Hepatocytes produce bile, transferring it to the gallbladder, where it is concentrated and stored until required by the digestive process.

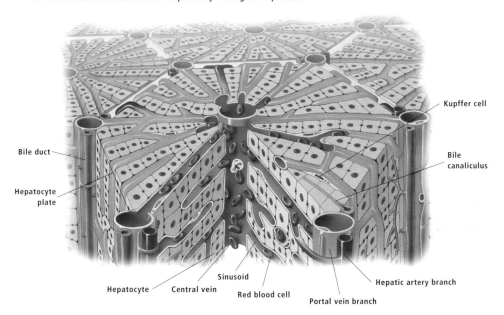

Kupffer cell

Bile duct

Bile canaliculus

Hepatocyte plate

Hepatocyte   Central vein   Sinusoid   Red blood cell   Hepatic artery branch   Portal vein branch

Microscopic structure of the liver

Tiny hexagonal structures, called lobules, composed of sheets of cells make up the composition of the liver. The cells that make up the sheets are hepatocytes. The hepatocytes store large amounts of glycogen, an energy storage chemical made from glucose, and are also responsible for the production of bile.

Between the sheets of cells are small blood vessels called sinusoids. It is here

that blood returning from the gastro-intestinal tract is processed and cleansed for transport back to the heart. Elements vital for efficient body metabolism are extracted by the hepatocytes, and waste products and ageing blood cells are eliminated by special macrophages in the sinusoids, called Kupffer cells.

Situated at each corner of the hexagonal structure are branches of the hepatic artery, branches of the portal vein, and the hepatic ducts, which are connected to a system of ductules running through the hepatocytes. The hepatic ducts eventually join up to form the bile duct, which carries bile to the gallbladder for concentration and storage. The blood returned to the liver from the gastrointestinal tract flows into the liver and through the sinusoids in the lobules. As it passes through the lobules, the hepatocytes extract the nutrients, bile salts, and any waste products for processing. The blood then makes its way to the central vein of each lobule, with these vessels eventually joining up and transferring blood to the hepatic veins, which ultimately drain back to the heart via the inferior vena cava.

Metabolic functions of the liver

The liver breaks down and converts many of the nutrients received from the gastrointestinal tract. Glucose is converted into glycogen for storage purposes, and can be converted back to glucose when required by the body. The liver produces and stores vitamin A; stores iron; produces albumin, an essential plasma protein in the blood; produces several blood clotting agents, including prothrombin and fibrinogen; and makes bile for the digestive process.

Liver lobule

The hexagonal structures that make up the liver are the liver lobules. Located at each corner of the hexagon are blood vessels and bile ducts. Blood filters through channels in the lobule called sinusoids, where nutrients are extracted, with the blood then flowing through to the central vein of the lobule. The central veins eventually join up, draining into the hepatic veins.

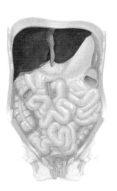

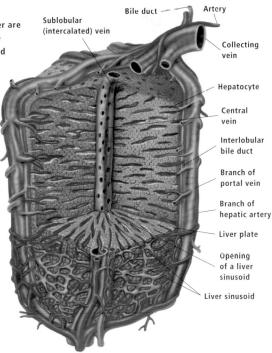

Bile duct

Artery

Sublobular (intercalated) vein

Collecting vein

Hepatocyte

Central vein

Interlobular bile duct

Branch of portal vein

Branch of hepatic artery

Liver plate

Opening of a liver sinusoid

Liver sinusoid

Gallbladder

The small sac-shaped organ of the gallbladder acts as a storage unit for bile received from the liver. When required by the digestive process, bile is transferred along the cystic duct to the bile duct, which connects to the duodenum.

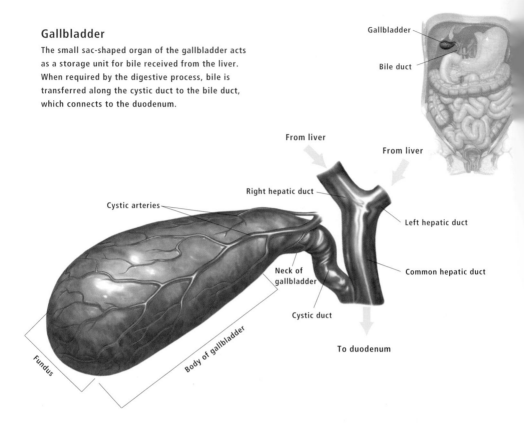

Gallbladder

Bile duct

From liver

From liver

Right hepatic duct

Left hepatic duct

Cystic arteries

Common hepatic duct

Neck of gallbladder

Cystic duct

To duodenum

Fundus

Body of gallbladder

Gallbladder

The gallbladder is tucked beneath the liver, under the cover of the right lobe. This small sac-shaped organ is attached to the liver by connective tissue, and a system of bile ducts connects the two organs. This system of ducts is known as the biliary tree.

Bile is produced by the hepatocytes in the liver; this clear yellow or orange fluid is an active ingredient of the digestive process. When bile comes into contact with digested food, it targets fats in food, breaking them down into smaller particles for easier digestion and absorption.

Bile produced by the liver is sent to the gallbladder through the bile ducts. The gallbladder stores and concentrates bile, ready for transport to the duodenum, via the cystic and bile ducts. Once digestion is complete, bile is extracted primarily by the terminal ileum (but also by the large intestine) and returned to the liver via the hepatic portal system where it is processed and recycled.

Pancreas

The pancreas is a member of two systems, the digestive system and the endocrine system. The predominant exocrine cells in its structure control its part in the digestive system.

The pennant-shaped pancreas lies behind the stomach, with its head nestled in the curve of the duodenum, and its tail meeting with the spleen. The digestive enzymes produced by the exocrine (acinar) cells are transported along a network of ducts that connect up to the main pancreatic duct, which feeds into the duodenum at the ampulla of Vater. Inactive until they reach the duodenum, the digestive enzymes produced by the pancreas aid in the digestion of fats, proteins, and starch, and have a slightly alkaline composition in order to neutralize the effects of the acid contents that arrive in the duodenum from the stomach via the pylorus.

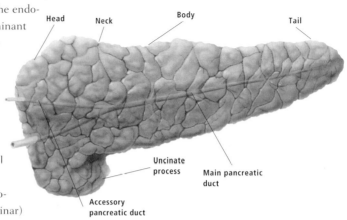

Head   Neck   Body   Tail

Uncinate process

Main pancreatic duct

Accessory pancreatic duct

Pancreas

Pancreas—digestive function

The digestive enzymes produced by the exocrine cells of the pancreas act on fats, proteins, and starches, breaking down these substances into smaller particles. Bicarbonate ions in the pancreatic secretions act to neutralize stomach acids in the chyme received by the duodenum.

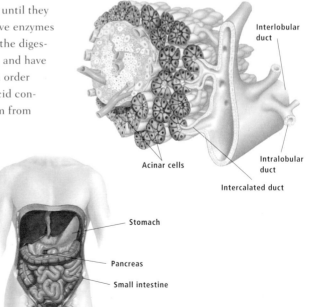

Interlobular duct

Intralobular duct

Intercalated duct

Acinar cells

Stomach

Pancreas

Small intestine

Peristalsis

Peristalsis describes the wave-like movements created by muscular contractions in the smooth muscle of various structures. Although several organs, such as the female uterus and fallopian tubes, are composed of smooth muscle, the action of peristalsis is perhaps best seen in the alimentary canal.

Food is firstly chewed and ground down by the teeth and mixed with saliva; it is formed into a ball (bolus) that slips easily into the pharynx and on into the esophagus. Here, the peristaltic movements begin. The smooth muscle of the esophagus pushes food down the hollow tube and into the stomach. Food arriving in the stomach is mixed with gastric juices by the peristaltic movements of the muscular layers of the stomach wall.

The churning action continues for some hours, until the combined food and gastric juices are converted to a semiliquid substance called chyme. Slower muscular contractions prompt the pyloric sphincter to relax and allow small amounts of chyme to enter the duodenum. The peristaltic action of the alimentary canal continues through the duodenum, jejunum, and ileum of the small intestine. As digested food passes through the small intestine, most of the nutrients and substances required for the well-being and upkeep of the body are absorbed. The digested food then passes through the ileocecal valve, which connects the ileum of the small intestine to the cecum of the large intestine. The wave-like contractions continue, pushing the digested material along the large intestine where any remaining water and bile salts are absorbed, before pushing the remaining material—feces—to the rectum and anus for defecation.

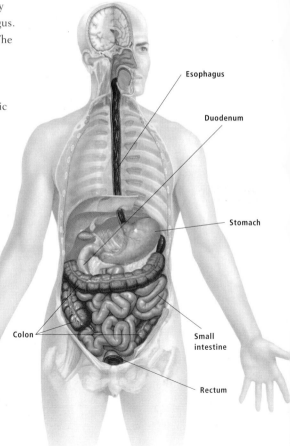

Esophagus

Duodenum

Stomach

Colon

Small intestine

Rectum

Peristalsis

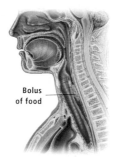

Bolus of food

Pharynx

gue

Esophagus

Trachea

(a) Chewed food is formed into a small ball, or bolus, and pushed to the back of the mouth.

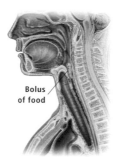

Bolus of food

(b) As the bolus enters the laryngopharynx, the epiglottis closes over, sealing off the airways. Food passes into the esophagus.

Bolus of food

(c) Muscular contractions push the bolus of food down the esophagus to the stomach.

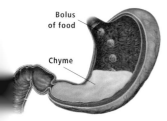

Bolus of food

Chyme

(d) Food enters the stomach, triggering the production of gastric juices.

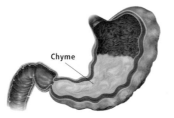

Chyme

(e) As the gastric juices begin to break down the food, muscular contractions of the stomach mix and churn the contents.

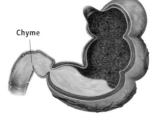

Chyme

(f) As the contractions slow, small amounts of chyme are released into the duodenum of the small intestine.

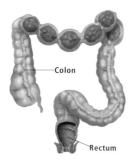

Colon

Rectum

(g) After nutrients have been absorbed by the small intestine, the food is transferred into the large intestine.

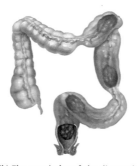

(h) The remainder of the digested food is pushed along the various sections of the colon, where any remaining water and bile salts are extracted.

(i) The remaining fecal matter is pushed towards the rectum and anus, to be periodically expelled.

Diseases and Disorders of the Digestive System

There are a wide range of diseases and disorders affecting one or all of the structures involved in the digestive system, ranging from gum disease to cancer of the major organs.

Discussed below are a few of the more commonly occurring problems.

Gingivitis

Tender, red, swollen gums that bleed easily usually indicate gingivitis. When food particles become trapped around and between the teeth, creating a breeding ground for bacteria, the subsequent proliferation of bacteria inflames the gums.

Improved dental hygiene and tooth brushing, coupled with flossing and regular visits to the dentist are recommended to keep infection at bay. If left untreated, gingivitis can lead to more serious gum disease, such as chronic gingivitis, periodontitis, or acute ulcerative necrotizing gingivitis, which can result in tooth loss.

Gastroenteritis

Generally associated with food poisoning by ingestion of disease-causing viruses or bacteria, gastroenteritis involves inflammation and infection of the stomach and intestines.

Symptoms include nausea, vomiting, and diarrhea, often with associated abdominal pain in the region affected.

Celiac disease

Celiac disease is intolerance and allergy to gluten. In celiac disease, the body develops an inability to cope with gluten in the diet due to the destruction of the nutrient-absorbing surface of the jejunum.

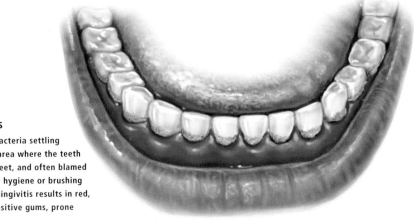

Gingivitis

Caused by bacteria settling around the area where the teeth and gums meet, and often blamed on poor oral hygiene or brushing technique, gingivitis results in red, swollen, sensitive gums, prone to bleeding.

Normal villi

Microvilli

Mucosa of jejunum

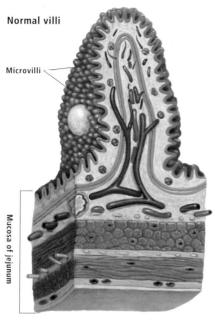

Jejunum

Celiac disease

Celiac disease is an inherited disorder in which an allergic intolerance to gluten in the diet causes changes in the small intestine. In celiac disease, the tiny villi that line the small intestine are flattened, resulting in a reduced surface area for nutrient absorption.

Flattened villi

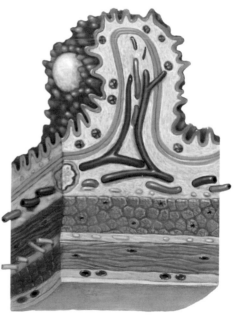

Normally rising to the body's defense, instead T lymphocytes attack gluten, causing the intestinal lining to become inflamed and resulting in poor absorption of nutrients.

Total elimination of gluten from the diet is essential, though challenging, as gluten is found in many everyday products, such as wheat and oats, as well as being found in many manufactured products.

Normal appendix

Inflamed appendix

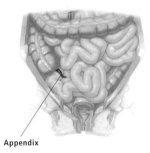

Appendix

Appendicitis

Inflammation of the appendix, known as appendicitis, can cause abdominal pain, fever, nausea, and vomiting, and urgent surgical removal is required.

Appendicitis

A blockage or obstruction of the tiny appendix is believed to cause appendicitis, bringing with it symptoms such as nausea, vomiting, and fever. Intermittent pain in the lower right side of the abdomen becomes more constant as infection takes hold. The inflammation can cause an abscess, which can burst, causing peritonitis, requiring immediate treatment. Appendicitis requires urgent surgery, followed by a recovery period of about one week, after which time there are usually no further complications.

Gallstones

Gallstones can range from tiny to large, vary in color from yellow to brown and black, and require different approaches dependent on their size and location. In most cases, gallstones are formed by cholesterol or, to a lesser extent, by bile pigments. In many cases, surgery is not required, but in cases where the gallstones lead to inflammation of the gallbladder, or when gallstones travel into the bile duct causing blockage, then surgery is often indicated. When gallstones block the bile duct, this can create further complications, as bile flow is interrupted, causing build-up of bile supplies in the liver and eventually causing jaundice.

Gallstones can lead to more serious complications, such as acute or chronic cholecystitis. Symptoms of these conditions manifest as abdominal pain, nausea, and vomiting. Where surgical removal of the gallbladder is indicated, this can either be by open abdominal operation or through a laparoscope; the latter option is generally preferred as the less invasive option offers shorter hospitalization and faster recovery.

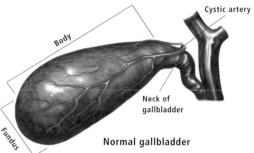

Normal gallbladder

Gallstones

A relatively common disorder of the gallbladder, gallstones are formed by cholesterol or bile pigments settling in the gallbladder. Stones can range in size from small $1/_2$–$1^1/_2$ inches (1–4 centimeters) in size to single stones large enough to fill the gallbladder.

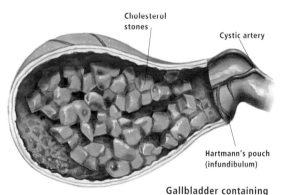

Gallbladder containing cholesterol stones

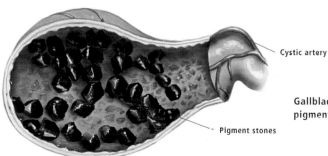

Gallbladder containing pigment stones

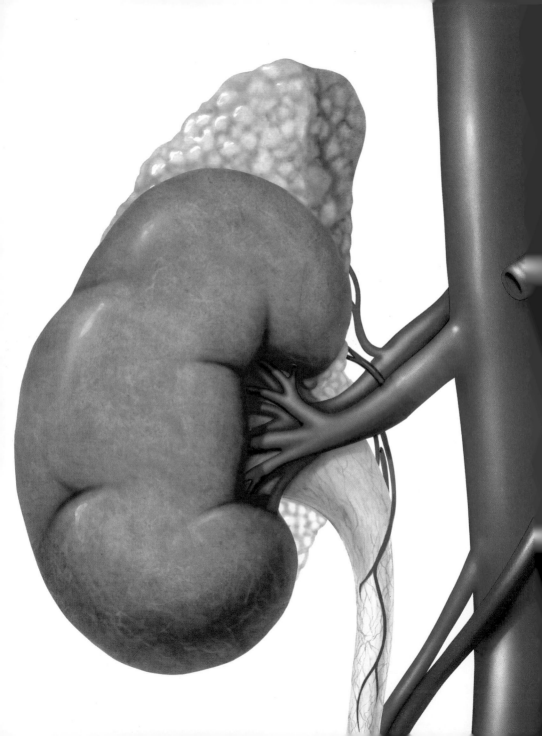

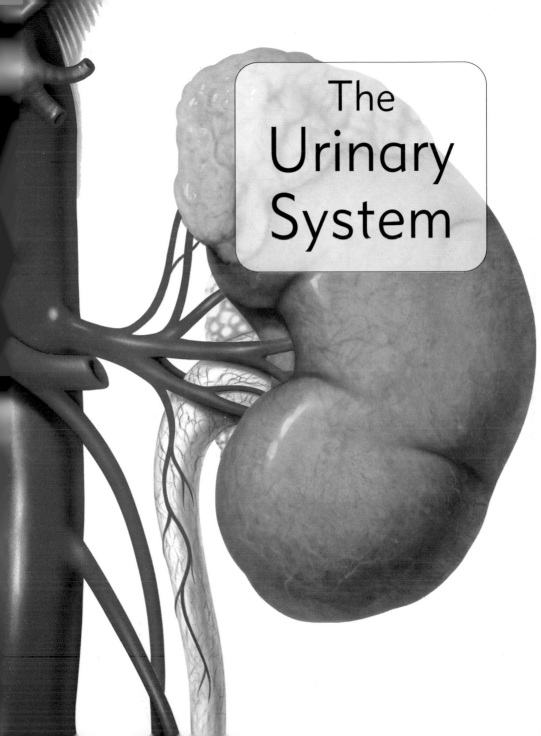

The Urinary System

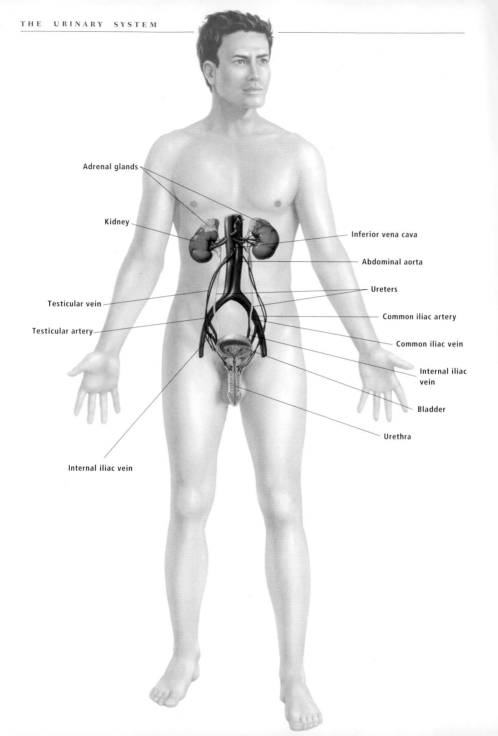

Adrenal glands

Kidney

Inferior vena cava

Abdominal aorta

Ureters

Testicular vein

Common iliac artery

Testicular artery

Common iliac vein

Internal iliac vein

Bladder

Urethra

Internal iliac vein

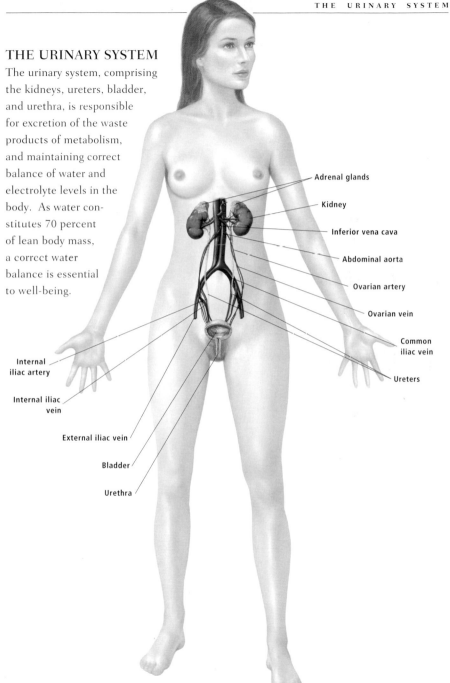

THE URINARY SYSTEM

The urinary system, comprising the kidneys, ureters, bladder, and urethra, is responsible for excretion of the waste products of metabolism, and maintaining correct balance of water and electrolyte levels in the body. As water constitutes 70 percent of lean body mass, a correct water balance is essential to well-being.

Adrenal glands

Kidney

Inferior vena cava

Abdominal aorta

Ovarian artery

Ovarian vein

Common iliac vein

Ureters

Internal iliac artery

Internal iliac vein

External iliac vein

Bladder

Urethra

Urinary Tract

The kidneys are one of the finely tuned filtering systems of the blood. Like the other major organs, the lungs, they are involved in maintaining the correct balances in the blood. The kidneys form part of the urinary system, which is the body's mechanism for elimination of waste extracted from the blood.

The blood is filtered by the kidneys through a massive network of capillaries, where useful components are reabsorbed. As a consequence, much of the content is reusable, with only a tiny percentage being required to be excreted.

The waste, comprised of water and substances such as urea and ammonia, is sent from collecting ducts in the kidneys to the ureters. The ureters transfer the urine, using muscular contractions, to the bladder, where it is stored. At the point where the ureters join the bladder is a slit-like opening.

The ureters run horizontally away from here before extending up to join the kidney. When the bladder is full, the pressure on the ureter near the entrance causes compression of the tube, preventing the backflow of urine to the kidneys.

Once the muscular sac of the bladder is full, the expulsion of urine is prompted. The muscles of the bladder contract, releasing urine into the urethra. The urethral sphincter muscle relaxes, allowing the urine to pass out of the body. An adult produces about 1–1½ quarts (1–1.5 liters) of urine per day. The volume and concentration of urine depends on fluid intake and fluid loss through perspiration, respiration, and elimination of feces.

Urinary tract

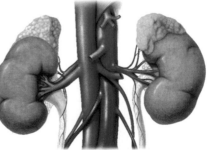

Kidneys

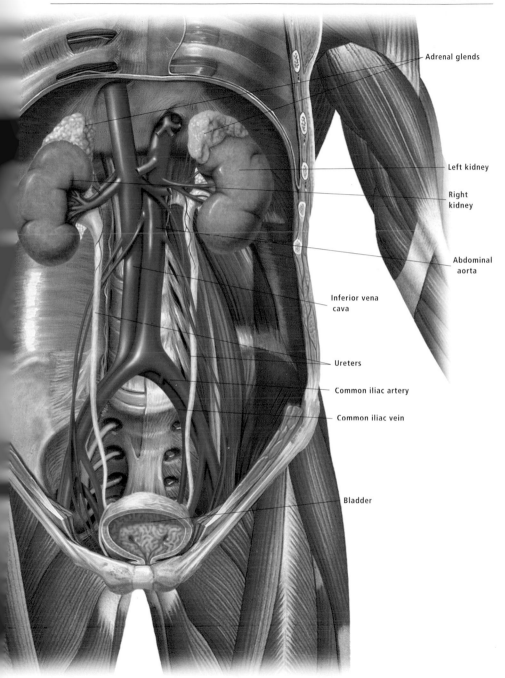

Adrenal glends

Left kidney

Right kidney

Abdominal aorta

Inferior vena cava

Ureters

Common iliac artery

Common iliac vein

Bladder

Kidneys

Lying on the back wall of the abdomen, the two bean-shaped kidneys are surrounded on all other sides by the organs of the abdominal cavity. The right kidney sits under the cover of the right lobe of the liver, contacting with the duodenum at its lower surface.

The left kidney, which usually sits higher in the abdomen than the right kidney, lies behind the stomach and pancreas and is in contact with the spleen and jejunum. Both kidneys are in contact with the rear surface of the transverse colon. Sitting on top of each kidney is an adrenal gland.

The two red-brown kidneys are enclosed in the protective membrane of the renal capsule. Around each renal capsule is a cushioning layer of adipose tissue, itself surrounded by a layer of connective tissue, the renal fascia, which holds each kidney in position, anchored to the back wall of the abdomen.

Each kidney weighs around 5 ounces (140 grams), and measures around 4 inches (10 centimeters) in length and 1 inch (2.5 centimeters) in width.

Within each kidney there is an outer layer, known as the cortex, surrounding the inner layer, the medulla, and a pelvis, a hollow inner structure that joins with the ureters, the tubes that conduct urine to the bladder. The blood vessels, nerves, and lymphatic vessels that serve each kidney enter and leave at the renal hilus, a small indentation in the kidney, and this is the point where the ureter exits the kidney.

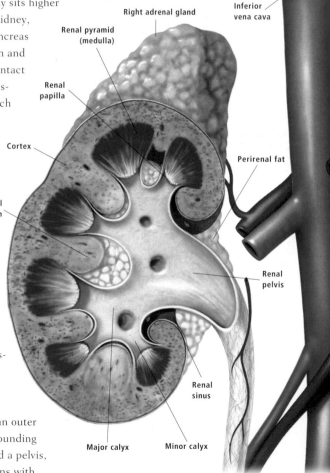

Right adrenal gland

Inferior vena cava

Renal pyramid (medulla)

Renal papilla

Cortex

Perirenal fat

Renal column

Renal pelvis

Renal sinus

Major calyx

Minor calyx

The cortex contains the filtration units of the kidneys, the glomeruli and tubules, while the medulla contains between 8 and 18 renal pyramids, positioned with their base facing outermost, and the tip pointing to the center of the kidney. The renal cortex fills the space between each of the pyramids; these areas are known as the renal columns.

Kidney

The kidneys are responsible for filtering the blood, maintaining the correct balance of water and electrolytes in the body, and eliminating waste in the form of urine.

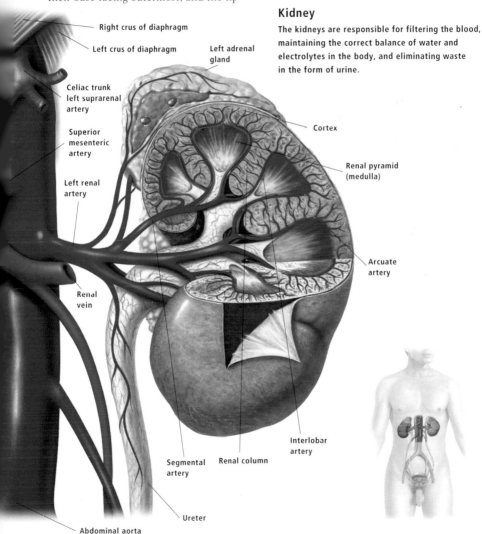

Right crus of diaphragm

Left crus of diaphragm

Left adrenal gland

Celiac trunk left suprarenal artery

Superior mesenteric artery

Left renal artery

Renal vein

Segmental artery

Ureter

Abdominal aorta

Cortex

Renal pyramid (medulla)

Arcuate artery

Interlobar artery

Renal column

Nephron

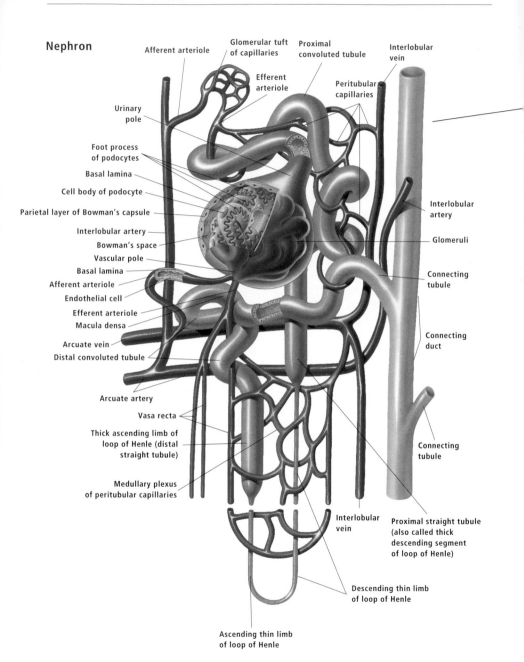

Afferent arteriole

Glomerular tuft of capillaries

Proximal convoluted tubule

Interlobular vein

Efferent arteriole

Peritubular capillaries

Urinary pole

Foot process of podocytes

Basal lamina

Cell body of podocyte

Parietal layer of Bowman's capsule

Interlobular artery

Bowman's space

Vascular pole

Basal lamina

Afferent arteriole

Endothelial cell

Efferent arteriole

Macula densa

Arcuate vein

Distal convoluted tubule

Arcuate artery

Vasa recta

Thick ascending limb of loop of Henle (distal straight tubule)

Medullary plexus of peritubular capillaries

Interlobular artery

Glomeruli

Connecting tubule

Connecting duct

Connecting tubule

Proximal straight tubule (also called thick descending segment of loop of Henle)

Descending thin limb of loop of Henle

Interlobular vein

Ascending thin limb of loop of Henle

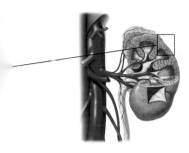

Kidney nephron

Each kidney contains over a million nephrons, the tiny filtering units responsible for filtering the blood. Water and electrolyte levels are kept stable by the partnership between the pituitary gland and the kidneys. The monitoring system of the pituitary gland influences the amount of water retained or excreted by the kidneys.

Nephron

There are over a million tiny filtration units, known as nephrons, in each kidney. Lying between the two layers of the kidney, each nephron is comprised of a renal corpuscle, which lies in the renal cortex, and a renal tubule, which runs through the renal pyramid.

The renal corpuscle is composed of an intertwined mass of capillaries called a glomerulus, enclosed within a double-walled filtration membrane called the glomerular or Bowman's capsule.

Blood passes through the renal corpuscle, where it is filtered, producing a liquid called filtrate. While purified blood is returned to the body, the filtrate passes through the various sections of the renal tubule. The renal tubule comprises the proximal convoluted tubule, the descending limb of the loop of Henle, the ascending limb of the loop of Henle, and the distal convoluted tubule.

Tiny peritubular capillaries serve the renal tubule, and as the filtrate passes through the various sections of the tubule, these capillaries reabsorb vitamins, electrolytes, and useful substances from the filtrate, and add additional wastes.

Reabsorption of much of the water, sodium, and potassium ions contained in the filtrate occurs in the section of the nephron known as the loop of Henle. The reabsorption process results in a large percentage (about 99 percent) of the filtrate being reabsorbed and returned to the body. The remaining 1 percent, amounting to about 1–1½ quarts (1–1¼ liters), which contains toxic wastes such as urea, is excreted by the body as urine.

The body's water requirements are monitored by the pituitary gland, which works in conjunction with the kidneys to maintain the correct balance. When the body requires more water, the capillaries absorb more. When the body has too much water, the kidneys excrete more urine. In a similar manner, the kidneys work in conjunction with the hypothalamus to control the water content in blood. Levels are corrected by the release of hormones, which alter the absorption rates by the kidneys to restore normal balance.

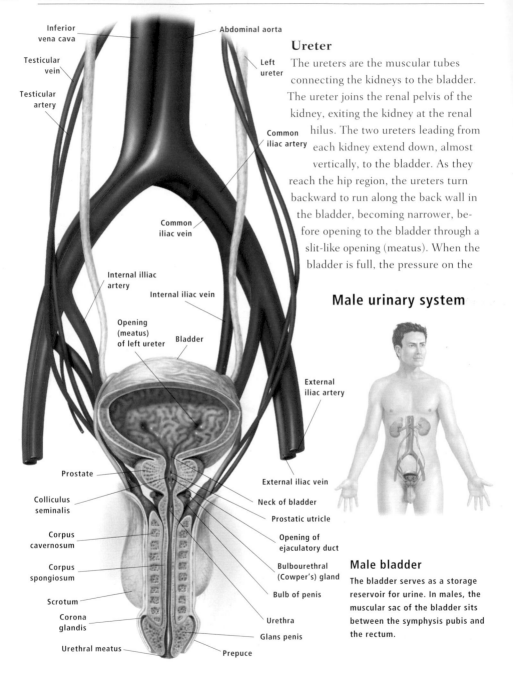

Inferior vena cava

Abdominal aorta

Testicular vein

Testicular artery

Left ureter

Common iliac artery

Common iliac vein

Internal iliac artery

Internal iliac vein

Opening (meatus) of left ureter

Bladder

External iliac artery

Prostate

Colliculus seminalis

Corpus cavernosum

Corpus spongiosum

Scrotum

Corona glandis

Urethral meatus

External iliac vein

Neck of bladder

Prostatic utricle

Opening of ejaculatory duct

Bulbourethral (Cowper's) gland

Bulb of penis

Urethra

Glans penis

Prepuce

Ureter

The ureters are the muscular tubes connecting the kidneys to the bladder. The ureter joins the renal pelvis of the kidney, exiting the kidney at the renal hilus. The two ureters leading from each kidney extend down, almost vertically, to the bladder. As they reach the hip region, the ureters turn backward to run along the back wall in the bladder, becoming narrower, before opening to the bladder through a slit-like opening (meatus). When the bladder is full, the pressure on the

Male urinary system

Male bladder

The bladder serves as a storage reservoir for urine. In males, the muscular sac of the bladder sits between the symphysis pubis and the rectum.

narrow section of the ureters running through the bladder wall is compressed, preventing backflow of urine toward the kidneys.

The ureters comprise three layers: an inner mucosal layer surrounded by two muscular layers, a longitudinal inner layer, and a circular outer layer. The smooth muscle of the ureters employs peristaltic movements to transfer urine from the kidneys to the bladder.

Bladder

The bladder is a muscular sac for the temporary storage of urine. The male and female bladders are positioned slightly differently.

The male bladder lies in front of the rectum, sits above the prostate, and is held in place by ligaments that attach it to the symphysis pubis and pelvic bones.

Similarly positioned ligaments hold the female bladder, which lies in front of the vagina, with the uterus lying above the bladder.

The full bladder can contain as much as 1 pint (475 milliliters) of urine. Urine release is stimulated when the bladder reaches capacity. When full, the bladder alerts the reflex center in the sacral part of the spinal cord.

Reflex contraction of the bladder muscle occurs, releasing the neck of the bladder, and allowing urine to pass into the urethra.

The bladder empties using peristaltic movements, creating wave-like contractions

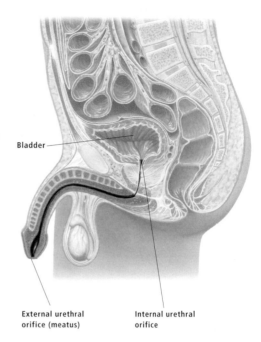

Bladder

External urethral orifice (meatus)

Internal urethral orifice

Male urethra

The male urethra is longer than that of the female. Extending from the neck of the bladder to the external urethral orifice at the tip of the penis, the male urethra measures around 8 inches (20 centimeters) in length.

to expel the contents. The detrusor muscle of the bladder comprises three layers. Between two layers of longitudinal muscle is a middle layer of circular muscle.

Beneath the muscular layers are the submucosa and mucosa. Within the bladder is the area known as the trigone, a zone demarcated by the openings of the ureters and the exit point of the urethra.

Urethra

The urethra is the passageway through which urine is expelled from the body. Major differences occur between the male and female urethras.

The male urethra is a shared passageway of the urinary and reproductive organs; it extends from the neck of the bladder, passing through the prostate, the membranous region of the pelvic floor, and the penis—a distance of about 8 inches (20 centimeters).

Female bladder

The female bladder lies between the symphysis pubis and the vagina. When full, the muscles of the bladder create peristaltic waves to empty the contents.

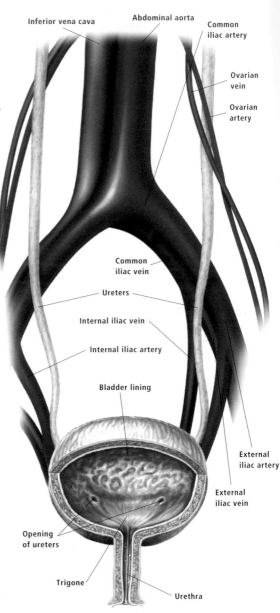

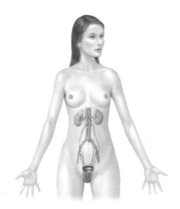

Female urinary system

Female urethra

The female urethra is much shorter than that of the male, extending from the neck of the bladder, through the pelvic floor to the external environment. The external urethral orifice lies in front of the vagina.

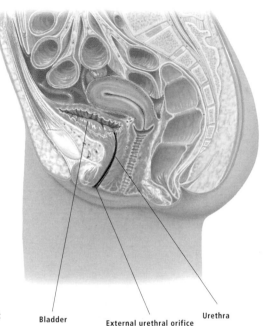

Bladder   External urethral orifice   Urethra

Sphincter muscle

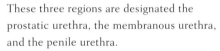

Sphincter muscles have fibers arranged in a circular pattern. These muscles are often present at the entrances and exits of passageways, such as the urethra.

These three regions are designated the prostatic urethra, the membranous urethra, and the penile urethra.

The female urethra is much shorter than that of the male—spanning about $1\frac{1}{2}$ inches (5 centimeters) from the bladder to the opening. Connected to the neck of the bladder, the female urethra passes through the pelvic floor to the external environment.

The internal urethral sphincter located at the neck of the bladder releases urine into the urethra; this movement is not under voluntary control; it is controlled by the micturition reflex.

The micturition reflex is prompted by nerve signals that detect bladder fullness. These signals prompt the contraction of the bladder and the conscious relaxation of the sphincter muscles.

The external urethral sphincter, which lies at the opening (meatus) to the external environment, is under voluntary control, and it is through relaxation of the external sphincter in the urinary tract that urine is expelled from the body.

Diseases and Disorders of the Urinary System

The structures of the urinary system are vulnerable to attack by bacteria, and the thin tubular components are susceptible to blockage. A variety of problems can affect the urinary system, and several of the more common problems are discussed below.

Kidney stones

While there may be an underlying cause to the formation of kidney stones, in many cases, there often is no explanation for their formation. Kidney stones are most frequently caused by a build-up of calcium salts, normally excreted in urine.

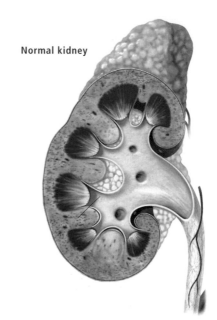

Normal kidney

Nephritis

Nephritis is inflammation of the kidney; various types of nephritis affect the various structures of the kidney.

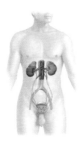

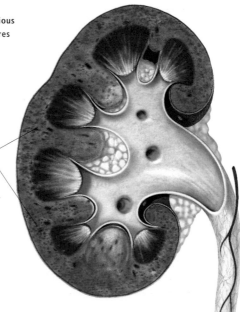

Inflamed cortical tissue

Glomerulonephritis

Glomerulonephritis is inflammation of the glomeruli, which disables their filtering capabilities, resulting in more serious complications.

The stones can form in the kidney tissue, causing damage and injury to the kidney itself. They can also form in the kidney, and as urine is passed to the ureters, the stones may be carried in the flow, creating a blockage in the ureter, causing pain, and sometimes preventing the passage of urine; this can cause problems in the renal pelvis and ultimately affect kidney function. Sometimes kidney stones can pass through the system without assistance, and sometimes they require surgical removal.

Nephritis

The many types of inflammation of the kidney fall under the general heading of nephritis. Glomerulonephritis is inflammation of the glomeruli, the blood filtering component of the nephron. When the glomeruli are compromised by inflammation, they are unable to properly filter the blood, resulting in serious kidney problems.

There are two types of glomerulonephritis, acute and chronic. Acute glomerulonephritis can strike following a streptococcal infection, with the common symptoms including discolored urine that is often smoky or slightly red, puffy eyes and ankles, nausea, vomiting, and headaches. Blood tests determining kidney function are necessary to determine the degree of inflammation, and treatment can range from bed rest to dialysis, depending on the results. Chronic glomerulonephritis can slowly establish a foothold in the kidneys over an extended period of time, often giving no indication of its presence. If symptoms are manifested

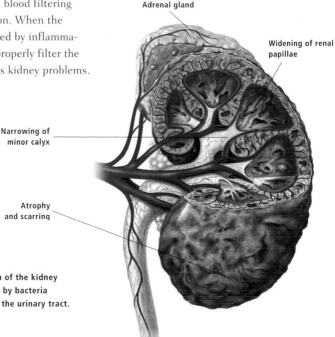

Adrenal gland

Widening of renal papillae

Narrowing of minor calyx

Atrophy and scarring

Pyelonephritis

Pyelonephritis is inflammation of the kidney and renal pelvis, often caused by bacteria spreading from other parts of the urinary tract.

Bladder stones

A build-up of salts, cholesterol, and some proteins are the main cause of the formation of bladder stones. These bladder stones can develop without symptoms, until they cause blockage to the outlet of the bladder. Surgery or ultrasound treatment is usually required. The bladder stones shown here are actual size.

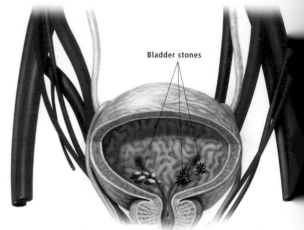

Bladder stones

they can include blood in the urine, fatigue, nausea, vomiting, muscle cramps, seizures, and confusion. Informed diagnosis can be made following blood tests. Glomerulonephritis is sometimes associated with deficiencies in the immune system; these can be addressed as a potential cause, however, treatment is very much dependent on the nature and extent of the inflammation.

Pyelonephritis affects the kidney and renal pelvis, and is most often caused by spread of infection from the bladder. Other kidney problems can precipitate the condition, which manifests in a variety of symptoms, including lower back pain, nausea, vomiting, fever with chills, frequent urination, or pain in passing urine. Generally

pyelonephritis can be addressed with antibiotics. Chronic pyelonephritis attacks and destroys kidney tissue, and is often associated with recurrent or untreated urinary tract infections. It too can often be addressed with antibiotics, although the underlying cause should be investigated and treated.

Bladder stones

Most bladder stones contain calcium, while some are composed of uric acid. Varying in size, they often accumulate and grow in the bladder without causing problems until they are large enough to block the urethral outlet, obstructing the flow of urine from the bladder. Bladder stones often require surgical removal, or alternatively they can be shattered or crushed using lithotripsy. Lithotripsy uses high-frequency sound waves to reduce the stone to small particles, which can then be passed with urine.

Cystitis

Cystitis, an inflammation of the bladder, often accompanies bladder stones. Because bacteria usually gain entry via the urethra, cystitis is more common in women than men due to their shorter urethra. Symptoms include frequent, painful urination, and there may be blood in the urine.

Treatment, based on confirmation by urine testing, usually involves the use of antibiotics.

Occasionally the bacteria from cystitis infection can spread to the kidneys, causing further complications.

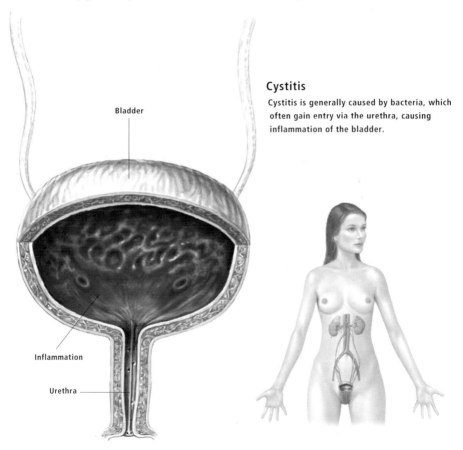

Cystitis

Cystitis is generally caused by bacteria, which often gain entry via the urethra, causing inflammation of the bladder.

Bladder

Inflammation

Urethra

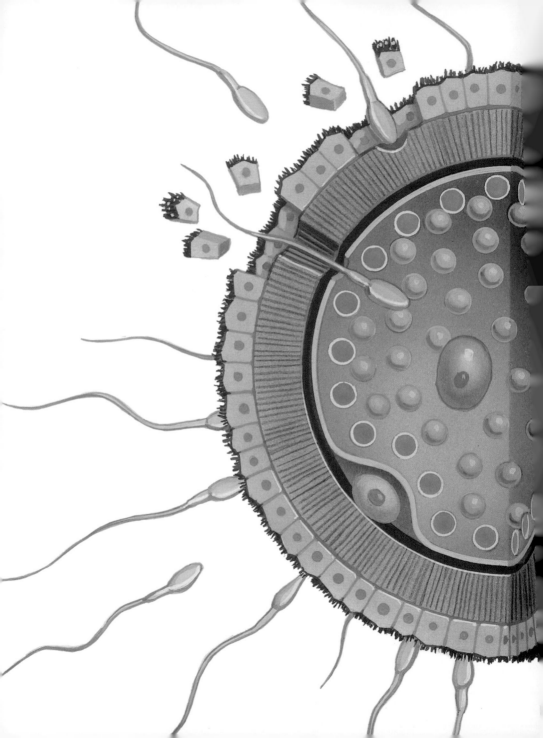

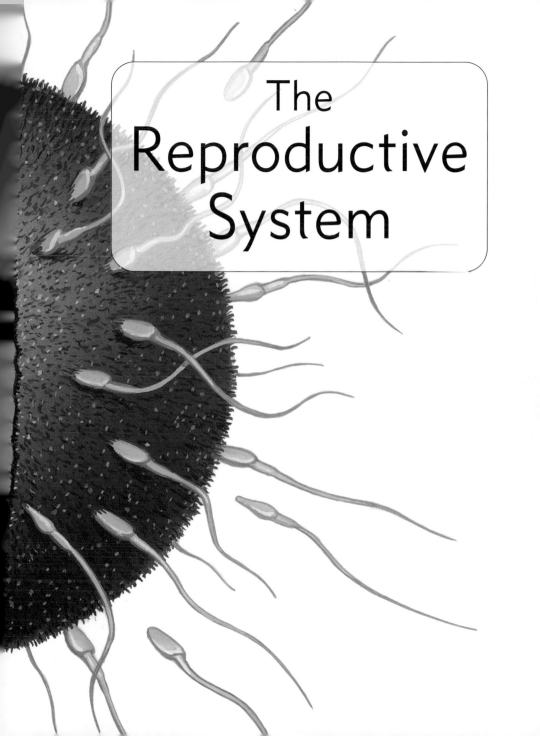

The
Reproductive
System

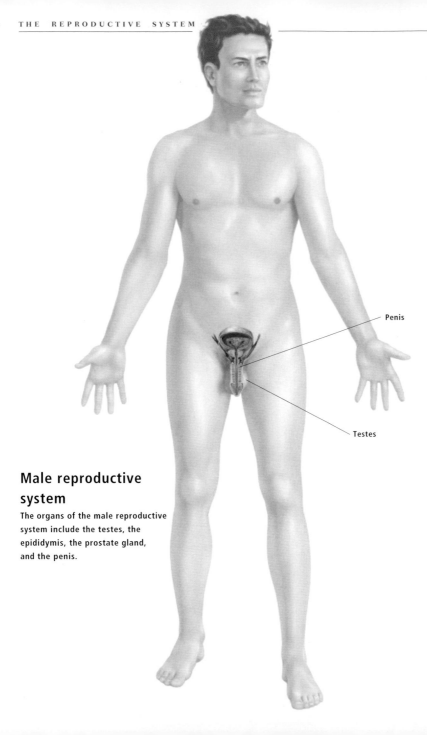

Penis

Testes

Male reproductive system

The organs of the male reproductive system include the testes, the epididymis, the prostate gland, and the penis.

THE REPRODUCTIVE SYSTEM

Both the male and female reproductive systems reach maturity during puberty. They are involved in the creation of the next generation and, during fetal development, are linked with the urinary system.

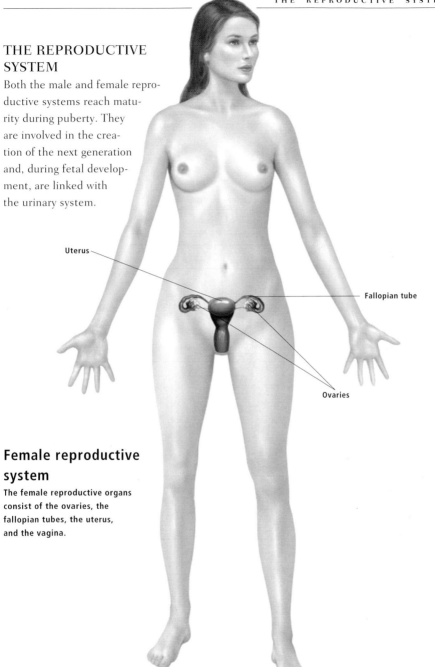

Uterus

Fallopian tube

Ovaries

Female reproductive system

The female reproductive organs consist of the ovaries, the fallopian tubes, the uterus, and the vagina.

THE MALE REPRODUCTIVE SYSTEM

Testes

The two testes are each held in a wrinkled sac called the scrotum. The scrotum is composed of skin and soft tissue overlying a thin layer of muscle. It is the contractions of this muscle that give the wrinkled appearance to the scrotum. Lying behind the penis, the testes are located outside the abdomen so that they remain at optimal temperature for hormone and sperm production.

Within the scrotum, the testes are separated by the scrotal septum, and commonly, the right one sits higher than the left one.

Each testis is surrounded by the membrane of the tunica vaginalis. Beneath the tunica vaginalis is the tunica albuginea, a tough fibrous tissue that extends through the testis to create several hundred compartments, known as lobules. The lobules are the site of sperm production, with each lobule connected to a network of ducts that merge into fewer, larger ducts and congregate to form the head of the epididymis. The coiled mass of ducts tapers to form the tail of the epididymis, which gives rise to the ductus deferens (or vas deferens).

Penis

The penis plays a role in both the reproductive and urinary systems. In the reproductive system, it is the male organ of

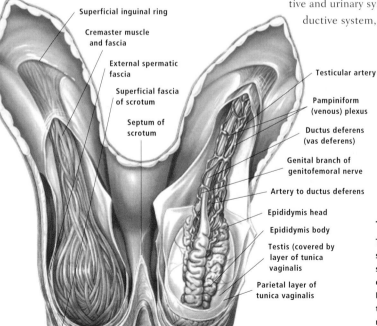

Superficial inguinal ring

Cremaster muscle and fascia

External spermatic fascia

Superficial fascia of scrotum

Septum of scrotum

Testicular artery

Pampiniform (venous) plexus

Ductus deferens (vas deferens)

Genital branch of genitofemoral nerve

Artery to ductus deferens

Epididymis head

Epididymis body

Testis (covered by layer of tunica vaginalis

Parietal layer of tunica vaginalis

Scrotal skin

Testes

The testes are the site of sperm production and, as such, are the main organs of reproduction in the male. Male hormones, primarily testosterone, are also manufactured by the testes.

Penis

The penis is the male urinary and reproductive organ, containing three cylinders of sponge-like vascular tissue that allow erection. The penis is attached to the pelvic region by connective tissue, and usually hangs flaccid unless stimulated.

copulation, and the passageway through which semen leaves the body. Running the length of the penis is the urethra, through which both semen and urine travel, and which opens out at the external urethral orifice. The urethra is within a cylinder of tissue known as the corpus spongiosum, which flares out at the tip of the penis to form the glans penis. Located at each side of the corpus spongiosum are the two cylinders of tissue known as the corpus cavernosum. All three cylinders then have an outer layer of skin; the retractable fold of skin overlying the sensitive region of the glans penis is known as the foreskin or prepuce.

The vascular tissue of the cylinders has a sponge-like, honeycombed structure, with blood coursing through its cavities. When this is in progress, the penis is flaccid; however, when triggered by the parasympathetic division of the autonomic nervous system, blood is prevented from leaving the venous system of the penis due to increased arterial blood flow, causing the blood to accumulate in the penis. This causes the tissue to become rigid, resulting in an erection.

Penis—urinary system

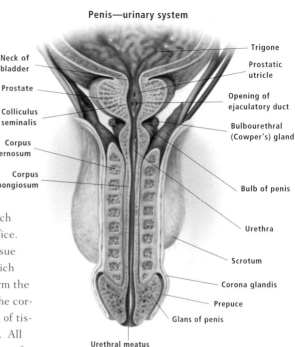

Labels: Neck of bladder; Prostate; Colliculus seminalis; Corpus cavernosum; Corpus spongiosum; Trigone; Prostatic utricle; Opening of ejaculatory duct; Bulbourethral (Cowper's) gland; Bulb of penis; Urethra; Scrotum; Corona glandis; Prepuce; Glans of penis; Urethral meatus

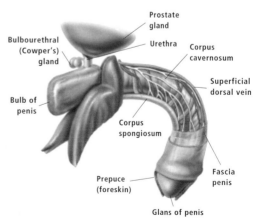

Labels: Bulbourethral (Cowper's) gland; Bulb of penis; Corpus spongiosum; Prepuce (foreskin); Glans of penis; Prostate gland; Urethra; Corpus cavernosum; Superficial dorsal vein; Fascia penis

Penis—reproductive system

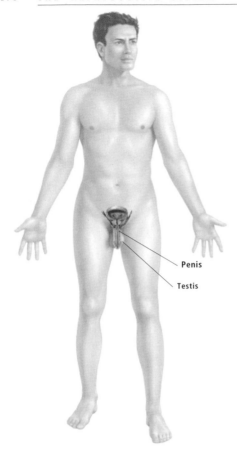

Penis

Testis

Glands of the Male Reproductive System

The prostate lies beneath the bladder, and comprises both glandular and muscular tissue. The prostate encircles both the neck of the bladder and the urethra, with the glands of the prostate secreting their fluids into the urethra.

Surrounded by muscular walls, the inner glandular tissue contains two major groups of glands. The secretions of these glands flow into the urethra, contributing approximately 25 percent of the content of semen. The urethra serves both the urinary and reproductive system, conveying urine from the bladder and semen from the reproductive organs.

Penetrating the top of the prostate is the ejaculatory duct. This duct carries the secretions from the seminal vesicle and

Glands of the male reproductive system

The accessory glands of the male reproductive system include the prostate, the seminal vesicle, the ductus deferens, and the bulbourethral glands.

Prostate gland

The prostate lies under the bladder, and is shaped like an inverted pyramid.

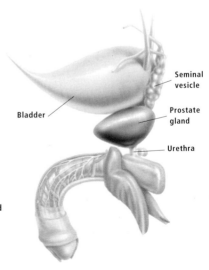

Seminal vesicle

Bladder

Prostate gland

Urethra

ductus deferens, whose individual ducts merge to form the ejaculatory duct. The seminal vesicle is a coiled tube lying above the prostate gland, and between the rear wall of the bladder and the rectum.

The ductus deferens is a component of the spermatic cord, which also incorporates the testicular artery, the pampiniform plexus, lymphatic vessels, and autonomic nerve fibers.

These components are wrapped in connective tissue and muscle, to form the spermatic cord. The ductus deferens originates from the tail section of the epididymis in the testis. As the spermatic cord, it runs up into the pelvic cavity and, as it nears the back of the prostate, the diameter of the cord expands, creating an ampulla. The ampulla joins with the duct from the seminal vesicle to form the ejaculatory duct.

Inactive sperm are produced in the testis, and are then transferred along the ductus deferens. When sperm from the ductus deferens meet with the secretions of the seminal vesicle, they are prompted into activity. The secretions of the seminal vesicle contribute a large percent (about 60 percent) of the content of semen.

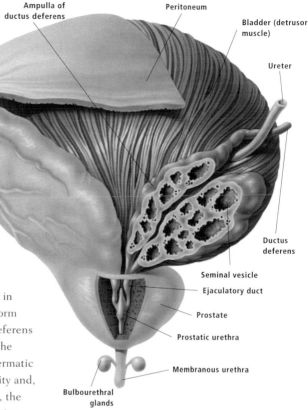

Ampulla of
ductus deferens

Peritoneum

Bladder (detrusor
muscle)

Ureter

Ductus
deferens

Seminal vesicle

Ejaculatory duct

Prostate

Prostatic urethra

Membranous urethra

Bulbourethral
glands

Male bladder

The urethra conveys urine from the bladder and semen from the reproductive organs.

Secretions from the prostate and seminal vesicle, and the sperm carried via the spermatic cord, all meet up in the prostatic urethra. The tiny bulbourethral glands are situated below the prostate gland, at the base of the penis. Secretions from these glands lubricate the urethra.

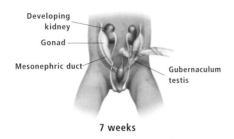

Developing kidney
Gonad
Mesonephric duct
Gubernaculum testis

7 weeks

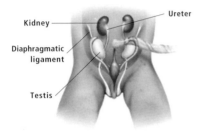

Kidney
Ureter
Diaphragmatic ligament
Testis

16 weeks

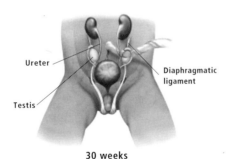

Ureter
Diaphragmatic ligament
Testis

30 weeks

Descent of the testes

The testes form in the embryo from the sexually undifferentiated gonad during the 7th week. Developing in the abdomen during much of the gestation period, they are usually fully developed around the 30th week. When they are fully developed, they begin their journey towards their final destination, gradually descending down the inguinal canal, and reaching the scrotum as the time of birth nears.

Descent of the Testes

In the developing fetus, the formation of the genital organs begins in the second month. The testes lie in the abdominal cavity, near the kidney on either side of the lumbar vertebrae. Around the seventh or eighth month, the testes are fully developed and their descent toward their final position begins.

Hormonal triggers start the process, with the testes gradually descending. They pass through the abdominal wall at the groin, taking with them an extraneous section of peritoneum, which is later pinched off into the tunica vaginalis. By full term, the testes have descended into the scrotum, each accompanied by a spermatic cord, which contains the testicular blood vessels, ducts, nerves, and lymph vessels.

Development of Male Genitalia

As the genital organs begin to develop, in the second month of gestation, there is no apparent difference initially between male and female. By the 7th week, the male and

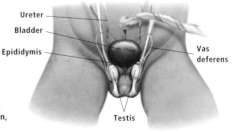

Ureter
Bladder
Epididymis
Vas deferens
Testis

Fully developed

Development of male external genitalia

The male and female sex organs in the developing fetus develop similarly during the initial weeks of pregnancy until about the 7th week. After this the male and female sex organs follow different paths of development. The external genitalia begin to show noticeable differences after the 12th week.

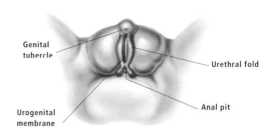

Undifferentiated

female sex glands follow different courses of development; this is determined by the chromosomes present. The presence of a Y chromosome triggers the production of male hormones in the testes, which in turn initiates the development of the male sexual organs.

The testes develop in the abdomen prior to the descent, commencing around the 7th month of gestation. They gradually move down the inguinal canal, coming to lie in the scrotum by the end of the 8th month. In the male, a pair of tubules form which join up to the testes, and then open into the urethra; pouches in the ducts become the seminal vesicles.

Similarly, male and female external genitalia are undifferentiated for the first 12 weeks of fetal development. After 12 weeks, the differences begin to appear; the male genital tubercle and urogenital folds develop into the penis, and the genital swellings fuse together to become the scrotum.

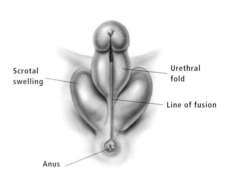

12 weeks

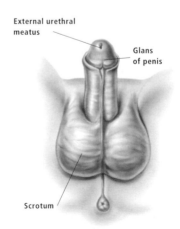

Fully developed

Testosterone

The development of secondary male characteristics, such as facial and pubic hair, enlargement of the penis and testes, enlargement of the larynx, and increase in muscle strength is initiated by testosterone. Testosterone is produced in the testes, with production ultimately controlled by the pituitary gland.

Sperm and Sperm Production

Within the lobules of the testes are the seminiferous tubules, where sperm production commences. The spermatogonia (germ cells) divide, becoming spermatids, which then mature into sperm, the male reproductive cells.

The sperm resembles a tadpole in shape; the head has an acrosomal membrane that releases enzymes to help the sperm penetrate the female ovum, while the tail helps the sperm move on its journey from the epididymis to the female reproductive organs. The chromosomes that will determine the sex of a fertilized ovum are carried in a nucleus in the head of the sperm; half of the sperm cells carry the Y chromosome and the other half carry the X chromosome.

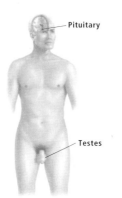

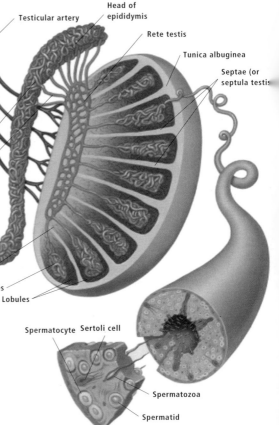

Testicular artery

Head of epididymis

Rete testis

Tunica albuginea

Septae (or septula testis

Pituitary

Ductus deferens

Body of epididymis

Efferent ductules

Tail of epididymis

Testes

Mediastinum testis

Seminiferous tubules

Lobules

Spermatocyte   Sertoli cell

Spermatozoa

Spermatid

Sperm production

Maturing from germ cells through various stages to become mature sperm, the production of sperm commences in the tiny tubules of the lobules in the testes.

Sperm structure

The nucleus in the head of the sperm contains the chromosomes that will determine the sex of a fertilized ovum. The acrosomal membrane surrounding the head releases enzymes to aid the sperm's penetration of the ovum. The tail provides mobility for the sperm, which employs a "swimming" motion to travel to the female reproductive organs.

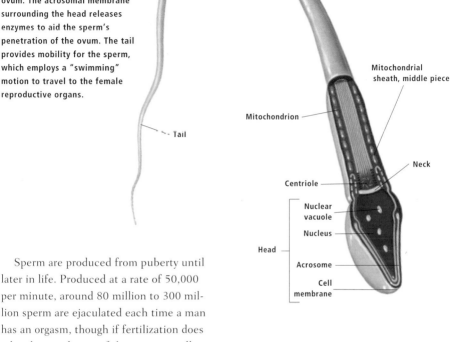

Mitochondrial
sheath, middle piece

Mitochondrion

Tail

Neck

Centriole

Nuclear
vacuole

Nucleus

Head

Acrosome

Cell
membrane

Sperm are produced from puberty until later in life. Produced at a rate of 50,000 per minute, around 80 million to 300 million sperm are ejaculated each time a man has an orgasm, though if fertilization does take place, only one of these sperm will penetrate and fertilize the ovum.

The journey of the sperm takes them from the testes to the epididymis, where they are stored. From the epididymis they travel along the ductus deferens where they are ejaculated. During ejaculation, muscles around the epididymis and ductus deferens contract, forcing semen into the urethra. The fluid is then expelled out of the body by spasmodic contractions of the bulbocavernosus muscle in the penis.

Semen

Semen is the liquid that contains the sperm and the secretions of the reproductive glands. The contributions of the glands such as the prostate and seminal vesicle ensure provision of the nutrients required to maintain the viability of the sperm.

Semen is predominantly composed of seminal fluids, with sperm and epididymal fluid accounting for only 5 to 10 percent of the content.

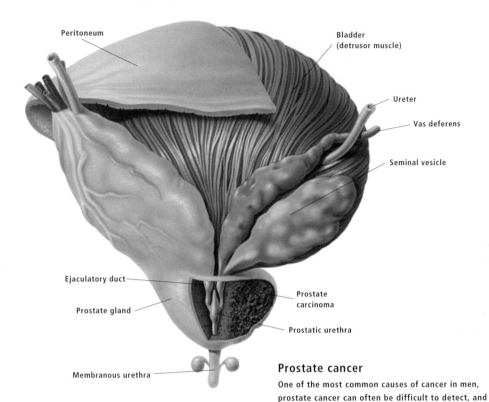

Peritoneum

Bladder
(detrusor muscle)

Ureter

Vas deferens

Seminal vesicle

Ejaculatory duct

Prostate
carcinoma

Prostate gland

Prostatic urethra

Membranous urethra

Prostate cancer

One of the most common causes of cancer in men,
prostate cancer can often be difficult to detect, and
is often well established, spreading to nearby
structures, before detection occurs.

Diseases and Disorders of the Male Reproductive System

Each of the different structures represented in the male reproductive system presents its own range of problems. Cancer of the prostate, for instance, is the one of the most common causes of cancer in men in the industrialized world.

The most serious problem of the testes is testicular cancer, which usually affects a younger age group (15–35) than prostate cancer, while the penis is most susceptible to sexually transmitted diseases. Some of the more common diseases and disorders are discussed below.

Prostate cancer

Generally affecting the over-60 age group, prostate cancer is the most common form of cancer in men. The exact cause of prostate cancer is unknown, though it may have genetic or hormonal links. Like many cancers it is often not detected until it spreads from its original site to other structures in the body, notably the bones of the spine and pelvis. Its detection is made more difficult as it is often masked by, and coexists with, other urinary tract problems, such as enlargement of the prostate—a common problem in older men.

Prostate cancer often begins in the outer portion of the prostate, so by the time urinary difficulties or symptoms present, affecting the ureter running through the central part of the prostate, it often means that the cancer has had an opportunity to establish and proliferate.

Treatment is very much dependent on the extent of spread. Hormonal controls, surgery, and radiation treatment are among the possible treatment options; however, early detection is important.

Hydrocele

Generally painless, hydroceles are the result of an accumulation of fluid in the tunica vaginalis, the membrane sac around the testes, usually caused by injury or inflammation. Treatment options include draining of the fluid or surgery.

Sexually transmitted diseases

There are a range of sexually transmitted diseases that generally affect the penis, including syphilis, which in its initial stages causes lesions on the skin of the penis. Gonorrhea affects the urethra, causing painful urination. Most sexually transmitted diseases respond well to appropriate antibiotics.

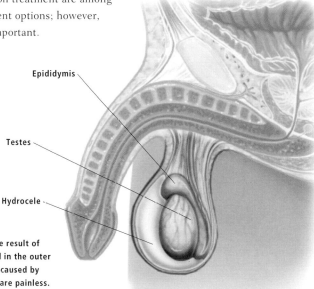

Epididymis

Testes

Hydrocele

Hydrocele

A hydrocele forms as the result of an accumulation of fluid in the outer sac of the testes. Often caused by injury, most hydroceles are painless.

THE FEMALE REPRODUCTIVE SYSTEM

Ovaries

Lying near the side walls of the pelvis, the two ovaries are oval-shaped organs about the size of almonds, held in place by ligaments. They are each connected to the top corner of the uterus by the ovarian ligament, and receive additional support from the broad ligament of the uterus.

Unlike sperm, which continues to be produced until late in life, the ovaries contain a set amount of cells—around 300,000—which lie dormant until puberty. At the onset of puberty, the ovaries are activated, with some 20 ova beginning to enlarge and develop at the beginning of each menstrual cycle. Each of these ova is contained in a sac-like structure, known as a follicle. Of the developing follicles, usually only one will go on to full maturity, with the others gradually degenerating during the ripening process. The developing follicle matures into a Graafian follicle, which contains the mature ovum, surrounded by follicular fluid, and encased in an outer follicular layer; this follicle is responsible for the production of estrogen, which prepares the uterus for pregnancy. During ovulation, the Graafian follicle bursts, releasing the ovum, which is gathered up by the fimbriae and conveyed along the fallopian tubes, where fertilization may take place.

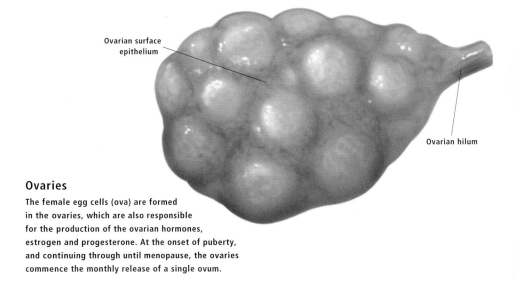

Ovarian surface epithelium

Ovarian hilum

Ovaries

The female egg cells (ova) are formed in the ovaries, which are also responsible for the production of the ovarian hormones, estrogen and progesterone. At the onset of puberty, and continuing through until menopause, the ovaries commence the monthly release of a single ovum.

Ovary cross section

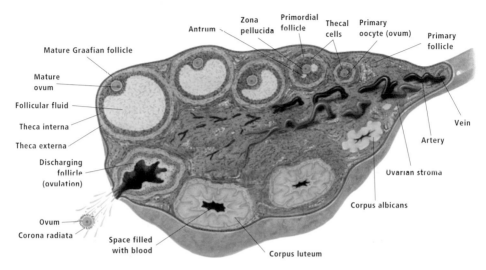

Once the ovum has been released, the ruptured follicle forms the corpus luteum, and produces the hormones progesterone and estradiol, further preparing the uterus for pregnancy. If the ovum remains unfertilized, the corpus luteum breaks down, progesterone and estrogen levels fall, and menstruation follows. The substance of the corpus luteum is gradually replaced by fibrous tissue, to become the corpus albicans.

The structure and function of the ovaries change as a woman ages. As menopause approaches, the number of follicles decreases. In addition, the amount of estrogen in circulation begins to decline, causing the ovaries to become smaller.

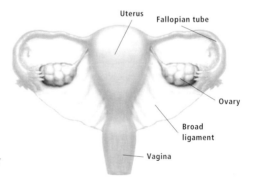

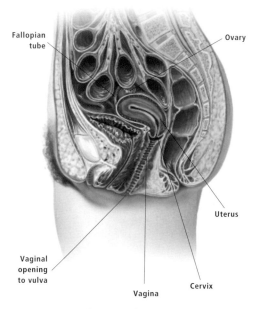

Fallopian tube

Ovary

Uterus

Vaginal opening to vulva

Cervix

Vagina

Female reproductive organs

Fallopian Tubes

The fallopian (uterine) tubes are joined to the top of the uterus by a narrow opening. From here they extend out like arms, with the end of the fallopian tube flaring out and forming feathery, finger-like projections called fimbriae, around the ovary. These fimbriae pick up the released ovum and from here it moves along the fallopian tube.

The ovum's movements along the fallopian tube are the result of muscular contractions, in conjunction with tiny cilia in the epithelial lining of the fallopian tubes, which assist in moving the ovum toward the uterus. If the egg is fertilized by a male sperm cell, this takes place in the outer third of the fallopian tubes.

Uterus

The two fallopian tubes join to each side of the upper part of the uterus, which lies between the bladder and the rectum, with the broad ligament holding the uterus to the side walls of the pelvis.

The non-pregnant uterus is somewhat pear-shaped, with its apex pointing downward and connecting to the vagina. The upper portion of the uterus is the body, while the lower third is the cervix. Three layers make up the walls of the uterus—the inner layer of endometrium, a central layer of myometrium, and an outer layer of peritoneum. The inner layer of endometrium undergoes cyclic changes in its structure, firstly preparing for a fertilized ovum, then, should the ovum remain unfertilized, the endometrial surface breaks down and is expelled as menstrual fluid.

The cervix is the lower portion of the uterus. This narrow canal has a muscular wall, with its internal, mucus-producing lining remaining intact during menstruation.

Vagina

Joining the cervix of the uterus to the external environment, the vagina is a muscular tube situated in the lower portion of the pelvis, between the bladder and the rectum.

Support is provided by the cervical ligaments and the pelvic floor muscles.

Generally, the walls of the vagina lie close together, and they are coated with mucus secretions from the cervix and plasma from the capillaries.

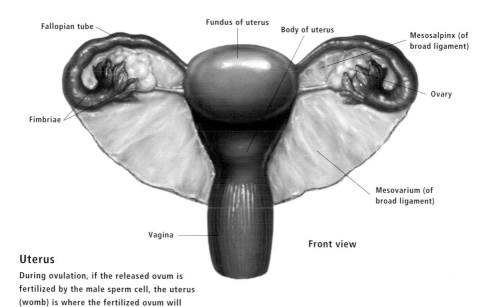

Fallopian tube

Fundus of uterus

Body of uterus

Mesosalpinx (of broad ligament)

Ovary

Fimbriae

Mesovarium (of broad ligament)

Vagina

Front view

Uterus

During ovulation, if the released ovum is fertilized by the male sperm cell, the uterus (womb) is where the fertilized ovum will establish and develop into an embryo and fetus.

Fallopian tube

Mesosalpinx (of broad ligament)

Ampulla

Infundibulum

Fimbriae

Ovary

Endometrium

Myometrium

Mesovarium (of broad ligament)

Cervix

Internal os

Vaginal fornix

Vagina

Back view

External os

External Genitalia

The female external genitalia are collectively known as the pudendum or vulva. Lying on each side of the vagina are the hairless folds of skin called the labia minora. At birth, the junction between the labia minora and the vagina is covered with a thin membrane called the hymen. Although an intact hymen is regarded by some cultures as an indication of virginity, this delicate membrane is prone to rupture or partial rupture, and often does so before puberty, thus allowing the unrestricted passage of menstrual blood. Located on either side of the vaginal opening (vestibule) are the Bartholin's glands, which exude mucus secretions into the vestibule. Outside the vaginal opening are the folds of the labia majora. These folds of skin cover a subcutaneous fat interior, while their exterior is covered in pubic hair.

Lying in front of the urethra is the erectile body of the clitoris. The tip of the clitoris has a glans or head, often hidden within a thin fold of skin known as the prepuce, formed by the junction of the labia minora.

Development of Female Genitalia

The genital organs of the developing fetus are undifferentiated for the first few weeks of gestation. Around the 7th week, the organs begin to develop differently; this development is determined by the chromosomes carried in the male sperm cell. If the

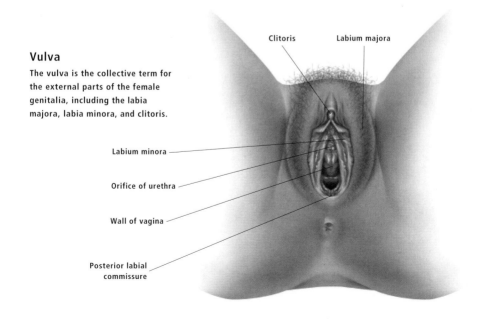

Vulva

The vulva is the collective term for the external parts of the female genitalia, including the labia majora, labia minora, and clitoris.

Clitoris

Labium majora

Labium minora

Orifice of urethra

Wall of vagina

Posterior labial commissure

ovum has been fertilized by an X chromosome sperm cell, a female baby will result. The ovaries begin to develop after the 7th week, gradually moving from their initial position in the abdomen to their lower permanent position. As the ovaries are developing, a pair of ducts develop at the same time, one end of which will fuse and ultimately become the uterus and vagina, while the other ends will come to lie next to each of the ovaries. No discernible outward difference in external genitalia is evident until around the 12th week, after which time the female genitalia begin to form, with the initial genital tubercle becoming the clitoris, and the folds and swellings forming the labia majora and labia minora of the vulva.

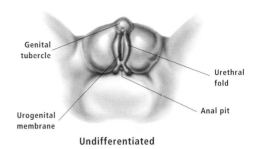

Undifferentiated

Development of female external genitalia

In fetal development, the external genitalia remain undifferentiated until about week 12, after which time the genitalia of the different sexes begin to develop. In females, the genital tubercle develops into the clitoris.

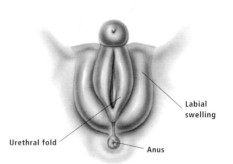

12 weeks

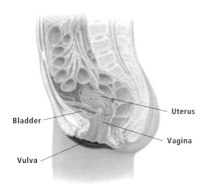

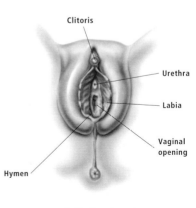

Fully developed

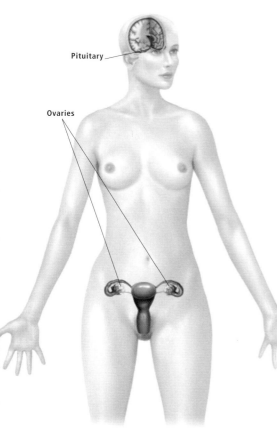

Pituitary

Ovaries

Estrogens

Estrogens are primarily produced in the ovaries, though other glands contribute small amounts to the overall supply. The production of estrogen is ultimately controlled by the pituitary gland. Estrogens control the development of secondary female sex characteristics and the reproductive system.

Ovarian Hormones

The two main hormones produced by the ovaries are estrogen and progesterone. These hormones have a unique role to play in ovulation, pregnancy, and the maturing female's development of secondary sexual characteristics.

Estrogen

Estrogens are a group of female sex hormones produced in the ovaries, and include estradiol and estrone. After release into the circulation, these hormones are broken down by the liver to form estriol.

Estrogens play a role in the development of secondary female characteristics, such as breast development, and follicle activity in the ovary. During the menstrual cycle, estrogens are released that initiate the preparation of the uterus for pregnancy, building up the inner endometrial lining to house the fertilized ovum.

Particularly prolific during pregnancy, estrogen production slows as menopause approaches; this can lead to various problems, including osteoporosis and vaginitis. Hormone replacement therapy can be an option to counteract the effects of reduced estrogen production.

Progesterone

After ovulation, the remains of the Graafian follicle form the corpus luteum in the ovary, and it is here that progesterone is produced. In anticipation and preparation for the eventuality of a fertilized ovum, progesterone is produced in order to ensure the ideal environment for implantation and nurturing of a fertilized ovum and the subsequent developing embryo.

If the ovum is not fertilized, progesterone production falls rapidly, the corpus luteum degenerates, and the lining of the uterus, no longer required for housing a potential fertilized ovum and embryo, falls away to be shed as menstrual blood.

If pregnancy occurs, progesterone is produced by the corpus luteum during the first trimester, after which time the role is fulfilled by the placenta. Shortly after birth, progesterone levels fall, while other hormonal levels rise, such as prolactin, a hormone involved in the production of breast milk.

Progesterone

After releasing its mature ovum, the ruptured follicle becomes a gland-like structure, called the corpus luteum, producing progesterone. Progesterone stimulates the uterus to prepare for an environment suitable for implantation by the fertilized ovum.

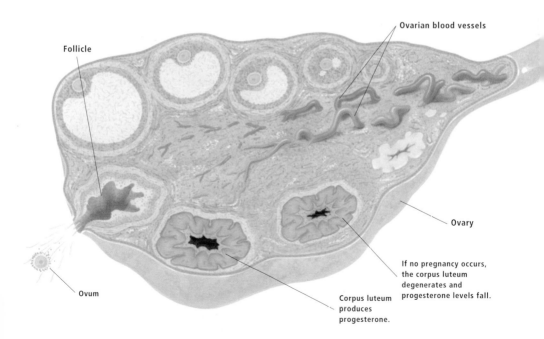

Follicle

Ovarian blood vessels

Ovum

Ovary

Corpus luteum produces progesterone.

If no pregnancy occurs, the corpus luteum degenerates and progesterone levels fall.

Menstruation and the Menstrual Cycle

The menstrual cycle is a 28-day cycle, which goes through a menstrual stage, a proliferative phase, ovulation, and a secretory phase. While the average cycle is 28 days, it can vary substantially.

Menstruation begins during puberty, generally somewhere between the ages of 10 to 16 years of age, and continues through to menopause.

The cycle begins with menstruation, with this stage lasting for around 5 days. During this stage (Days 1–6), the build-up of uterine lining (the endometrium) is sloughed away, leaving a thin layer of lining necessary for regeneration during the next cycle.

Falling progesterone levels trigger this shedding stage, as blood vessels cut supply to the majority of the endometrium, causing tissue breakdown and bleeding leading to menstruation, when this discharge of tissue and blood is forced down through the cervix and vagina by the muscular contractions of the uterus.

Days 7–13 signify the proliferative phase. During this phase, the endometrium begins to rebuild. Simultaneous with this, several egg cells in the ovary are maturing ready for release, with most degenerating before ovulation, leaving one dominant egg ready for the next phase.

Ovulation takes place on or around Day 14. The dominant follicle, now known as a Graafian follicle, releases its mature ovum into the fallopian tube.

Menstrual cycle

The 28-day menstrual cycle goes through several stages: menstruation, the proliferative phase, ovulation, and the secretory phase.

Days 15–28 encompass the secretory phase. During this phase, the ovum travels along the fallopian tube. The remnants of the Graafian follicle form the corpus luteum, which initiates production of progesterone.

Progesterone is released until about Day 26 of the menstrual cycle, during which time it stimulates secretions in the endometrium, which will establish a nourishing base for the fertilized ovum.

After Day 26, if the ovum remains unfertilized, progesterone production levels drop markedly, causing the uterine lining to gradually break down, leading to the onset of menstruation on or around Day 28, marking the beginning of the menstrual cycle once more.

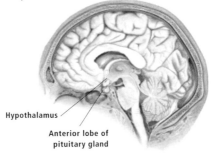

Hypothalamus

Anterior lobe of pituitary gland

Cycle regulation

The menstrual cycle is regulated by the hypothalamus and the pituitary gland in the brain, which stimulate the ovaries to release estrogen and progesterone.

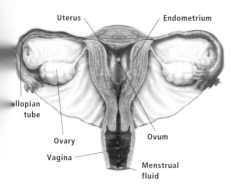

a **Days 1-6**

Menstruation—the rich vascular lining of the endometrium and its associated blood vessels are shed over a 5 day period.

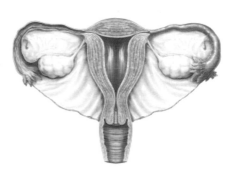

b **Days 7–13**

Proliferative phase—at the start of the proliferative phase, the endometrium begins to repair and rebuild.

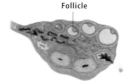

c During the proliferative stage, several ova in the follicles of the ovary begin to mature.

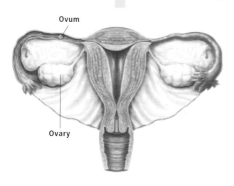

e **Days 15–28**

Secretory phase—as the ovum makes its way along the fallopian tube, the remains of the ruptured follicle forms the corpus luteum, performing gland-like activity, and triggering the production of progesterone. Progesterone release causes the endometrium to thicken in readiness to support a fertilized ovum. If fertilization does not take place, hormone levels fall, causing the endometrium to gradually break down.

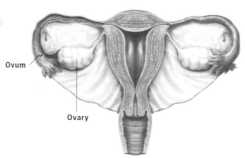

d **Day 14**

Ovulation—by Day 14 all but one of the follicles have degenerated. The one remaining follicle ruptures, releasing the mature ovum into the fallopian tube.

Diseases and Disorders of the Female Reproductive System

Disorders of the organs of the female reproductive system are common and often require specialist gynecological management. Some of the more common problems are discussed below.

Endometriosis

Endometriosis can cause a range of problems including painful and abnormal periods, abdominal pain before and during menstruation, lower back pain, pain during intercourse, fatigue, and bloating.

Conversely, some women with endometriosis do not experience any pain associated with the condition.

The condition is caused by endometrial tissue establishing in other parts of the body, such as the ovaries, cervix, pelvic cavity, and bladder. Once established, it performs the same actions it would if it were in the uterus, thickening its tissue prior to menstruation, as if to prepare for nourishing a fertilized egg, and then bleeding. The result is scarring and adhesions (clusters of endometrial cells), which may implant in the ovaries and fallopian tubes and obstruct the passage of the ovum.

Treatment varies depending on the location of the endometrial tissue, and can range from treatment with painkillers for mild symptoms, to surgical options in more serious cases.

Salpingitis

The most common disorder of the fallopian tubes is salpingitis due to pelvic infection. Left untreated, the infection can cause blockage of the fallopian tubes, either preventing sperm cells from reaching the ovum, or preventing the fertilized ovum to travel to the uterus. Salpingitis accounts for 20 percent of the cases of infertility.

Endometriosis

Endometriosis occurs when endometrial tissue forms in other parts of the body, such as the pelvic cavity and other reproductive organs. It continues to perform its functions of increasing and thickening its tissue base, before breaking down prior to menstruation. This often causes pain and a variety of other symptoms, resulting in scarring and adhesions.

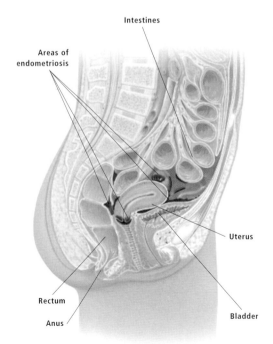

Intestines

Areas of endometriosis

Uterus

Rectum

Anus

Bladder

Cancer of the cervix

While the incidence of cervical cancer has decreased significantly in recent years due to the widespread use of the Pap smear, cervical cancer is one of the leading cancers affecting women. Early detection, using the Pap smear test, is crucial to monitor any changes in the cells of the cervix. Early detection enables cervical cancer to be treated before it becomes invasive.

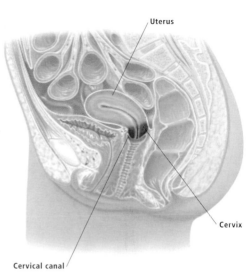

Uterus

Cervix

Cervical canal

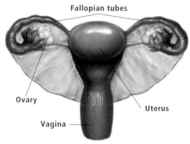

Fallopian tubes

Ovary

Uterus

Vagina

Normal fallopian tubes

Cervix

The cervix is the lower portion of the uterus, and is the region of the reproductive system most prone to cancer attack, accounting for almost 10 percent of cancers in females in developed countries.

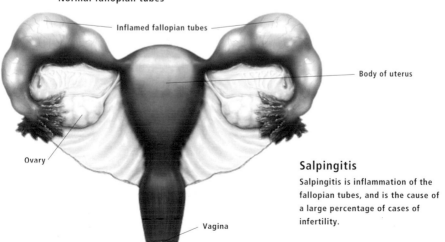

Inflamed fallopian tubes

Body of uterus

Ovary

Vagina

Salpingitis

Salpingitis is inflammation of the fallopian tubes, and is the cause of a large percentage of cases of infertility.

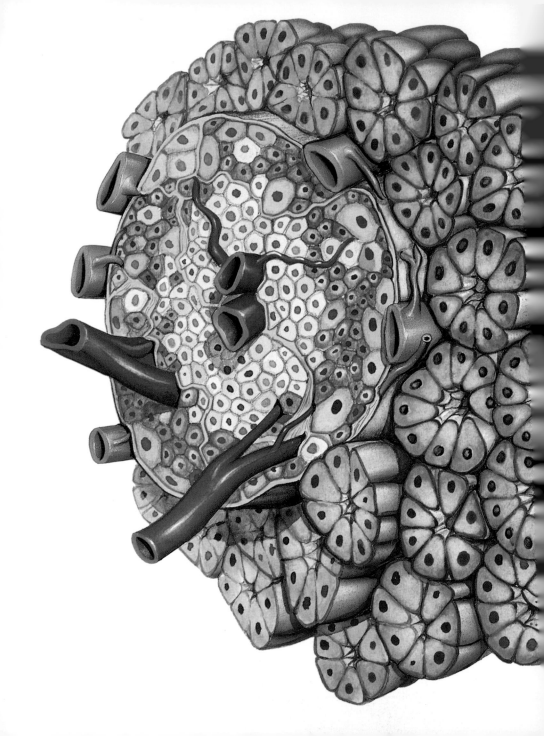

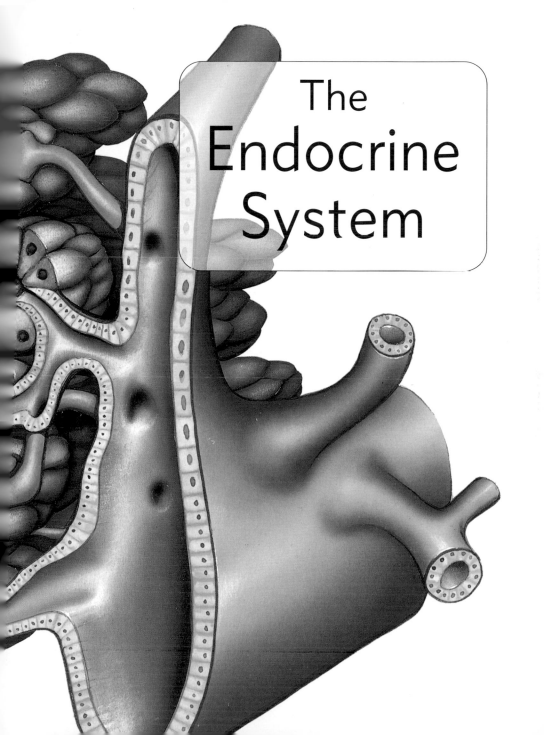

The
Endocrine
System

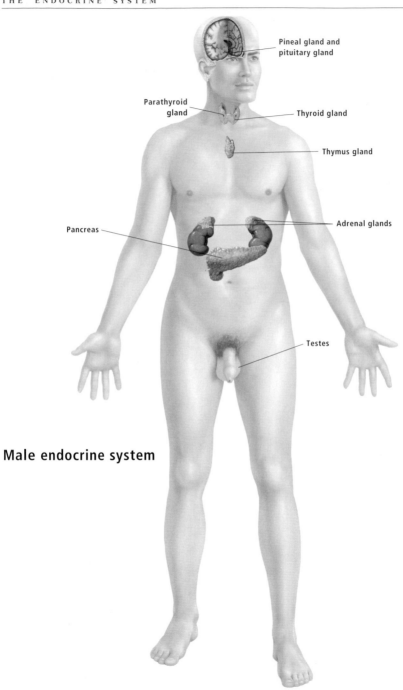

Pineal gland and pituitary gland

Parathyroid gland

Thyroid gland

Thymus gland

Pancreas

Adrenal glands

Testes

Male endocrine system

THE ENDOCRINE SYSTEM

The endocrine system controls hormone production, and is responsible for the slow or long-term changes in the body, such as growth, and the gradual changes experienced during puberty, as secondary sexual characteristics develop. The endocrine glands include the pineal, thymus, thyroid, parathyroid, adrenals, pancreatic islets, ovaries, and testes. The placenta, which develops during pregnancy, also has an endocrine function.

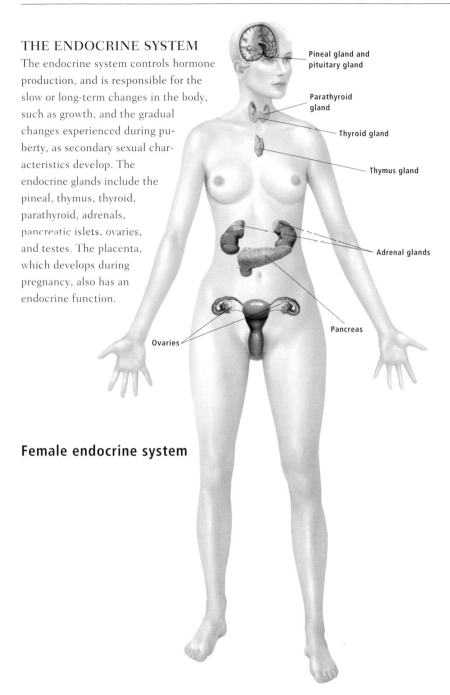

Pineal gland and pituitary gland

Parathyroid gland

Thyroid gland

Thymus gland

Adrenal glands

Pancreas

Ovaries

Female endocrine system

Pituitary Gland

The activities of the glands in the endocrine system are regulated by the pituitary gland, which is in turn closely controlled by the hypothalamus that lies directly behind it in the brain. This tiny gland is divided into two lobes: the anterior pituitary or adenohypophysis, and the posterior pituitary or neurohypophysis. Cells in the anterior lobe produce growth hormone, adrenocorticotropic hormone, follicle-stimulating hormone, luteinizing hormone, melanocyte-stimulating hormone, prolactin, and thyroid-stimulating hormone. The posterior lobe is responsible for oxytocin and antidiuretic hormones,

which it receives from the hypothalamus. The hypothalamus and pituitary gland are closely linked, working in harmony to provide the hormones required by the body for efficient operation. These hormones are either supplied directly by the pituitary or by initiating hormone production in other glands of the endocrine system.

Of the hormones produced by the pituitary gland, each has a specific task. Essential during childhood and into adolescence, growth hormone (GH) promotes muscle and long bone growth. Adrenocorticotropic hormone (ACTH) stimulates the adrenal gland to produce corticosteroid hormones.

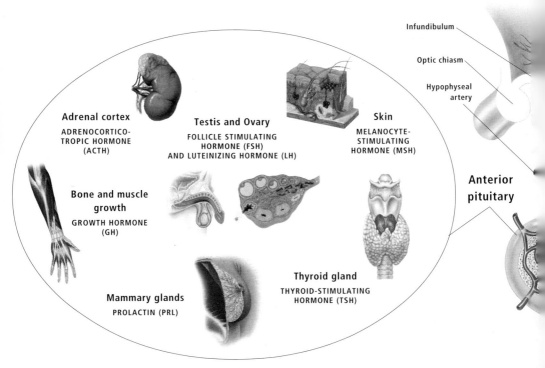

Infundibulum

Optic chiasm

Hypophyseal artery

Adrenal cortex
ADRENOCORTICO-TROPIC HORMONE (ACTH)

Testis and Ovary
FOLLICLE STIMULATING HORMONE (FSH) AND LUTEINIZING HORMONE (LH)

Skin
MELANOCYTE-STIMULATING HORMONE (MSH)

Anterior pituitary

Bone and muscle growth
GROWTH HORMONE (GH)

Mammary glands
PROLACTIN (PRL)

Thyroid gland
THYROID-STIMULATING HORMONE (TSH)

Follicle-stimulating hormone (FSH) affects both the male and female reproductive systems, stimulating the maturation of eggs in the ovaries of the female, and sperm production in the testes of the male. Similarly, luteinizing hormone (LH) also affects both the male and female reproductive systems, initiating the release of a mature egg from the ovary (ovulation) and prompting the production of progesterone in the female system, and stimulating the production of testosterone in the male system.

Melanocyte-stimulating hormone (MSH) influences the amount of melanin produced in the skin. Melanin is responsible for skin pigmentation and hair color. Prolactin (PRL), or lactogenic hormone, stimulates the production and continued supply of breast milk. Thyroid-stimulating hormone (TSH) activates the thyroid gland, prompting it to produce and release thyroid hormone.

Oxytocin (OT) acts on the female reproductive system, triggering the smooth muscle contractions associated with childbirth, and stimulating breast milk production. Antidiuretic hormone (ADH) triggers increased absorption of water by the kidneys, keeping water and salt levels in the body constant.

Neurosecretory cells

Hypothalamus

Mamillary body

Axon

Hypophyseal portal system

Pituitary stalk

Posterior pituitary

Hypophyseal artery

Vein

Pituitary gland

The pituitary gland, working with the neighboring hypothalamus, produces hormones and influences activity in other glands of the endocrine system.

Mammary glands
OXYTOCIN (OT)

Kidney tubules
ANTIDIURETIC HORMONE (ADH)

Uterus smooth muscle
OXYTOCIN (OT)

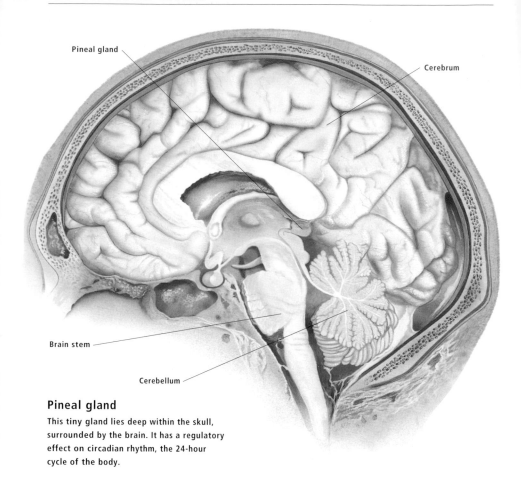

Pineal gland

Cerebrum

Brain stem

Cerebellum

Pineal gland

This tiny gland lies deep within the skull,
surrounded by the brain. It has a regulatory
effect on circadian rhythm, the 24-hour
cycle of the body.

Pineal Gland

The tiny pineal gland is located deep inside
the skull cavity, surrounded by the brain.
The pineal gland is responsible for the production of melatonin. Strongly linked to
the body's sleep–wake clock, or circadian
rhythm, melatonin production is at its
greatest during our sleeping hours, with
production falling off when we are awake.

While the exact role of the pineal gland
is not fully understood, it is thought to
have an influence on the male and female
gonads, and also on mood.

Thymus

From birth through to puberty, the thymus
gland grows steadily in size, eventually

reaching about 1 ounce (28 grams) in weight; it is located behind the sternum, close to the heart. After puberty, the thymus gradually degenerates, and by adulthood the thymus has reduced to about half that of its maximum size, weighing about ½ ounce (14 grams). As well as its role as a lymphatic organ, the thymus also serves a gland-like function in the endocrine system, producing hormones essential to the manufacture of the special lymphocytes, T lymphocytes, generated by the thymus. Thymosin, secreted by the thymus, encourages the development of the T lymphocytes, which play a vital role in the body's immune system.

Thymus microstructure

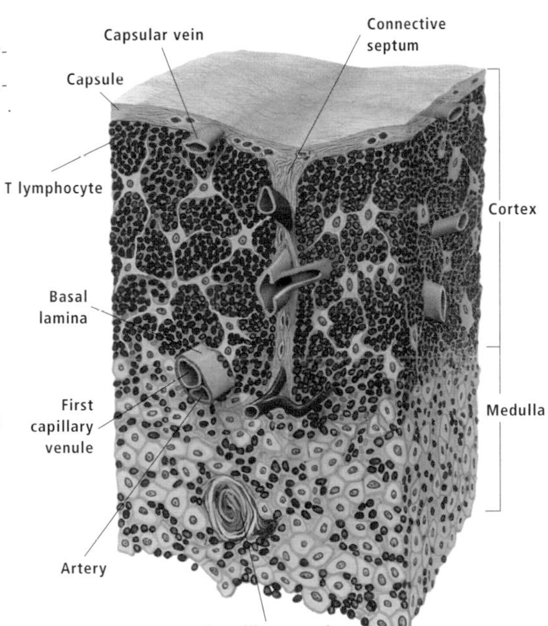

Capsular vein
Connective septum
Capsule
T lymphocyte
Cortex
Basal lamina
Medulla
First capillary venule
Artery
Hassall's corpuscle

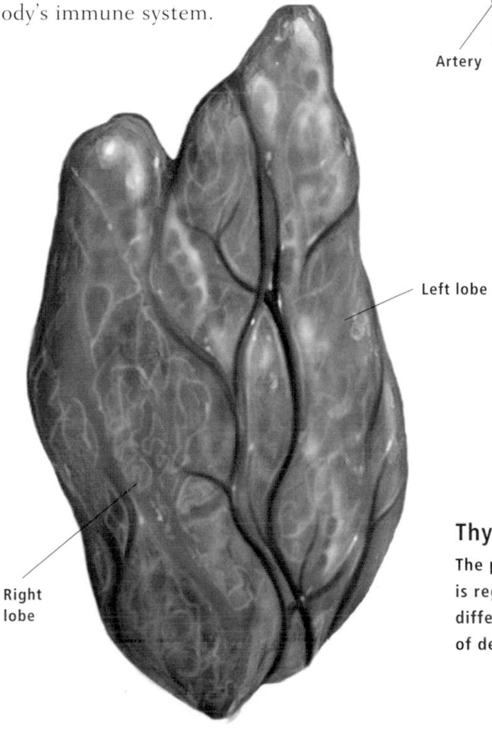

Left lobe
Right lobe

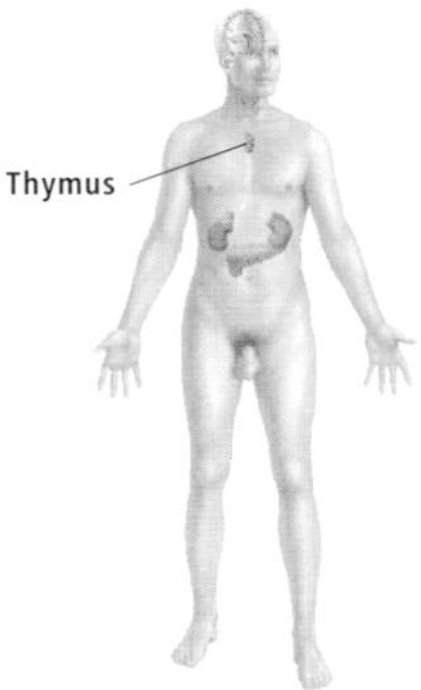

Thymus

Thymus

The production of T lymphocytes in the thymus is regulated by self-produced hormones. Several different hormones play a role in the various stages of development of the T lymphocytes.

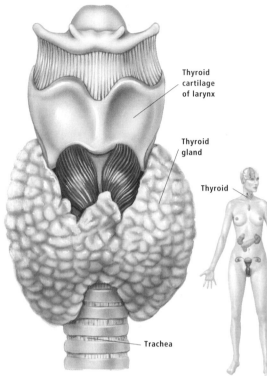

Thyroid
cartilage
of larynx

Thyroid
gland

Thyroid

Trachea

Thyroid

The thyroid is the largest of the endocrine glands, and lies at the front of the neck. This butterfly-shaped gland lies in front of the junction between the larynx and trachea.

Thyroid microstructure

The internal structure of the thyroid gland comprises numerous tiny follicles, with parafollicular cells between the follicles. The follicular cells produce thyroid hormone, while the parafollicular cells produce calcitonin.

Thyroid

Sitting in front of the larynx and trachea, and forming two lobes joined by a narrow bridge, the thyroid gland is one of the endocrine glands.

When the pituitary releases thyroid-stimulating hormone (TSH), the thyroid is activated into production. The hormones produced by the thyroid include thyroid hormone and calcitonin. The thyroid is made up of many follicles, each arranged around a cavity, or lumen, which contains a gel-like substance called colloid. Thyroid hormone is produced by the cells of the thyroid follicles, and stored in the lumen until required. Between the thyroid follicles are the parafollicular cells which produce calcitonin.

Thyroid hormone is composed of two different substances: thyroxine (also called T4, or tetraiodothyronine) and triiodothyronine (T3), which determine metabolic rate and energy production. Correct levels of production must be

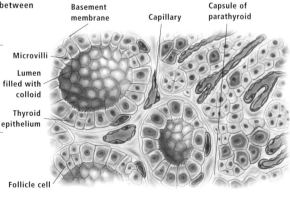

Basement
membrane

Capillary

Capsule of
parathyroid

Follicle

Microvilli

Lumen
filled with
colloid

Thyroid
epithelium

Follicle cell

maintained, as too much causes metabolism to speed up and too little causes it to slow down. Thyroid hormone is essential during fetal development and throughout the childhood years, with stable levels of the hormone being crucial to normal growth and development.

Calcitonin, produced in the parafollicular cells, is released by the thyroid gland when calcium levels in the blood

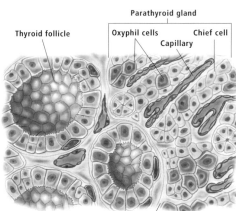

Parathyroid gland microstructure
The chief cells of the parathyroid gland manufacture parathyroid hormone, which is released when blood calcium levels fall.

are too high; calcitonin reduces and stabilizes the calcium content in the blood.

Parathyroid Glands
Generally there are four, though occasionally three, tiny pea-sized parathyroid glands, which lie on the rear surface of the thyroid gland, and are sometimes embedded within the thyroid. Each parathyroid has an outer capsule surrounding the cells within; two types of cell make up the parathyroid glands: chief cells and oxyphil cells. The parathyroid gland monitors calcium levels in the blood. When calcium levels fall, the chief cells release parathyroid hormone, acting on the bone cells, osteoclasts, to increase activity, resulting in an increase of calcium in the blood.

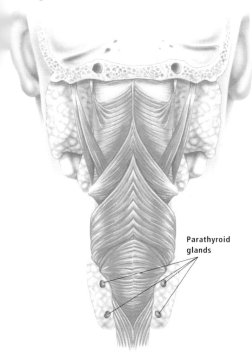

Parathyroid glands
The tiny parathyroid glands lie on the rear surface of the thyroid gland, with the outer capsule of the parathyroid gland forming a barrier between the cells of the two glands.

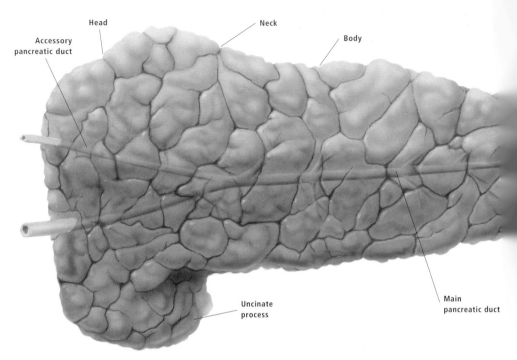

Head

Neck

Body

Accessory
pancreatic duct

Uncinate
process

Main
pancreatic duct

Pancreas

The pancreas is part of two systems, the
endocrine system and the digestive system.
While the majority of the pancreatic cells are
involved in producing enzymes for its digestive function, the pancreas plays an equally
important role in the endocrine system because it produces insulin and glucagon.

The pennant-shaped organ of the pancreas lies behind the stomach, with its head
lying in the curve of the duodenum and its
tail contacting the spleen.

The endocrine function of the pancreas
sees the production of its hormones by
select cells in the pancreatic structure,

grouped together in clusters, called the
islets of Langerhans. The cells of the islets
of Langerhans include alpha cells, beta
cells, and delta cells. Alpha cells produce
glucagon, which increases blood glucose
levels. Beta cells produce insulin. Release
of insulin into the bloodstream is effected
in response to increased blood sugar levels.
The delta cells secrete somatostatin, which
inhibits the release of both glucagon and
insulin. In this way, the different cells of
the islets of Langerhans work in harmony,
and are able to regulate blood sugar content
and keep levels stable.

Pancreas—endocrine function

The organs of the endocrine system are primarily
concerned with control of the body's metabolic
functions. The endocrine function of the pancreas
is to maintain stable blood sugar levels.

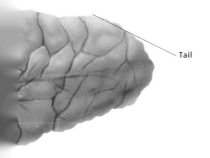

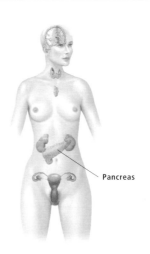

Tail

Pancreas

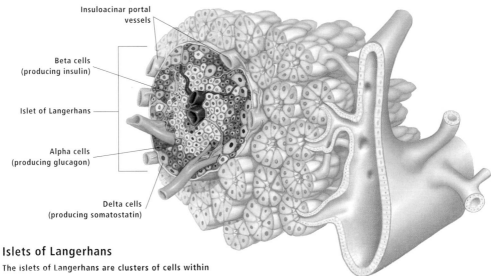

Insuloacinar portal
vessels

Beta cells
(producing insulin)

Islet of Langerhans

Alpha cells
(producing glucagon)

Delta cells
(producing somatostatin)

Islets of Langerhans

The islets of Langerhans are clusters of cells within
the pancreas. These islets are comprised of three
types of cells: alpha cells, beta cells, and delta cells.
Hormones produced by the alpha and beta cells raise
and lower blood sugar levels respectively, while the
delta cells produce somatostatin, which can inhibit
the release of both glucagon and insulin.

Adrenal Glands

The pyramidal adrenal glands lie on top of each kidney. Their small size belies their importance in the smooth operation of the body. The outer part of the adrenal gland, the cortex, surrounds the inner component of the medulla, with each part contributing different hormones required by the body. Hormone types produced in the adrenal cortex include glucocorticoids, mineralocorticoids, and sex steroids, while the adrenal medulla produces epinephrine and norepinephrine (adrenaline and noradrenaline).

The adrenal cortex is divided into three separate zones, with each zone producing the different types of cortical hormones. The outer zone, the zona glomerulosa, produces mineralocorticoids, with aldosterone being the principal mineralocorticoid. Mineralocorticoids, particularly aldosterone, influence the reabsorption of sodium in the kidneys, which in turn controls blood volume. If there is a decrease in blood flow or blood volume in the kidney, the kidney releases renin, which in turn acts on the hormone angiotensin, to increase aldosterone levels. This helps conserve sodium and hence to restore blood volume. The middle zone, zona fasciculata, produces

Adrenal gland

The adrenal glands secrete steroids, which are involved with glucose metabolism, keep the volume of blood at the correct level, and, in males, contribute to the development of sexual characteristics.

Left adrenal gland

Left adrenal cortex

Left adrenal medulla

Left suprarenal artery

Left kidney

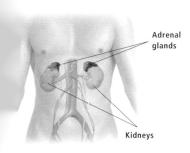

Adrenal glands

Kidneys

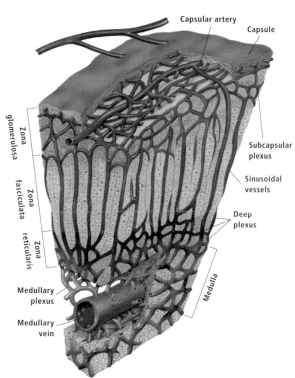

Capsular artery

Capsule

Zona glomerulosa

Zona fasciculata

Zona reticularis

Subcapsular plexus

Sinusoidal vessels

Deep plexus

Medulla

Medullary plexus

Medullary vein

Adrenal gland microstructure

Each zone of the adrenal cortex produces hormones essential to metabolic stability. The hormones epinephrine and norepinephrine are produced in the medulla.

glucocorticoids. Secretion of glucocorticoids is controlled by the pituitary gland, which secretes adrenocorticotropic (ACTH) hormone, which in turn stimulates the release of glucocorticoids. These glucocorticoids play a role in the metabolism of carbohydrates, fat, and protein, and are particularly involved in glucose metabolism. The inner zone, zona reticularis, is where the androgens are produced. Dehydroepiandrosterone is the principal androgen produced, which contributes to the development of male sexual characteristics.

The hormones produced in the adrenal medulla, epinephrine and norepinephrine (adrenaline and noradrenaline), are allied to the sympathetic division of the autonomic nervous system.

Released in response to stress or anxiety, these hormones aid in the body's response to emergency situations. When released into the bloodstream, the hormones act to increase heart rate, blood pressure, and flow of blood to muscles; they also initiate glucose release by the liver, providing the body with energy.

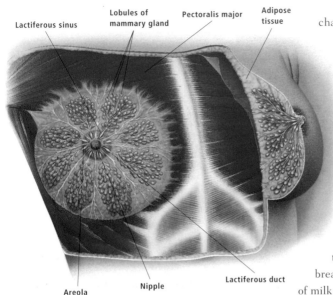

Lactiferous sinus

Lobules of mammary gland

Pectoralis major

Adipose tissue

Lactiferous duct

Areola

Nipple

Mammary glands

The mammary glands are stimulated into milk production and continued supply by hormones released by the pituitary gland, prolactin, and oxytocin.

Mammary Glands

The foundations of the mammary glands are laid down during fetal development, developing along the milk lines, which extend from the armpit (axilla) to the groin. The mammary glands are composed of fatty tissue and the glandular structures called lobules, which are connected by a network of ducts to openings at the nipple.

The female breasts develop during puberty, increasing in fat and fibrous and glandular tissue content; the nipple changes in pigmentation and is elevated during this development. The mammary glands are stimulated by hormones during pregnancy, which sees structural changes within the breast and a huge increase in the amount of glandular components. Stimulated by the pituitary gland, which releases prolactin (PRL), the breasts produce milk for breast-feeding, while the release of milk is prompted by another hormone released by the pituitary gland, oxytocin (OT). After weaning, most of the glandular tissue is once again replaced by fat and fibrous tissue.

Ovaries

The ovaries are responsible for the production of estrogen and progesterone. These two hormones play an important role in the female reproductive system, and undergo cyclic fluctuations during the 28-day menstrual cycle.

Follicle-stimulating hormone (FSH) and luteinizing hormone ((LH) released by the pituitary stimulates the production of these hormones. Both hormones are involved in creating a viable environment for a fertilized ovum, while estrogen is involved in the development of the breasts and reproductive organs, and progesterone is primarily responsible for preparing the lining of the uterus in readiness for a fertilized ovum.

Testes

The testes produces testosterone, the hormone responsible for the development of secondary male sexual characteristics, including facial and pubic hair, enlargement of the penis and testes, enlargement of the larynx and the resultant deepening of the voice, and increase in muscle bulk. The production of testosterone is influenced by follicle-stimulating hormone (FSH) and luteinizing hormone (LH) released by the pituitary gland.

Placenta

During pregnancy, the placenta acts as an endocrine organ, producing human chorionic gonadotrophin (HCG), estrogen, progesterone, growth hormone, relaxin, and placental prolactin.

Production of HCG encourages the corpus luteum in the ovary to continue provision of progesterone and estrogen, until such time as the placenta is able to take over the role—usually toward the end of the first trimester.

Placenta

Connecting mother and baby via the umbilical cord, the placenta transfers nutrients and oxygen to the developing fetus. It also acts as an endocrine organ during pregnancy, producing hormones such as human chorionic gonadotrophin, placental prolactin, relaxin, estrogen, and progesterone.

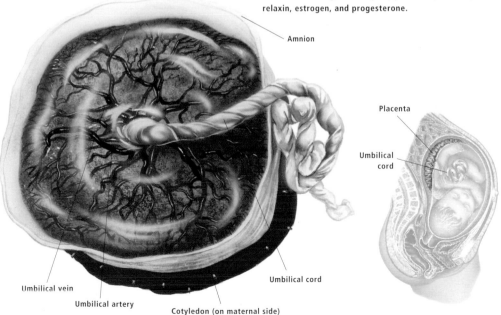

Amnion

Placenta

Umbilical cord

Umbilical vein

Umbilical artery

Cotyledon (on maternal side)

Umbilical cord

Diseases and Disorders of the Endocrine System

Hormonal imbalances can cause a range of problems. Too much or too little of a particular hormone can have far-reaching effects on the body. The body has a very efficient mechanism to maintain correct levels—negative feedback. By implementing this mechanism, the body organizes an appropriate response to an excess or deficiency of a particular hormone, by stimulating a counteraction or by inhibition or increase of hormone production. In this way, hormone levels are kept stable, and metabolic functions are kept operating efficiently.

However, if hormone imbalances are not corrected, problems can arise. Some of the more common problems associated with the endocrine system are discussed below.

Goiter

Commonly the result of a lack of iodine in the diet, goiter is caused by an imbalance of one of the hormones produced by the thyroid, thyroxine. An incorrect level of thyroxine can result in a goiter, which is an enlargement of the thyroid gland—associated with high, low, and sometimes even normal levels of thyroxine. The production of thyroxine is stimulated by thyroid-stimulating hormone (TSH), produced by the pituitary

Goiter

Goiter is enlargement of the thyroid gland. This enlargement can cause abnormal swelling in the front of the neck.

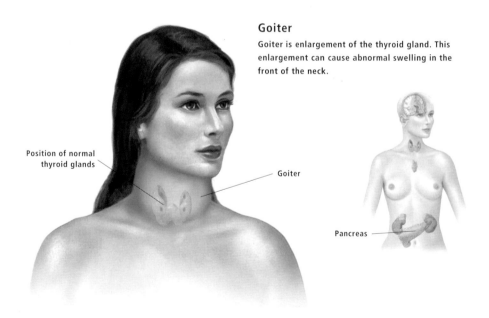

Position of normal thyroid glands

Goiter

Pancreas

gland. Overproduction of TSH can result in goiter, and problems can arise if the swelling of the thyroid causes interference with breathing or swallowing.

Graves' disease

The most common cause of hyperthyroidism—an overactive thyroid gland—Graves' disease is an autoimmune disease, whereby the body's own immune system attacks the thyroid gland. The overactivity of the thyroid gland causes problems such as rapid heartbeat, increased sweating, weakness, tremor, and weight loss despite increased appetite. Exophthalmos is a condition affecting the eyes. It causes protruding eyeballs due to fluid accumulation around the eye, which can often develop as a result of Graves' disease.

Treatment options include drugs to counteract the overactivity of the gland; in more severe cases, surgery to remove part of the thyroid gland may be necessary.

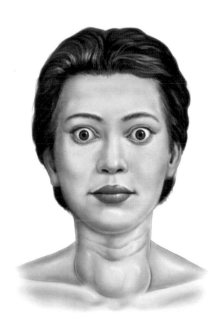

Graves' disease
One of the problems associated with Graves' disease is abnormal protrusion of the eyes (exophthalmos).

Hypoparathyroidism

Hormones released by the parathyroid gland help to regulate blood calcium levels. In persons suffering from hypoparathyroidism, the parathyroid gland fails to produce sufficient hormone, and in some cases, the absence of parathyroid glands means that no parathyroid hormone is produced for blood calcium regulation. Without parathyroid hormone to counteract falling blood calcium levels, conditions such as tetany can occur.

Hypoparathyroidism can be a congenital condition, and those born with the condition can suffer from dry skin, hair loss, and vulnerability to yeast (*Candida*) infections.

Treatment options include calcium and vitamin D supplements, which need to be taken for life.

Tumors of the islet cells

The islets cells of the pancreas can be a site for tumors. Tumors known as insulinomas can result in the beta cells producing too much insulin. The overproduction of insulin causes typical low blood sugar symptoms such as memory lapses, palpitations, sweating, unpredictable behavior, and unconsciousness.

Acute treatment usually involves drugs to restore normal insulin quantities. Definitive therapy involves surgical resection.

Addison's disease

Another autoimmune disease, where the body's immune system attacks its own tissue, Addison's disease is characterized by the destruction of the outer layer, the cortex, of the adrenal gland. This destruction in turn affects the adrenal gland's ability to produce hydrocortisone, which is vital to the body's metabolism.

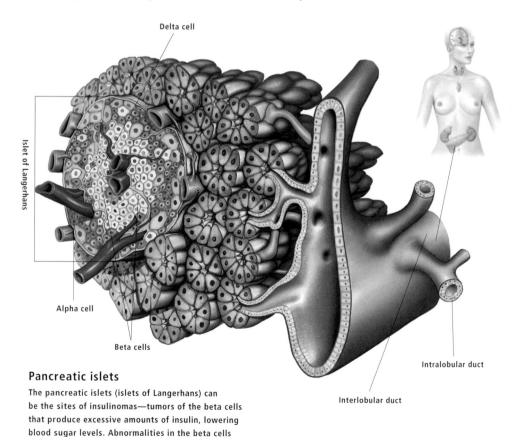

Delta cell

Islet of Langerhans

Alpha cell

Beta cells

Intralobular duct

Interlobular duct

Pancreatic islets

The pancreatic islets (islets of Langerhans) can be the sites of insulinomas—tumors of the beta cells that produce excessive amounts of insulin, lowering blood sugar levels. Abnormalities in the beta cells can also cause diabetes mellitus.

Symptoms include weight loss, loss of appetite, fatigue, and weakness, with the skin color often darkening.

Blood tests can confirm the presence of the condition, and treatment with drugs to replace the necessary hydrocortisone is extremely effective.

Cushing's syndrome

Excessive cortisol secretions, or prolonged use of cortisol-based drugs, can result in Cushing's syndrome. Indications of this condition include weight gain, a "moon" face, facial hair, muscle weakness, and high blood pressure. It can be caused by a tumor in the pituitary gland, which causes excess secretions of ACTH, in turn stimulating overproduction of cortisol by the adrenal gland. Testing of blood and urine samples can confirm the existence of the condition. Treatment options are varied, dependent on the exact nature and cause of the condition.

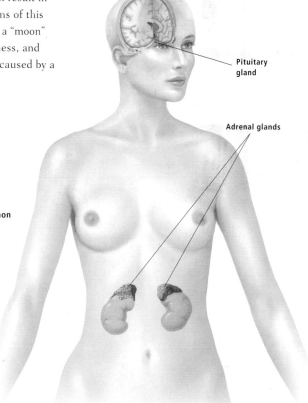

Pituitary gland

Adrenal glands

Cushing's syndrome

Weight gain and fatty deposits in the face and trunk areas are common symptoms associated with Cushing's syndrome, caused by excess cortisol in the body. This can be caused by overproduction, or prolonged use of cortisol drugs.

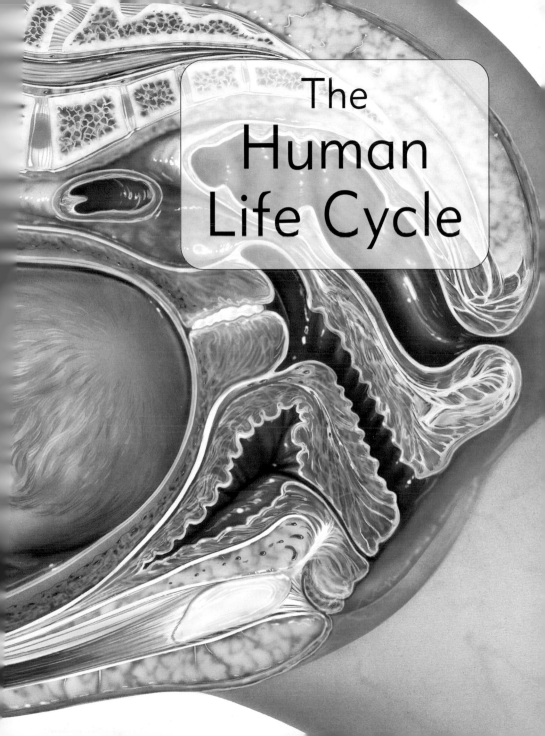

The
Human
Life Cycle

THE HUMAN LIFE CYCLE

Ageing

The human body goes through various stages of development, particularly during the formative years, progressing to adulthood and on through life to old age, and ultimately death.

While each individual is different, and there is no set time for many of the changes our bodies experience, there are definite periods of the human life cycle that have unique characteristics, signified by changes in outward appearance and in the function of body systems.

Following birth, the first stage of the human life cycle is infancy. More growth and development takes place during the first 12 months of life than at any other time. This first phase of life sees many achievements, with gains in weight and height, the eruption of the first teeth,

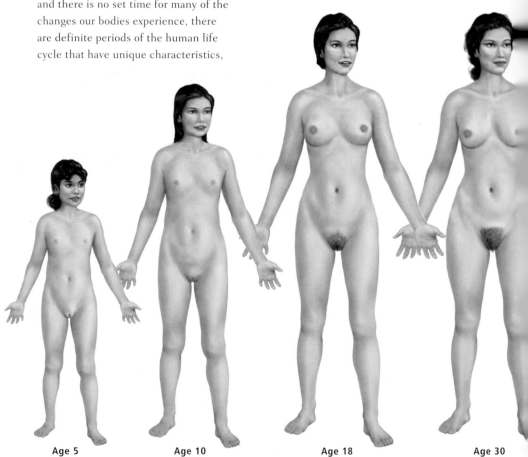

Age 5   Age 10   Age 18   Age 30

the first steps and first words, along with increased sensory discrimination.

Following infancy is the period of childhood, which lasts until the beginning of puberty. During these years, bones complete much of their growth, and the brain develops greatly during this time, allowing improved motor skills, and increased perception and understanding.

The onset of puberty signifies the beginning of the changes that will carry the individual into adulthood.

The male and female reproductive systems mature, and outward physical characteristics change markedly.

Males gradually develop facial and pubic hair, acquire the deep voice associated with manhood, and increase in muscle strength and bulk.

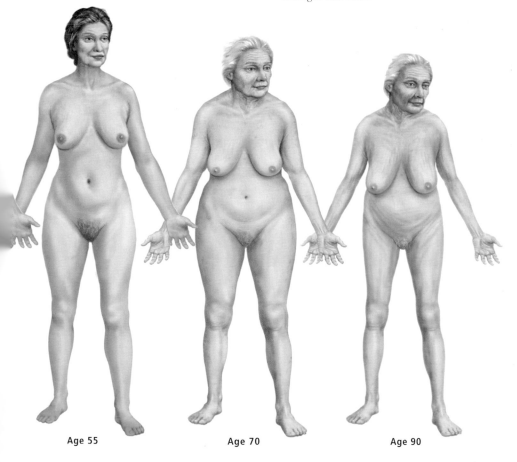

Age 55   **Age 70**   **Age 90**

Sperm production begins as the testes grow in size, and the penis increases in length and girth.

The changes for females include the commencement of menstruation (menarche), as the reproductive organs reach maturity. The breasts develop, the external genitalia mature, and pubic hair develops.

Both sexes experience growth spurts, with girls usually beginning earlier than boys. Acne can pose skin problems for both sexes, due to overactive sweat and sebaceous glands.

Adulthood is attained when full anatomical, physiological, and sexual maturity has been reached. From a legal standpoint, most Western societies have designated adulthood as being 18 or 21 years of age. Adulthood signifies the completion of bone growth and the full establishment of the body systems.

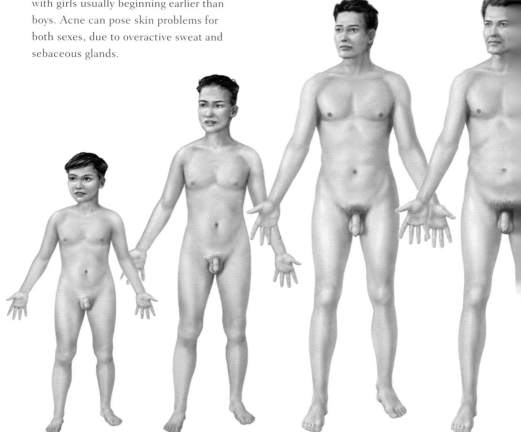

Age 5   Age 10   Age 18   Age 30

As we get older, the body systems begin to deteriorate in a very gradual process—marked by some specific changes along the way. Organs become less efficient, senses are less heightened, and the body's response to infection and disease is reduced as the immune system declines as we age. The male reproductive system declines in productivity, though continues to produce sperm in lesser quantities until old age. The female reproductive system slows down from the mid-thirties, until menopause in the 5th or 6th decade, when menstruation eventually ceases, bringing with it physical changes in outward appearance.

As we enter the senior years, the body systems are gradually less effective, and brain function is affected, often resulting in poor memory capabilities.

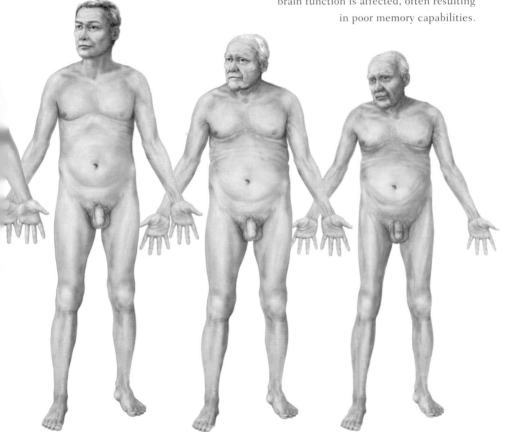

Age 55   **Age 70**   **Age 90**

Fertility

Fertility—the ability to conceive a child—is governed by several factors in the function of both the male and female reproductive systems.

In males, the number of active sperm in semen must be of sufficient quantity to complete the journey to the ovum. Sperm can survive for about 5 days in the female reproductive tract.

In women, the production of viable eggs and regular ovulation is important, as is a favorable environment in the vagina, cervix, and uterus to facilitate the movement of the sperm along the female reproductive tract.

A woman is fertile for about 36 hours in every menstrual cycle.

Fertilization

Penetration of a female ovum by a male sperm is the final event in fertilization, and the beginning of a new life.

Conditions must be right for fertilization to take place. The ovary releases the ovum around day 14 of the menstrual cycle. This ovum has two protective coatings. Immediately around the ovum is the zona pellucida, a thin, noncellular membrane, and around this are some of the cells from the outer wall of the ovarian follicle.

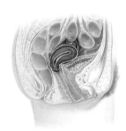

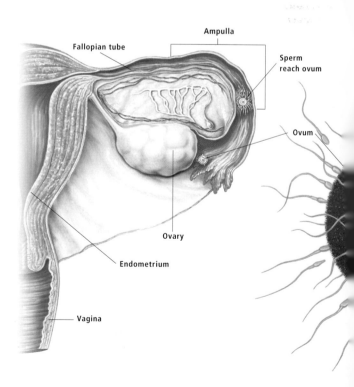

Ampulla

Fallopian tube

Sperm reach ovum

Ovum

Ovary

Endometrium

Vagina

Female fertility

Female fertility is dependent on egg production and regular ovulation, coupled with a favorable uterine environment. Once an ovum has been released it travels along the fallopian tube. When it reaches the region known as the ampulla, it is primed for fertilization.

Between 60 million and 100 million sperm are contained in each ejaculation deposited into the vagina. Only around 200 of these reach the ampulla of the fallopian tube where the ovum is located. Of the remaining 200 sperm, only one will fertilize the ovum. The acrosomal membrane surrounding the head of the sperm aids the breakdown of the outer layers of the ovum, until the ovum is finally penetrated by one of the sperm. The fusion of the sperm and the ovum forms a zygote.

Male fertility

Male fertility is dependent on the number of viable sperm produced. Active motile sperm are required in large quantities, as survival rates on the journey to the ovum are low, with only a few hundred of the original 60–100 million surviving the journey. Of these, only one will successfully penetrate the ovum.

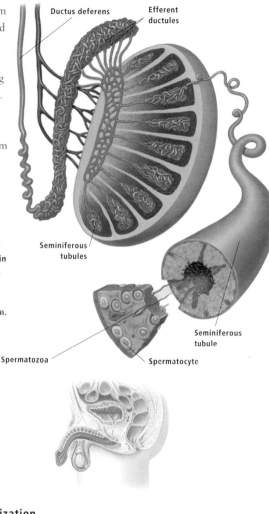

Ductus deferens

Efferent ductules

Seminiferous tubules

Seminiferous tubule

Spermatozoa

Spermatocyte

Zona pellucida

Fertilization

Optimal conditions for fertilization include a healthy ovum and a plentiful supply of active sperm. When the two meet, and the protective coating of the ovum is eventually broken down by one successful sperm, the fusion of these two cells results in a zygote.

Pregnancy

The fusion of a male sperm cell and a female egg cell, forming a zygote, signals the beginning of a pregnancy. Pregnancy is the state of having a developing fetus in the uterus, and is usually calculated either from the first day of the last menstrual period, amounting to 280 days (10 lunar months), or from conception, amounting to 266 days, although only a small percentage of babies are born on the estimated due date.

Pregnancy is usually confirmed by blood or urine tests. The progress of pregnancy is marked by three trimesters.

The first trimester sees various changes including enlarged and tender breasts, enlarged nipples, nausea—often referred to as "morning sickness," but which can occur at any time—increased frequency of urination, tiredness, moodiness, and skin changes.

Many of these symptoms have usually passed by the second trimester, and it is during this trimester that the mother becomes aware of fetal movement in the womb.

In the last trimester as the due birth date approaches, various conditions can arise. These include Braxton-Hicks contractions, which are false contractions by the uterus that emulate the contractions of childbirth. It is during this final trimester that the mother can suffer from various symptoms of discomfort.

Many of the symptoms, such as backache, varicose veins, and fluid retention can be alleviated

Placenta

Small intestine

L2   L3   L4

Pregnancy

The course of pregnancy is broken up into three sections, known as trimesters. During this time, the fetus develops from a single-celled zygote into a fully developed fetus.

with rest. In the case of hemorrhoids, preparatory creams may be prescribed.

Stretch marks often occur, due to stretching and rupturing of the elastic fibers of the skin; very little can be done to prevent this, though in time they will usually fade.

Antenatal care, including regular check-ups by qualified health care providers and information about a healthy lifestyle, has reduced many of the problems that women previously encountered during pregnancy. With modern obstetrics, the risks to mother and baby have been greatly reduced.

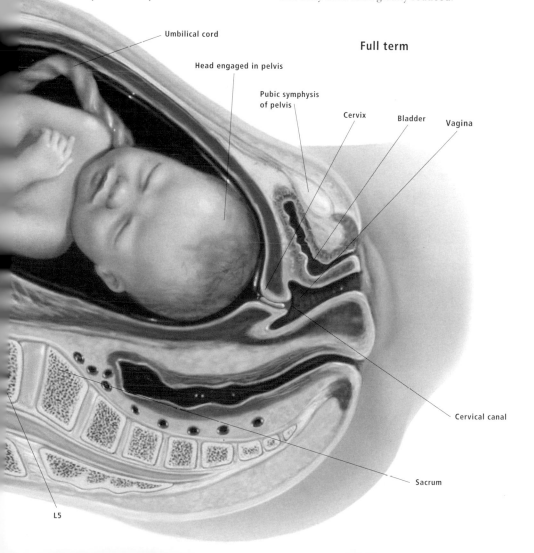

Full term

Umbilical cord

Head engaged in pelvis

Pubic symphysis of pelvis

Cervix

Bladder

Vagina

Cervical canal

Sacrum

L5

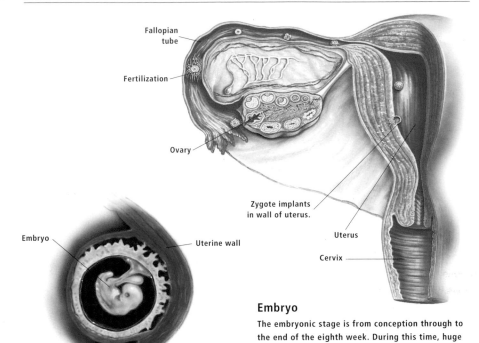

Fallopian tube

Fertilization

Ovary

Zygote implants in wall of uterus.

Uterus

Cervix

Embryo

Uterine wall

5 weeks

Embryo

The embryonic stage is from conception through to the end of the eighth week. During this time, huge developments take place, transforming the mass of cells of the blastocyst into the embryo with distinct human characteristics.

Embryo

The female ovum that has been fertilized by a male sperm cell is known as a zygote. This zygote immediately begins to divide (mitosis) and multiply, and reaches the uterus some 4–5 days after conception. The mass of cells, now known as a blastocyst, develops a cavity; the inner mass of cells on one side of the cavity will become the embryo and the embryonic membranes, while the outer cells will become the placenta. The blastocyst implants in the wall of the uterus, most commonly in the upper posterior wall, around day 7.

After implantation, the cells are rearranged once more. The single-celled zygote of conception will become the embryo, and over the next 8 weeks the foundations of the body systems, organs, limbs, and facial features will be laid down, and the amnion, a fluid-filled sac, will form around the developing embryo.

Placenta

About 2 weeks after implantation, the placenta begins to develop. The outer cells of the blastocyst merge into the surface of the endometrium, with the endometrium healing over the blastocyst. Those cells of the blastocyst, which will become the placenta, create a folded surface, which invades further into the endometrial lining of the uterus, resulting in maternal blood filling the cavernous areas of the folds. Eventually a circulatory supply is set up that feeds maternal blood through arteries into the spaces, drained by a similar venous system. Meanwhile, running through the folds in the placenta are the fetal blood vessels that connect to the fetus via the umbilical cord. In this way, the maternal and fetal circulations do not come into direct contact, but by diffusion through the capillaries, nutrients and oxygen from the maternal blood are delivered to the fetus via the placenta, and waste products from the fetus are returned to the maternal circulation through the placenta.

Umbilical Cord

Containing two umbilical arteries and one umbilical vein, the umbilical cord unites the fetus to the placenta. Variable in length, the diameter of the umbilical cord is usually around $\frac{1}{5}$–1 inch (1–2 centimeters). Through the blood vessels of the umbilical cord, the fetus receives nutrients and oxygen from the placenta, and returns deoxygenated blood back to the placenta.

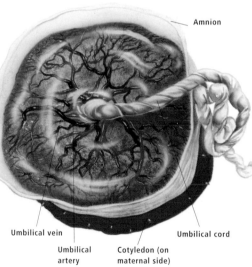

Amnion

Umbilical cord   Umbilical arteries

Umbilical
vein

Area filled with
maternal blood

Syncytial trophoblast

Chorionic
villi

Placenta

Endometrium

Maternal blood vessels

Myometrium

Umbilical vein

Umbilical
artery

Cotyledon (on
maternal side)

Umbilical cord

Placenta

Umbilical cord

Placenta

The placenta will provide nutrients and oxygen to the developing fetus for its entire time in the womb. The placenta acts as an intermediary between maternal and fetal blood, without allowing the two circulations to come into direct contact.

Fetal development cycle

Fetal development is a gradual process, but certain events take place at particular times. At 12 weeks, the facial features begin to form.

5 weeks

11 weeks

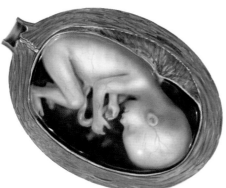

20 weeks

Fetal Development

The embryo is known as a fetus from the 8th week onward. During the embryonic stage, the foundations of all the major organs, systems, limbs, and facial features have been laid down, ready for further development. The cartilage, which will ultimately ossify and become bone, is also laid in place during the embryonic stage, and by birth, much of the ossification has commenced.

Also during the embryonic stage, the heart and blood vessels are established, with a discernible heartbeat able to be detected using ultrasound, by around week 5.

During the second month, the sex glands begin to develop. These are determined by the initial fusion of the sperm and egg cells, and depending on whether an X chromosome- or Y chromosome-carrying sperm cell fused with the ovum, will determine the sex of the fetus.

Merging with an X-chromosome ovum, the possibilities are XX, in which case a girl will result, or XY, in which case a boy will result. So, during the second

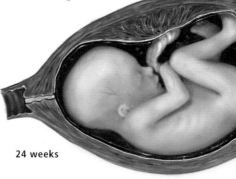

24 weeks

month, the predetermined sex organs and external genitalia begin to develop.

At 12 weeks, the external genitalia will differentiate, the facial features begin to take form, nails begin to grow on the fingers and toes, the ears and eyelids form (though the eyelids remain closed until around the 6th month), and the formation of the buds of the permanent teeth takes place.

During the 4th month, the mother can feel fetal movement, as the fetus responds to noise and sudden movement.

During the 5th and 6th months, the body grows larger, and the fetus develops a covering of fine hairs, called lanugo; this downy covering is shed prior to birth. Prior to this time, the head has been disproportionately large compared to the rest of the body to accommodate the growth of the brain.

The 7th month sees an increase in weight and the gradual maturation of the major organs, in readiness for birth. Vernix, a sebum-based secretion produced by the glands in the skin, provides a protective layer over the delicate fetal skin.

At full term (40 weeks), the fetus weighs 6–9 pounds (2.5–4 kilograms), and is ready to be born.

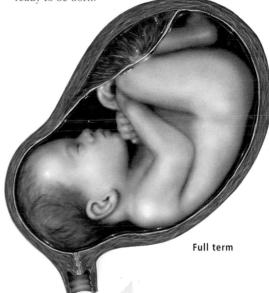

Full term

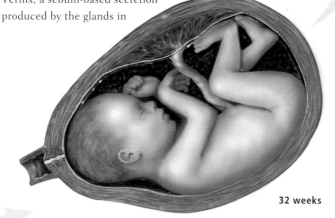

32 weeks

Labor and Childbirth

At around 40 weeks, the baby is ready to be born, and this readiness can be indicated by various signs. The waters can break, signaling that the amniotic sac has ruptured; the mucus plug at the cervix can be expelled, known as the show, though this can happen in the weeks leading up to actual birth; and contractions can commence. Three stages are involved in labor—actual labor, birth, and expulsion of the placenta.

During the first stage, the cervix begins to dilate, and hormones are produced to stimulate contractions of the uterus, pushing the baby's head down on the cervix and encouraging further dilation. The cervix must be fully dilated to allow the passage of the baby down the birth canal.

The second stage of labor, which is considerably shorter than the first, culminates in the birth of the baby. At the beginning of the second stage, the urge to bear down increases, and the mother usually gives four or five pushes to move the baby along the birth canal, resulting in the baby's head presenting at the opening to the vagina; this is known as the "crowning." Once this has occurred, the head passes beneath the pelvic bone, followed by the shoulders, and the baby is then born.

A surge of the hormone oxytocin, which stimulates the placenta to separate away from the uterus, triggers the third stage, the expulsion of the placenta.

Oxytocin is also responsible for stimulating the release of breast milk, although actual milk production is governed by the hormone prolactin. As the uterus continues to contract, the placenta separates from the uterus, sliding naturally along the birth canal, and is expelled.

Childbirth

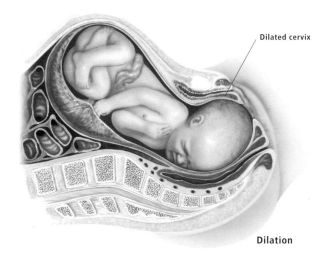

Dilated cervix

Dilation

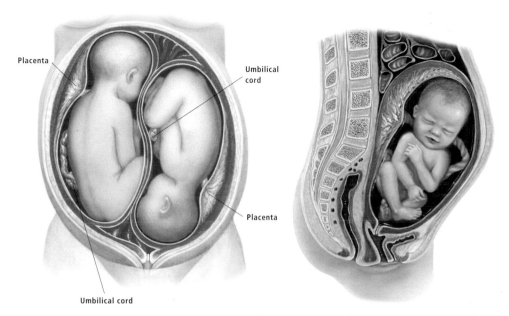

Placenta

Umbilical
cord

Placenta

Umbilical cord

Twins

Fraternal twins are produced by the fertilization of
two eggs; identical twins result when a single egg
divides in two.

Breech birth

Six percent of babies are in breech position,
whereby the fetus faces the mother's pubis, and
the legs and buttocks enter the birth canal first.

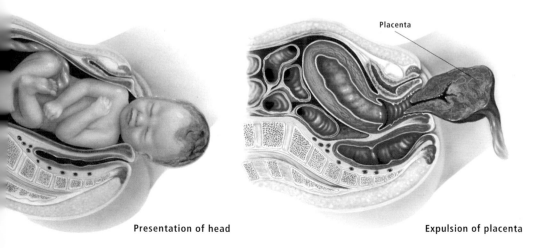

Placenta

Presentation of head

Expulsion of placenta

4 months

Infancy

Infants develop at their own pace, but there is a certain order of developmental changes. For example, babies must master the muscular skill of head control before they can sit up.

8 months

12 months

Infancy

The first 12 months of life are known as infancy. During this time, rapid growth and development can be observed, as baby's senses heighten, weight is gained, teething begins, the first steps are often taken, and the first words are spoken. While there is no fixed time for any of these formative milestones, there is an order of development, and each baby will develop and grow at its own pace.

As a general rule, the following developments can be expected during the first year. After birth, baby's first movements are reflex movements, such as grasping, sucking, and kicking. Within the first 3 months, sensory awareness is heightened, and infants are able to distinguish color and form sounds.

Some coordination is established by the 4th month, with baby able to grasp objects within reach. Muscular control of the head allows baby to sit unsupported for short periods of time.

Between the 5th and 12th month, the first tooth will appear, usually one of the lower incisors. Crawling usually begins around 7 months. By 10 months baby can sit unsupported for longer periods of time, by 12 months many babies can stand and some can walk by this time too. At 12 months of age, many babies have uttered their first word. The soft spots in the skull, the fontanelles, have usually closed over by 12 months of age.

During the first year, babies will generally triple their birth weight and increase in height by 50 percent. Some babies will walk earlier than others, others will talk sooner than others—each is an individual and develops at their own rate.

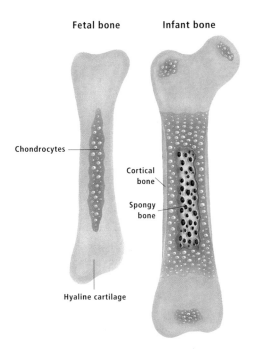

Fetal bone   **Infant bone**

Chondrocytes

Cortical bone

Spongy bone

Hyaline cartilage

Bone development

Bone growth begins very early in fetal development. At 6 weeks, chondrocytes in the center of the cartilage begin to enlarge—the first stage in bone development. By the time the baby is born, bone has been laid down in the center of the shaft and in a collar around the middle of the shaft.

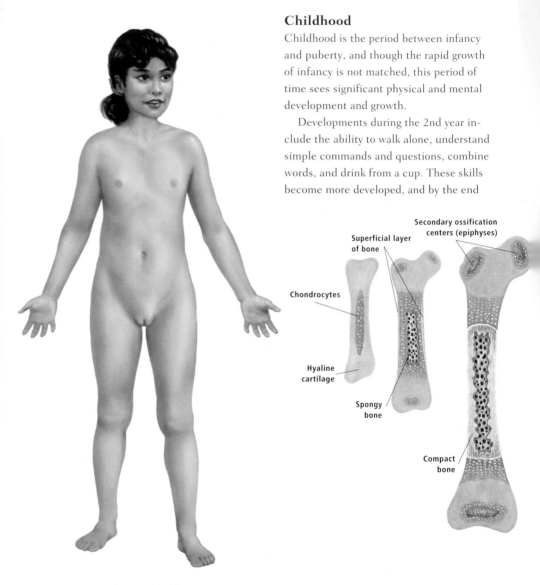

Childhood

Childhood is the period between infancy and puberty, and though the rapid growth of infancy is not matched, this period of time sees significant physical and mental development and growth.

Developments during the 2nd year include the ability to walk alone, understand simple commands and questions, combine words, and drink from a cup. These skills become more developed, and by the end

Secondary ossification centers (epiphyses)

Superficial layer of bone

Chondrocytes

Hyaline cartilage

Spongy bone

Compact bone

5-year-old girl

of the 3rd or 4th year, the child is able to walk, jump, and run, and use simple sentences. The vocabulary slowly builds, and by age 6, the child has a concept of grammar, and knows how to think, remember, recognize, and perceive. Also by age 6, the primary teeth begin to be replaced by the permanent teeth.

Growth occurs steadily, though interspersed with growth spurts, again to an individual timetable. Long bone growth is often not completed until early adulthood.

Spongy bone

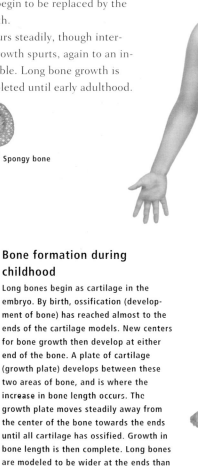

Bone formation during childhood

Long bones begin as cartilage in the embryo. By birth, ossification (development of bone) has reached almost to the ends of the cartilage models. New centers for bone growth then develop at either end of the bone. A plate of cartilage (growth plate) develops between these two areas of bone, and is where the increase in bone length occurs. The growth plate moves steadily away from the center of the bone towards the ends until all cartilage has ossified. Growth in bone length is then complete. Long bones are modeled to be wider at the ends than the middle, providing extra support at the joints where it is needed.

5-year-old boy

Puberty

Puberty signifies the change from childhood to adulthood, when physical and hormonal changes take place. Puberty can encompass various periods of time, depending on the individual, and can take between 2–6 years, with females generally entering and completing puberty before males. The changes experienced during male and female puberty are discussed below.

Male puberty

Male puberty is a time of great change, though most of the changes are gradual.

It generally commences around 13 years of age and continues through to around 18 years, although height can continue to increase, with continued long bone growth, into the early twenties.

During puberty, the reproductive organs gradually reach maturity and the secondary sexual characteristics develop. At this time, rapid growth of the testes and penis takes place. The larynx also experiences rapid growth, with this growth elongating the

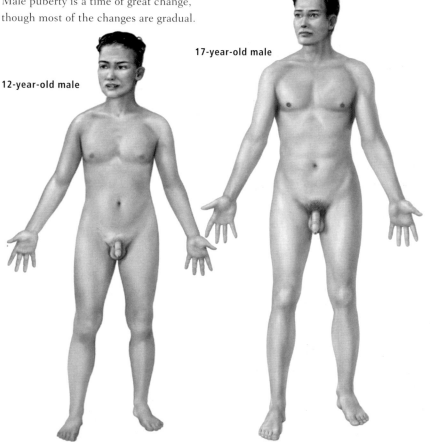

17-year-old male

12-year-old male

Male puberty

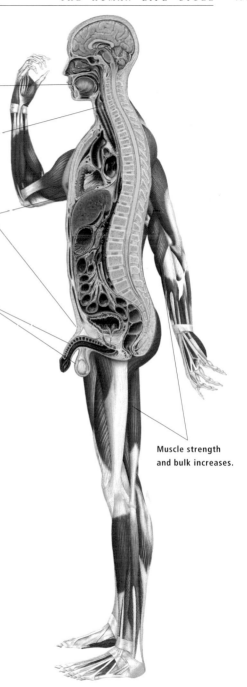

Appearance of facial hair.

Larynx grows rapidly, elongating vocal ligaments. Voice deepens in pitch. Adam's apple becomes more prominent.

Appearance of underarm and pubic hair.

Penis and testes increase in size. Sperm production begins, stimulated by the male hormone testosterone.

Muscle strength and bulk increases.

vocal ligaments, causing the voice to deepen in pitch, and creating the protuberance of the Adam's apple. Facial, pubic, and underarm hair begin to appear, and there is increased glandular activity in these areas. Generally there is an overall increase in body hair, muscle bulk, and height.

The hormone testosterone initiates sperm production in the testes, and usually about a year or so after the penis begins enlarging, the first ejaculation occurs.

The sebaceous glands of the skin tend to increase their production of sebum, which can cause acne.

The hormonal changes experienced by the body can have an effect on emotional and behavioral states.

Female puberty

Female puberty involves more physical changes than that of the male. Females also usually begin puberty earlier than males, with puberty generally commencing around age 11, and continuing through to around age 16, though this is a very general time frame. Much depends on inherited tendencies, nutrition, and environment.

Many of the changes experienced during female puberty are triggered by the sex hormones estrogen and progesterone. Development of the breasts is triggered by the estrogen, which promotes the laying down of fat deposits in the breasts. The shape and size of breasts is usually determined by inherited factors.

Underarm and pubic hair begins to appear and the hips widen. With the maturation of the reproductive organs, the

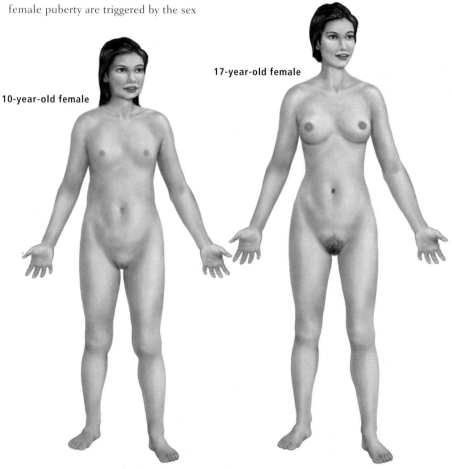

17-year-old female

10-year-old female

Female puberty

Initiated by hormones, fatty deposits accumulate in the breasts.

Underarm and pubic hair appears.

menstrual cycle begins, with the first menstruation called the menarche. Menstruation usually commences about 2 years after the onset of puberty, and occurs somewhat irregularly before establishing a regular pattern.

Growth spurts in girls usually occur around 10 to 11 years of age, with many girls being taller than boys of the same age, although this pattern is reversed over the next few years, when boys begin to commence their growth spurt.

Increased glandular activity occurs during puberty. The apocrine glands, located in the underarms, genitals, anus, and breasts, become more active, as do the sebaceous glands in the skin, causing the common problem experienced by many teenagers—acne.

Hips widen.

Uterus is now fully developed. Ovaries begin to produce sex hormones. Egg production and menstruation begin.

Body shapes

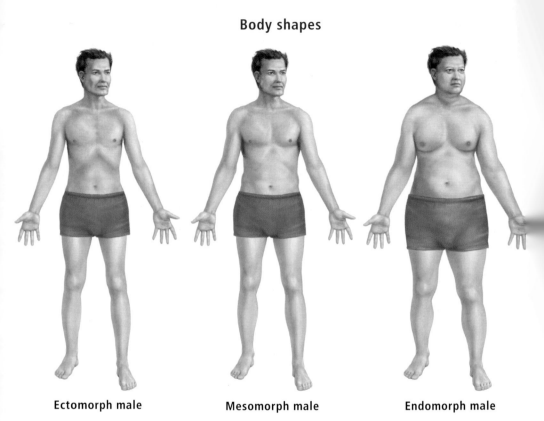

Ectomorph male   **Mesomorph male**   **Endomorph male**

Adulthood and Middle Age

On reaching adulthood, all of our body systems are fully matured, and our physical growth is complete.

Our final body shape is often determined by inherited factors, falling into one of three body shape categories (somatotypes): ectomorphic, mesomorphic, and endomorphic. Each somatotype has unique characteristics: ectomorph types are lean and low in fat, and angular and delicate in appearance; mesomorph types are evenly proportioned, with a strong, muscular frame; and endomorph types are more rounded, with a propensity for fat storage.

As we age, our bodies still have to perform repairs, fight disease and infection, and maintain a fully functioning operation, but as we get older, this does not happen with such rapid response or such effectiveness as when we are younger.

Menopause

The advent of middle age usually signals more physical and hormonal changes; this is particularly true for females, who go through the climacteric period, or menopause. As with many other body changes, there is no fixed time for menopause to occur, but it more commonly takes place between the mid-40s and mid-50s.

The climacteric period signals the end of the reproductive period of life, as the ovaries cease egg production, and as a result, production of the sex hormones estrogen and progesterone also diminishes or ceases. One of the first symptoms of menopause is irregular menstrual periods, often coupled with heavier or lighter bleeding than normal. This irregular cycle can persist for several years before the final menstrual period.

Some of the more common symptoms that might be experienced during menopause include: night sweats, hot flashes, anxiety and depression, palpitations, headaches, sleep problems (often due to hot flashes), vaginal dryness, and diminished or loss of interest in sexual intercourse.

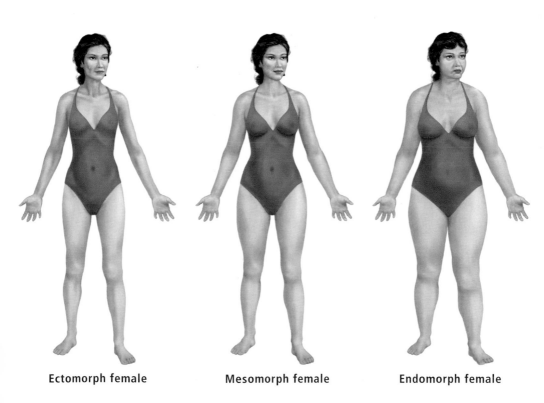

Ectomorph female   **Mesomorph female**   **Endomorph female**

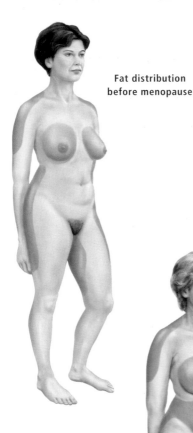

**Fat distribution
before menopause**

**Fat distribution
after menopause**

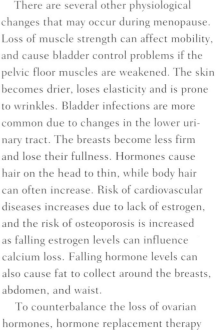

There are several other physiological changes that may occur during menopause. Loss of muscle strength can affect mobility, and cause bladder control problems if the pelvic floor muscles are weakened. The skin becomes drier, loses elasticity and is prone to wrinkles. Bladder infections are more common due to changes in the lower urinary tract. The breasts become less firm and lose their fullness. Hormones cause hair on the head to thin, while body hair can often increase. Risk of cardiovascular diseases increases due to lack of estrogen, and the risk of osteoporosis is increased as falling estrogen levels can influence calcium loss. Falling hormone levels can also cause fat to collect around the breasts, abdomen, and waist.

To counterbalance the loss of ovarian hormones, hormone replacement therapy (HRT) is an option for some women.

Technically, there is no official male menopause condition, though there are some medical authorities who consider there is. Certainly there are hormonal, and often physical changes that occur, though perhaps they are less radical than that of the female, since the reproductive organs continue to operate and produce sperm, albeit at reduced levels. Hormone levels can affect hair regeneration in men over 50, with hair replacement unable to keep pace with hair loss, resulting in baldness. Prostate enlargement becomes increasingly more common in men over 50. Weight may increase, and the skin and muscles become softer.

During a hot flash

Pores open and sweat emerges.

Hairs flatten.

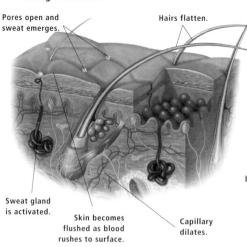

Sweat gland is activated.

Skin becomes flushed as blood rushes to surface.

Capillary dilates.

Hot flash

Hormone imbalances during menopause cause the capillaries of the skin to dilate, sending a sudden flood of warm blood to the surface. After the flash has passed, the capillaries rapidly constrict, the skin becomes pale and cold as blood drains away, and hairs rise to provide an insulating layer.

After a hot flash

Hairs rise for insulation.

Skin becomes pale.

Pores close.

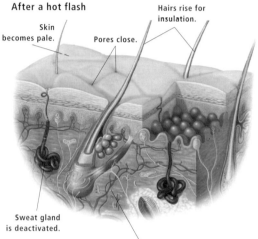

Sweat gland is deactivated.

Capillary contracts and blood drains from skin.

Male pattern baldness

There are several forms of baldness, with some diseases and conditions causing temporary hair loss. Male pattern baldness, however, is typified by a characteristic loss of hair from the temple area first, creating a receding hairline, and often leaving hair growth at the sides of the head only. Hair loss is often age related, and allied to hormonal changes in the body as hormone production declines.

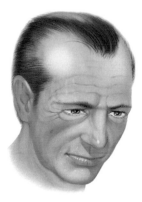

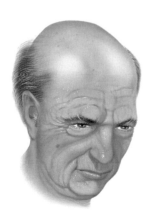

Senescence

It is an inevitable fact that we will age, and with the ageing process come physical and sometimes mental changes. There is no way to delay or halt the ageing process, but a healthy lifestyle, exercise, and good nutrition can help to maintain good health and physical abilities well into old age.

However, as a general rule, there is a definite decline in the efficiency of most of the systems; major organs become less efficient and mobility can be affected.

The senses often decline as we get older. Sight and hearing deteriorate, with visual impairment affecting many basic tasks, such as reading and driving. Cataracts of the eyes are common, although these are relatively simple to treat. Hearing, too, is often affected, with difficulty detecting higher frequencies.

Senescence

As a general rule, people are living longer, and the ageing population makes up a large proportion of the populations of the industrialized countries. While there are often many health problems associated with ageing, an active lifestyle and healthy diet can contribute to good health and well-being.

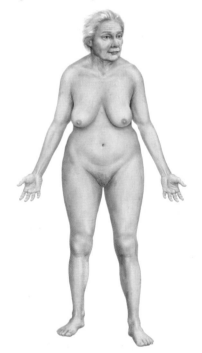

70-year-old woman

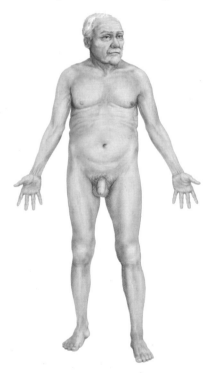

70-year-old man

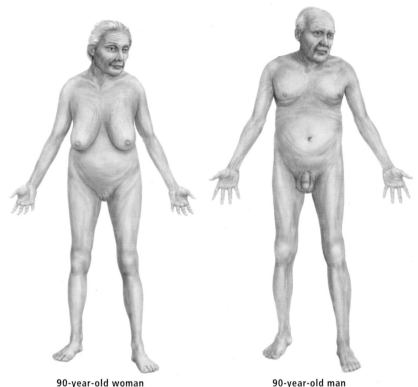

90-year-old woman   **90-year-old man**

The skin loses its elasticity, causing wrinkling. Reduced melanocyte production results in graying hair and paler skin. Loss of bone density can cause osteoporosis, which can cause further complications such as increased risk of fractures, and general loss of height as bone structures in the spine partially collapse.

Circulatory problems can increase with age, as blood flow to the major organs decreases, affecting both their efficiency and operation.

Lowered disease resistance, due to a less effective immune system, renders the older person more vulnerable to disease and infections, and this lowered resistance can provide an ideal environment for more aggressive diseases.

Mental impairment affects many elderly persons. Conditions such as Alzheimer's disease and stroke result in reduced mental ability. Other diseases common in the elderly include fracture, heart attack, and incontinence.

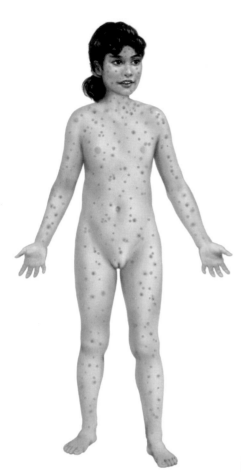

Childhood Diseases

There are a variety of diseases that are associated with the childhood years, although many previously dangerous or life-threatening diseases have now been all but eradicated due to widespread vaccination.

Some of the more common childhood diseases are discussed below.

Chickenpox

Highly contagious, chickenpox is an airborne viral disease. Although it can occur at any age, it is more common in school-age children. Chickenpox is also known as varicella.

The incubation period for the disease is around 13–17 days from contact, with a rash developing firstly on the trunk and then spreading to the rest of the body. The rash begins as small red bumps that can be extremely itchy. Headache and flu-like symptoms may precede the rash.

The bumps then develop into blisters that eventually scab over. Since this is a

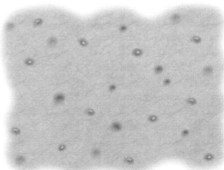

Chickenpox

Indications of chickenpox include a red itchy rash that develops blisters before crusting over. The rash first develops on the trunk and then spreads to the rest of the body. Other symptoms include headache and mild fever or cold-like symptoms.

particularly irritating and itchy disease, efforts should be made to reduce scratching, as this can result in scarring. Soothing oatmeal baths or the use of calamine lotion can reduce the itchiness.

While one exposure usually results in resistance to further attack, the virus can remain dormant in the system, causing shingles at a later date.

Measles

Measles is another viral disease, usually associated with childhood, although it can attack later in life. The incubation period is 10–14 days, after which time a fever develops, along with flu-like symptoms such as a dry cough, nasal discharge, and watery eyes.

Before the measles rash breaks out over the body, small white spots may appear in the mouth. The body rash begins on the face and gradually spreads down the body, leaving enlarged lymph glands and inflamed eyes in its wake. Eventually the rash fades, though it may leave brown marks.

Complications associated with measles include middle ear infection, bronchitis, and pneumonia. More serious complications include encephalitis and meningitis.

Little can be done to treat measles beyond alleviation of the symptoms of fever, though the more serious complications will need medical attention.

Immunization against measles is now available, and attacks of this virus are now far less common.

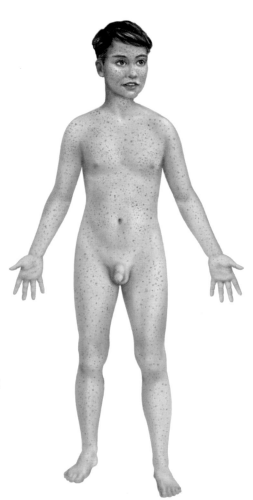

Measles

Measles begins with a rash on the face and behind the ears, which gradually spreads over the entire body. Measles is usually accompanied by flu-like symptoms, headache, and fever.

Rubella

Rubella, or German measles, is a highly infectious viral infection. Relatively mild in nature, rubella has an incubation period of 2–3 weeks, after which a light rash appears, usually beginning on the face and spreading to the rest of the body. Symptoms include irritability and mild fever. German measles usually only lasts for a few days, although complications can arise in a small number of cases, where patients can develop encephalopathy, causing inflammation and damage to the brain.

Immunization against rubella is now available, usually administered in conjunction with measles and mumps vaccines.

Rubella can cause problems for pregnant women, should they become infected, and it is advisable to check rubella antibody presence 3 months before starting a pregnancy, at which time a vaccination can be administered if rubella antibodies are not present.

Mumps

Mumps is a viral infection commonly affecting the large salivary glands, the parotid glands. The incubation period is 2–3 weeks, and an infected person is contagious for a day or two before swelling occurs and for 3–4 days thereafter. Sometimes only one gland is affected, and sometimes all of the salivary glands become swollen.

While mumps is usually a childhood disease, it can also affect adults, and is generally more severe in their case. Adult males can develop swelling of the testes, which can occasionally cause sterility. Adult females can suffer painful swelling of the ovaries.

Less common, but more serious complication include encephalitis, inflammation of the kidney, pancreas, or thyroid, thrombocytopenia, and deafness.

Outbreaks of mumps are far less common nowadays as a vaccination is available against the disease.

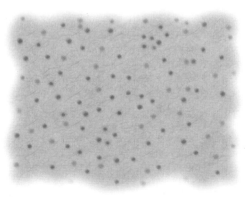

Rubella

A relatively mild disease, rubella is spread by droplets. Following droplet inhalation, the disease incubates for 2–3 weeks, before producing a mild rash, sometimes accompanied by swelling of the lymph glands of the head and neck.

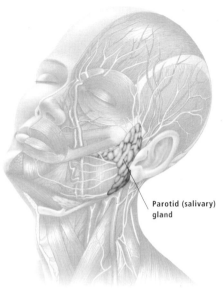

Parotid (salivary) gland

Mumps

Mumps is a viral infection spread by droplets. Inhaled droplets incubate for several weeks before producing the characteristics swelling of the salivary glands. Adult cases are often more severe, producing associated symptoms such as swelling of the testes in men, and swelling of the ovaries in women.

Whooping cough

Whooping cough (pertussis) is a highly infectious respiratory tract infection, which can lead to even more serious complications, such as pneumonia.

Initial symptoms are cold-like—runny nose, slight fever, and cough. The cough becomes persistent, accompanied by the characteristic "whoop."

The incubation period for whooping cough is 1–2 weeks following exposure.

The illness can last for several months, and during the first month, the sufferer is infectious. Full recovery also can take several months.

Although far more common in young children, it can also be contracted by older children and adults, although the symptoms are often milder.

Immunization against whooping cough, in conjunction with tetanus and diphtheria, is available.

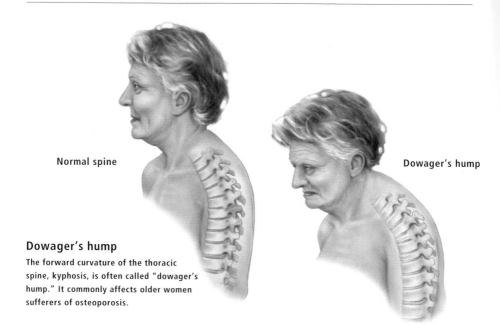

Normal spine

Dowager's hump

Dowager's hump
The forward curvature of the thoracic spine, kyphosis, is often called "dowager's hump." It commonly affects older women sufferers of osteoporosis.

Geriatric Diseases and Disorders

Age brings with it many diseases and disorders most often brought about by the declining efficiency of the body systems and major organs. All of the systems can be affected to some degree, each condition bringing with it its own set of problems, symptoms, and treatments. Several of the more common problems encountered by the elderly are discussed below.

Kyphosis

Advanced cases of osteoporosis can manifest as kyphosis ("dowager's hump"). The progressive loss of bone density can cause one or more of the vertebrae to compress, causing the forward curvature of the spine to be exaggerated, creating the stooped posture of Dowager's hump. This condition is seen almost exclusively in women, with osteoporosis itself more common in women due to falling estrogen levels following menopause.

Macular degeneration

A relatively common problem in the elderly, macular degeneration involves the breakdown of cells in the eye that are responsible for perception of fine detail. As peripheral vision remains, the disease often goes unnoticed by the sufferer until it is well advanced.

Early detection sometimes allows the use of laser treatment to slow the progress of the disease. Once the disease is well established, there is no treatment available.

Macular degeneration is one of the leading causes of blindness in the industrialized countries of the world.

Parkinson's disease

Generally occurring after age 50, Parkinson's disease attacks the central nervous system, manifesting in muscle stiffness and tremors. These symptoms result in reduced mobility and clumsiness, and make simple tasks difficult.

Stress and fatigue accentuate the symptoms. As the disease progresses, it can cause mental deterioration and dementia.

Parkinson's disease is incurable, but some of the symptoms can be alleviated to improve quality of life. While the exact cause of Parkinson's disease is uncertain, reduced levels of the neurotransmitter dopamine are implicated, and treatment to increase these levels can improve some of the symptoms.

Macular degeneration

Macular degeneration is the result of the slow breakdown of cells in the eye. The area affected involves the cells responsible for central vision, while cells responsible for peripheral vision are unaffected. Its gradual progress makes it difficult to detect until considerable irreversible degeneration has taken place.

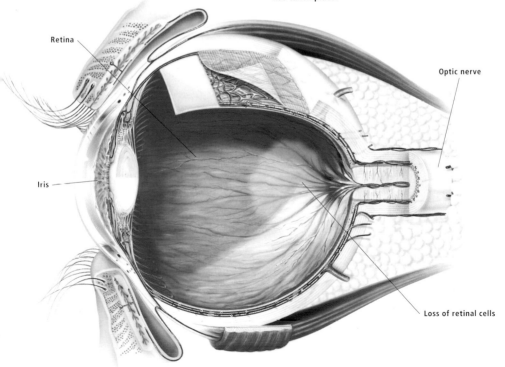

Retina

Optic nerve

Iris

Loss of retinal cells

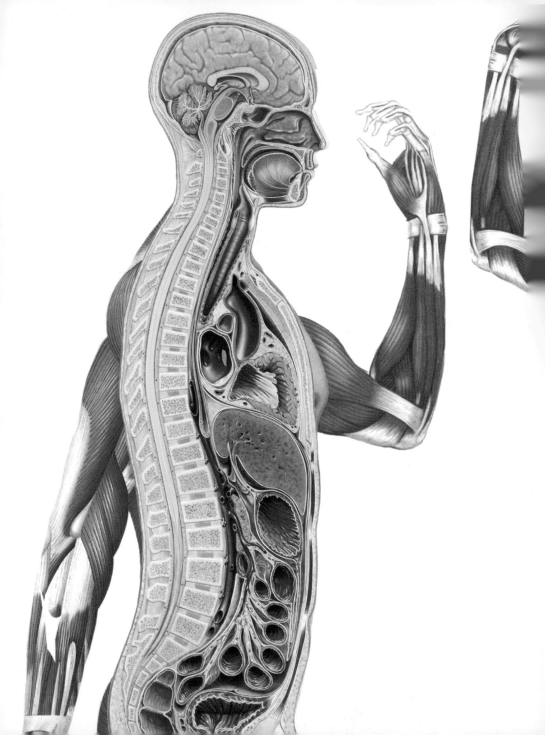

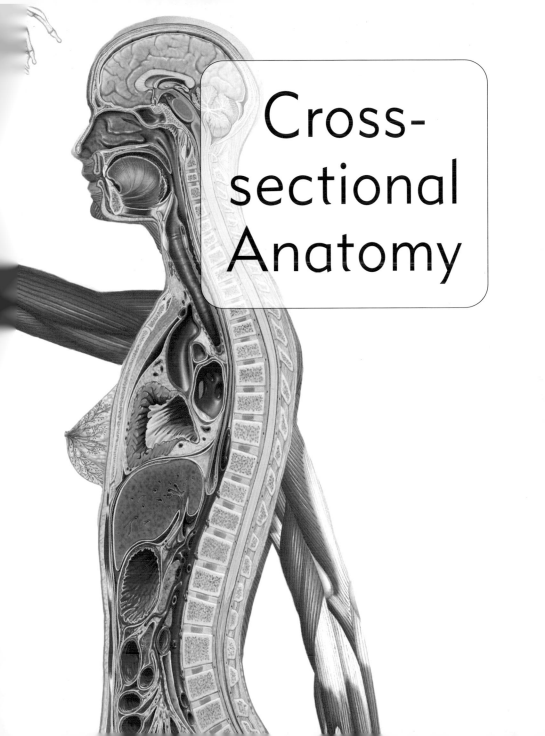

Cross-
sectional
Anatomy

Cross section of the neck

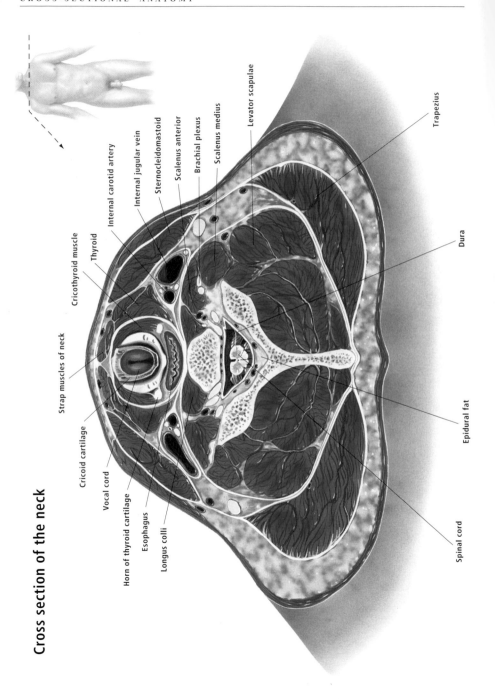

Internal carotid artery

Internal jugular vein

Sternocleidomastoid

Scalenus anterior

Brachial plexus

Scalenus medius

Levator scapulae

Trapezius

Cricothyroid muscle

Thyroid

Dura

Strap muscles of neck

Epidural fat

Cricoid cartilage

Vocal cord

Horn of thyroid cartilage

Esophagus

Longus colli

Spinal cord

Cross section of the heart and lungs

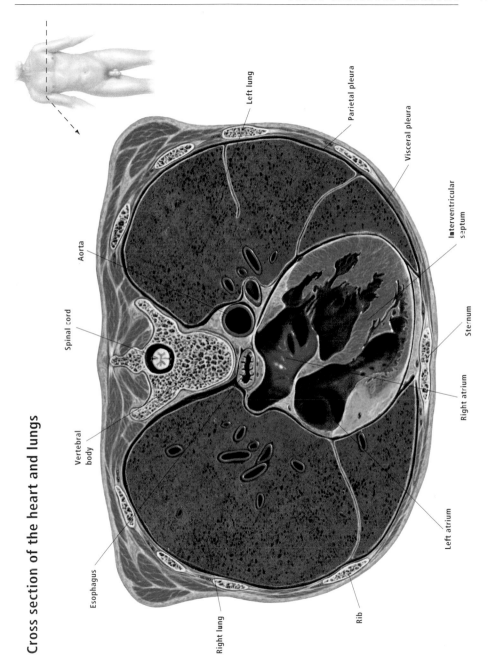

Left lung

Parietal pleura

Visceral pleura

Interventricular septum

Aorta

Spinal cord

Sternum

Vertebral body

Right atrium

Esophagus

Left atrium

Right lung

Rib

Cross section of the abdomen

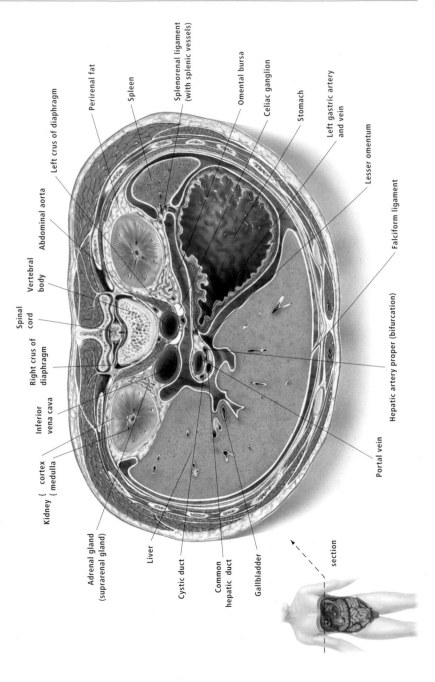

Perirenal fat

Spleen

Splenorenal ligament (with splenic vessels)

Omental bursa

Celiac ganglion

Stomach

Left gastric artery and vein

Lesser omentum

Left crus of diaphragm

Abdominal aorta

Falciform ligament

Vertebral body

Spinal cord

Right crus of diaphragm

Inferior vena cava

Kidney { cortex { medulla

Hepatic artery proper (bifurcation)

Portal vein

Adrenal gland (suprarenal gland)

Liver

Cystic duct

Common hepatic duct

Gallbladder

section

Cross section of the intestines

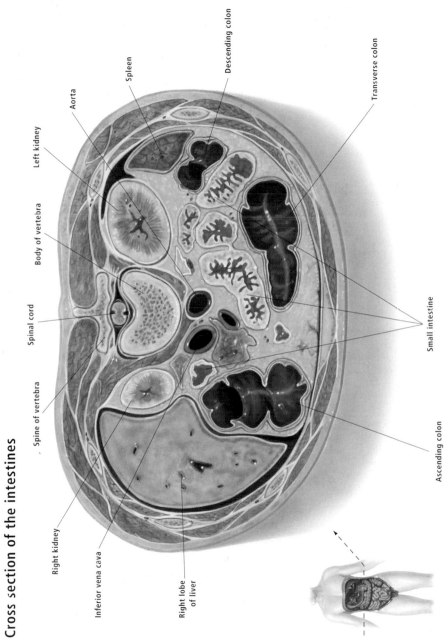

Spleen

Aorta

Descending colon

Transverse colon

Left kidney

Body of vertebra

Spinal cord

Spine of vertebra

Right kidney

Inferior vena cava

Right lobe of liver

Small intestine

Ascending colon

Cross section of the arm

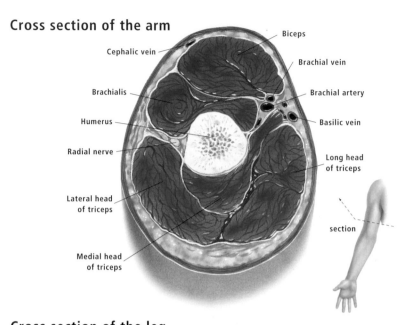

Cephalic vein

Biceps

Brachial vein

Brachialis

Brachial artery

Humerus

Basilic vein

Radial nerve

Long head
of triceps

Lateral head
of triceps

section

Medial head
of triceps

Cross section of the leg

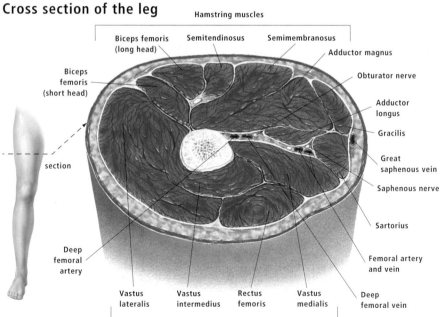

Hamstring muscles

Biceps femoris
(long head)

Semitendinosus

Semimembranosus

Adductor magnus

Biceps
femoris
(short head)

Obturator nerve

Adductor
longus

Gracilis

section

Great
saphenous vein

Saphenous nerve

Sartorius

Deep
femoral
artery

Femoral artery
and vein

Vastus
lateralis

Vastus
intermedius

Rectus
femoris

Vastus
medialis

Deep
femoral vein

Quadriceps muscles

Symptoms table

This section provides information on common symptoms and associated diseases. You may find it useful to refer to this symptoms table if you would like to find out more information on a particular symptom, and what it may mean.

This symptoms table does not replace your doctor, and should not be used as a guide to diagnosing medical conditions in yourself or a family member. Refer to your doctor or other appropriately qualified health professionals if you believe you or another person may have one of the diseases described.

How to use this table

Symptoms are listed alphabetically, and usually contain the name of the affected body part. For example, if you would like information on pain in the eye, refer to *Eye problems* to find entries on pain and other problems relating to the eye.

Each entry includes a list of descriptions of various ways in which the symptom may be experienced. Look down the *Symptom characteristics* column until you find a description that best matches the symptom in which you are interested.

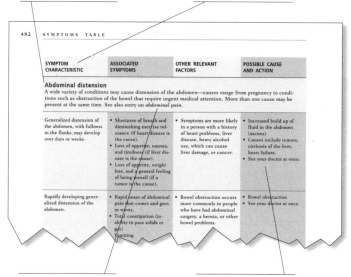

There is also a list of other symptoms that may be associated with the main symptom. Symptoms listed in the *Associated symptoms* column may not all occur at once, and not all of them must be present for the medical problem suggested under *Possible cause* to be a likely explanation for the symptoms.

Diseases mentioned in the *Possible cause* column suggest problems that may cause the symptoms described, but do not represent definitive diagnoses. Not all disease possibilities are listed, especially those that are less common. The symptoms appear in alphabetical order and are listed at right.

LIST OF SYMPTOM ENTRIES

Abdominal distension
Abdominal pain
Arm and shoulder problems
Back deformity
Back pain
Blackouts,
   including fainting
Bones, fracture/brittleness
Bowel, bleeding from
Breast lump
Breathing problems
Chest pain
Cough
Diarrhea
Ear pain (earache)
Eye problems
Eyesight problems
Fever
Foot and ankle problems
Genital pain
"Glands," swollen
Hand and wrist problems
Headache
Hearing loss (deafness)
Heartburn
Hip problems
Knee problems
Leg problems
Menstrual problems
Nausea and vomiting
Neck problems
Numbness and tingling
Palpitations
Seizures (known as "fits,"
   including convulsions)
Sexual function problems
Skin problems
Urination, frequent
Urination, painful
Vaginal problems
Weakness and paralysis
Wounds or sores,
   non-healing

SYMPTOM CHARACTERISTIC	ASSOCIATED SYMPTOMS	OTHER RELEVANT FACTORS	POSSIBLE CAUSE AND ACTION

Abdominal distension

A wide variety of conditions may cause distension of the abdomen—causes range from pregnancy to conditions such as obstruction of the bowel that require urgent medical attention. More than one cause may be present at the same time. See also entry on *abdominal pain*.

SYMPTOM CHARACTERISTIC	ASSOCIATED SYMPTOMS	OTHER RELEVANT FACTORS	POSSIBLE CAUSE AND ACTION
Generalized distension of the abdomen, with fullness in the flanks; may develop over days or weeks.	• Shortness of breath and diminishing exercise tolerance (if heart disease is the cause). • Loss of appetite, nausea, and tiredness (if liver disease is the cause). • Loss of appetite, weight loss, and a general feeling of being unwell (if a tumor is the cause).	• Symptoms are more likely in a person with a history of heart problems, liver disease, heavy alcohol use, which can cause liver damage, or cancer.	• Increased build up of fluid in the abdomen (ascites) • Causes include tumors, cirrhosis of the liver, heart failure. • See your doctor at once.
Rapidly developing generalized distension of the abdomen.	• Rapid onset of abdominal pain that comes and goes in waves. • Total constipation (inability to pass solids or gas). • Vomiting	• Bowel obstruction occurs more commonly in people who have had abdominal surgery, a hernia, or other bowel problems.	• Bowel obstruction • See your doctor at once.
Distension in the lower area of the abdomen or pelvic area	• Difficulty passing urine, vague abdominal pain.	• Ovarian cancer is more common with age and occasionally occurs in women under 35 years of age.	• Pregnancy, ovarian cyst, tumor of the ovary or uterus, bladder distension due to retention of urine. • See your doctor at once.
Lump next to the umbilicus, near a surgical scar, or (in men) in the groin.	• Painless • Appears when straining or coughing or when you stand up (i.e. when pressure in the abdomen rises); lump then disappears spontaneously or can be pushed back.	• Weakness in the abdominal wall may be caused by a congenital defect in the abdominal wall, a surgical incision, or muscle weakness due to obesity, pregnancy, or wasting.	• Hernia or inguinal hernia • See your doctor.
Lump that remains visible whether or not you are straining, coughing, or standing.	• Associated symptoms, if present, will depend on the cause.		• Enlargement of an abdominal organ (e.g. liver or spleen), tumor, hernia that has become trapped. • See your doctor at once.

SYMPTOM CHARACTERISTIC	ASSOCIATED SYMPTOMS	OTHER RELEVANT FACTORS	POSSIBLE CAUSE AND ACTION

Abdominal pain

Abdominal pain may be related to any abdominal organ, and the type of pain will vary widely, depending on the cause. See your doctor or emergency room immediately if abdominal pain is severe or of sudden onset, since many common causes require immediate surgical intervention. See also entry on *abdominal distension*.

SYMPTOM CHARACTERISTIC	ASSOCIATED SYMPTOMS	OTHER RELEVANT FACTORS	POSSIBLE CAUSE AND ACTION
Episodes of pain in the upper abdomen, typically central and characterized as a burning sensation or heartburn.	• Bitter fluid may return into the mouth from the stomach. • Symptoms may be brought on by stooping, straining, or lying down.	• Symptoms may be alleviated by antacids.	• Esophagitis caused by reflux. • See your doctor.
Episodes of pain in the upper abdomen, typically central but may also be on the right or left; may be characterized as a burning or gnawing sensation.	• Pain is typically related to food, and may either occur before meals and be relieved by eating, antacids, or vomiting, or may occur shortly after meals, sometimes making the person reluctant to eat.	• The incidence of peptic ulcer is highest among people aged 30–50 years; may occur in teenagers and rarely in children. • Ulcers are most frequently caused by a *Helicobacter pylori* infection or by the use of nonsteroidal antiinflammatory drugs.	• Peptic ulcer • See your doctor.
Episodes of colicky or constant pain in the upper abdomen, typically central or on the right side and of rapid onset; pain may radiate around the right side of the body into the shoulder and typically lasts for hours at a time.	• Nausea • Sometimes vomiting • Sometimes yellowness of the skin may ensue (jaundice), possibly with darkened urine and pale stools. • This pain is not relieved by vomiting or antacids.	• An episode may follow eating a fatty meal. • Occurs in adults; women on oral contraceptives are at higher risk. • May have a family history of gallstones.	• Gallstones • See your doctor.
Upper abdominal pain, gradually worsening and typically constant over hours or days before subsiding; pain may radiate directly through to the back.	• Nausea, vomiting • May be aggravated by food, drink, or vomiting. • Typically aggravated by lying down.	• Symptoms may follow a bout of heavy drinking. • May have background of alcohol use or gallstones. • May be alleviated by sitting up and leaning forward.	• Sudden-onset (acute) pancreatitis • See your doctor at once.
Persistent upper abdominal pain, typically characterized as a constant background pain; pain may radiate directly through to the back.	• May later develop weight loss or pale stools. • Pain may be aggravated by lying down.	• May have background of alcohol use or gallstones. • Pain may be alleviated by sitting up and leaning forward.	• Long-term (chronic) pancreatitis or pancreatic tumor • See your doctor.

SYMPTOM CHARACTERISTIC	ASSOCIATED SYMPTOMS	OTHER RELEVANT FACTORS	POSSIBLE CAUSE AND ACTION
Abdominal pain (cont.)			
Constant discomfort on the right side of upper abdomen, typically felt as a dull ache.	• Nausea, loss of appetite, tiredness, fever, eventually jaundice • Other symptoms depend on the cause. • Shortness of breath, diminishing exercise tolerance, and ankle swelling may occur with a heart problem. • Loss of appetite, weight loss, and a general feeling of being unwell may occur with a tumor.	• Liver symptoms are more likely to occur in people with a history of heart disease, heavy alcohol users, cancer, recent intravenous drug use or blood transfusion, overseas travel, recent contact with a person with a similar illness.	• Liver pain due to heart failure (causing swelling of the liver), viral hepatitis, hepatitis caused by heavy alcohol use, liver tumor • See your doctor.
Rapid onset of diffuse cramping pains, typically centered around the central abdomen.	• Diarrhea, nausea, vomiting, fever	• Symptoms may follow overseas travel or eating carelessly prepared food.	• Gastroenteritis or food poisoning • Drink plenty of fluids. • See your doctor if you are worried or if symptoms persist.
Rapid onset of pain on the right side of the lower abdomen; the pain may either be constant or occur in waves; pain typically starts around the middle of the abdomen, then moves to the right lower abdomen.	• Loss of appetite, nausea, vomiting, low fever	• Most common in children or young adults.	• Appendicitis • This is a medical emergency. • Go to the emergency room or call an ambulance immediately.
Pains in the lower abdomen that may be cramping, occur in waves, or vague discomfort.	• Change in bowel habit. • May have blood or mucus in the stools. • Pain may be worse before defecation and alleviated by defecation or passing wind.	• Cancer is a common cause in patients over 50, but can also occur in younger people. • In older patients (>70 years of age), bowel obstruction is one of the most common causes of abdominal pain requiring surgical intervention.	• Inflammatory bowel disease or tumor of the colon • See your doctor. • If severe pain, abdominal distension and total constipation (inability to pass solids or gas) occurs, this is a medical emergency. • Go to the emergency room or call an ambulance immediately.

SYMPTOM CHARACTERISTIC	ASSOCIATED SYMPTOMS	OTHER RELEVANT FACTORS	POSSIBLE CAUSE AND ACTION
Sudden severe pain on one flank, typically radiating from the loins around to the front of the abdomen and down to the groin.	• May have a change in the color of your urine or blood in the urine. • May be associated with nausea.	• May have a predisposing factor such as gout. • Passing a stone in the urine may cause severe pain.	• Kidney stone • This is a medical emergency. • Go to the emergency room or call an ambulance immediately.
Loin pain, typically a dull ache	• Fever • Passing urine more frequently. • Painful urination	• May be a predisposing condition like an abnormality in the urinary tract or a disorder that impairs immune defenses. • In women and infants, may occur without a predisposing factor.	• Pyelonephritis (kidney infection) • See your doctor at once.
Severe, sharp, well localized abdominal pain; pain may be of abrupt onset, but sometimes is more gradual.	• Pain aggravated by movement or coughing.	• Predisposing conditions include peptic ulcer, abdominal aneurysm, hernia.	• Peritonitis caused by appendicitis, perforation of a peptic ulcer, complication of abdominal aneurysm, strangulation or twisting of the bowel. • This is a medical emergency. Go to the emergency room or call an ambulance immediately.
Bouts of severe abdominal pain in an infant	• Screaming and drawing up the legs during the bouts of pain, vomiting, red stools.	• Intussusception is most common during the first two years of life.	• Intussusception • This is a medical emergency. Go to the emergency room or call an ambulance immediately.

Arm and shoulder problems

Symptoms in the arm or shoulder often result from overuse, injury, and falls. Sporting activities and occupational tasks are common causes of problems in this area. Pain in the arm can also be a symptom of a heart attack, so see your doctor immediately if you experience unexplained pain. See also entry on *chest pain*.

Pain in the shoulder with movement, especially lifting the arm over the head; may occur following an injury or activity involving repetitive movements.	• Tenderness and swelling in the area, weakness of the arm. • Symptoms are aggravated by further movement. • Pushing and twisting movements (e.g. reaching behind the back, dressing) are often painful.	• Lifting, twisting and sporting activities often cause these problems.	• Tendinitis • Rest and apply ice. • See your doctor if you are worried or if pain persists.

SYMPTOM CHARACTERISTIC	ASSOCIATED SYMPTOMS	OTHER RELEVANT FACTORS	POSSIBLE CAUSE AND ACTION
Arm and shoulder problems (cont.)			
Sudden pain on the top of the shoulder; may be worse upon waking.	• May be some redness and tenderness, limited movement. • Symptoms are usually aggravated by raising arm out to the side or twisting movement.	• Can occur with overuse, infections, arthritis, injuries.	• Bursitis • See your doctor for advice.
Gradually worsening elbow pain that occurs with repeated movements.	• Pain may spread down into the muscles in the forearm. • May be tender to touch. • Symptoms are aggravated by further movement and alleviated by rest.	• More likely to occur in those with weak wrist and arm muscles, and in tennis players using an incorrect backhand stroke.	• Tennis elbow • Rest and apply ice. • See your doctor if you are worried or if pain persists.
Increasing shoulder pain and stiffness that has worsened over several weeks.	• Limited movement in all directions • Symptoms are aggravated by movement in most directions, especially away from the body.	• Often occurs in older people, especially following an injury or shoulder/chest surgery.	• Frozen shoulder • See your doctor.
Intense arm pain following a fall or direct blow to the arm.	• Arm or shoulder may be deformed. • Symptoms are aggravated by an attempt at movement.		• Fracture or dislocation • See your doctor at once.
Persistent, severe pain radiating from the chest down the left arm.	• Shortness of breath, nausea, sweating • Feeling of chest being squeezed. • Pain does not change with arm movements.	• Increased likelihood in a person who has previously had a heart attack or with a history of coronary heart disease.	• Myocardial infarction (heart attack) • This is a medical emergency. Go to the emergency room or call an ambulance immediately.

SYMPTOM CHARACTERISTIC	ASSOCIATED SYMPTOMS	OTHER RELEVANT FACTORS	POSSIBLE CAUSE AND ACTION

Back deformity

Abnormalities of the spine may occur due to an acquired disease, as a developmental problem in growing children, or as part of one of the syndromes of multiple anatomical abnormalities that are caused by genetic defects. Any unusual appearance of the spine should be checked by a doctor.

SYMPTOM CHARACTERISTIC	ASSOCIATED SYMPTOMS	OTHER RELEVANT FACTORS	POSSIBLE CAUSE AND ACTION
Sideways curvature of the spine (scoliosis) in an adolescent, visible while sitting erect and straight; spine is S-shaped or corkscrew shaped.	• Postural problems, pain, fatigue of back muscles, visible disfigurement in severe cases	• The cause of idiopathic scoliosis is unknown, but it is thought to be due to complex inherited causes. • The condition is twice as common in females as males, and usually appears at around puberty.	• Idiopathic scoliosis
Exaggeration of the normal outward curve of the spine at the level of the rib cage (kyphosis).	• Poor posture, fatigue or pain in back muscles	• Abnormal curvature may first appear in a young growing child.	• Exaggerated kyphosis ("hunchback"), due to injury or a developmental abnormality. • See your doctor.
Exaggeration of the normal inward curve of the spine at the lower back (lordosis).	• Poor posture, fatigue or pain in back muscles.	• Abnormal curvature may first appear in a young growing child.	• Exaggerated lordosis ("swayback"), due to injury or a developmental abnormality • See your doctor.
Sideways curvature of the spine (scoliosis) which worsens over time.	• Difficulty running and jumping, muscle weakness, enlarged calf muscles, breathing problems. • Children use their arms to pull themselves up from the ground after sitting or lying down. • Spinal curvature may become painful over time.	• Duchenne muscular dystrophy is a genetic disorder affecting only boys. • The condition is present at birth, but signs only appear between 3–5 years of age.	• Duchenne muscular dystrophy • See your doctor.
Gradual development of a exaggerated outwards curvature (kyphosis) or sideways curvature of the spine (scoliosis) in an elderly person.	• Multiple fractures of the vertebrae may result in "dowager's hump." • New vertebral fractures may be very painful, but pain usually subsides within weeks to months.	• Osteoporosis is more common in women than men.	• Osteoporosis • See your doctor.

SYMPTOM CHARACTERISTIC	ASSOCIATED SYMPTOMS	OTHER RELEVANT FACTORS	POSSIBLE CAUSE AND ACTION

Back pain

The majority of cases of backache relate to the lower back. Low back pain is one of the most common symptoms that people experience. In adults, a majority of cases are based on a mechanical problem relating to posture, muscle strain, arthritis, or disk problems. Sometimes, no physical cause is found.

In some circumstances, prompt and adequate early treatment of acute mechanical back pain may help to avoid progression to ongoing long-term pain, so do not hesitate to see your doctor. Back pain that is progressive, severe, persistent, or associated with neurological symptoms such as weakness, numbness, or tingling sensations always warrants a visit to a doctor. See also entry on *numbness and tingling*.

SYMPTOM CHARACTERISTIC	ASSOCIATED SYMPTOMS	OTHER RELEVANT FACTORS	POSSIBLE CAUSE AND ACTION
Sudden episodes of pain, usually in the lower back; often described as sharp and/or shooting in character; often radiates down into the buttock and leg.	• Neurological symptoms may develop, such as tingling or some loss of sensation in the buttock or part of the leg. • Pain aggravated by bending (especially to the side), coughing/ sneezing, laughing.	• Pain may have been precipitated by trauma or an unusual movement. • The pain may spontaneously go away for a time. • Pain may be alleviated by lying in the fetal position.	• Intervertebral disk problem (e.g. herniated disk or disk protrusion) • See your doctor at once.
Gradual onset of pain, usually in the lower back	• Similar symptoms in other joints, stiffness in the mornings • Stiffness tends to gradually ease as the day progresses, pain tends to be worse with activity and eased by resting.	• Not precipitated by physical activity or trauma. • Osteoarthritis is more common with age; other forms of arthritis may occur in younger people.	• Arthritis (e.g. osteoarthritis or rheumatoid arthritis) • See your doctor.
Dull aching pain; may radiate to one side; most commonly in the middle or lower spine.	• Worse with exertion, posture, movement, coughing.	• May result from severe trauma, or minor or unnoticed trauma in a person with osteoporosis or a bone abnormality.	• Vertebral fracture • See your doctor at once.
Lower back ache, sometimes throbbing in nature.	• May have stiffness, rash, other joints affected, or symptoms from the gastrointestinal, urinary, or genital tract.	• Often in young adults or people with inflammatory disease of the bowel. • Characterized by being worse with rest, and alleviated by activity; this contrasts with mechanical causes of back pain.	• An inflammatory joint disorder • See your doctor.
Unremitting continuous and possibly progressive pain, usually of gradual onset; may be deep, boring pain.	• Weight loss, loss of appetite, malaise, fever, chills, nervous system problems • Posture or motion typically have little effect on the pain.	• Pathological lesions are rarely the cause of low back pain in adults; they are a more common underlying cause of pain higher up in the spine or low back pain in a child.	• Cancer metastasis, cancer of the blood (e.g. leukemia), inflammation, or infection • See your doctor at once.

SYMPTOM CHARACTERISTIC	ASSOCIATED SYMPTOMS	OTHER RELEVANT FACTORS	POSSIBLE CAUSE AND ACTION

Blackouts, including fainting

Fainting results from a transient loss of consciousness. Lapses in awareness or falling to the ground suddenly without warning may also occur without loss of consciousness. Faints are more common in elderly people, who are also at greatest risk of injury from the resulting falls. In about 50 percent of cases, no cause is found.

SYMPTOM CHARACTERISTIC	ASSOCIATED SYMPTOMS	OTHER RELEVANT FACTORS	POSSIBLE CAUSE AND ACTION
Transient loss of consciousness often without warning; may occur when upright or when lying down.	• Lightheadedness, palpitations, chest pain, or breathlessness preceding loss of consciousness. • While unconscious, the person may appear pallid. • On waking, the person is usually well oriented and quickly feels well again.	• In some cases, fainting is precipitated by exercise. • The risk is increased if the person is also dehydrated.	• Heart disease • See your doctor at once.
Falling to the ground suddenly without warning; may occur with or without loss of consciousness.	• Sudden onset of persistent nervous system problems (e.g. blindness, weakness, pins and needles, numbness, difficulties with speech, lack of muscle coordination)	• More common with advancing age. • Risk increased in people with hypertension, lipid abnormalities (e.g. high cholesterol), diabetes.	• Stroke • This is a medical emergency. • Go to the emergency room or call an ambulance at once.
Sudden transient loss of consciousness, or falling to the ground, without warning.	• There may be a sudden onset of transient neurological problems such as loss of vision, weakness, pins and needles, numbness, difficulties with speech, lack of muscle coordination, vertigo. • These symptoms spontaneously resolve.	• May have a history of cardiovascular disease or osteoarthritis of the neck. • May have been precipitated by a particular movement of the head. • More common with advancing age • Risk increased in people with hypertension, lipid abnormalities (e.g. high cholesterol), diabetes.	• Transient ischemic attack • This is a medical emergency. • Go to the emergency room or call an ambulance at once.
Loss of consciousness, generally prolonged	• Before losing consciousness may experience confusion, sweating, possibly headache, faintness, and weakness.	• Person on medication to treat diabetes. • May have missed a meal, exercised more than anticipated, or recently altered their diabetic medication. • Loss of consciousness can usually be prevented by eating glucose candy or jelly beans and taking sugary drinks (e.g. fruit juice) as soon as the attack is recognized.	• Hypoglycemia (low blood sugar) • This is a medical emergency. • Go to the emergency room or call an ambulance at once.

SYMPTOM CHARACTERISTIC	ASSOCIATED SYMPTOMS	OTHER RELEVANT FACTORS	POSSIBLE CAUSE AND ACTION

Bones, fracture/brittleness

Suspected fractures always require immediate medical attention, especially when there is substantial blood loss, symptoms of nerve damage, or any suggestion of damage to internal organs. If the injury may affect the spinal column—for example following a direct blow to the neck or back, falls from a height, or high-speed accidents—great care must be taken to immobilize the spine until an ambulance arrives to transport the person to hospital.

SYMPTOM CHARACTERISTIC	ASSOCIATED SYMPTOMS	OTHER RELEVANT FACTORS	POSSIBLE CAUSE AND ACTION
Pain with loss of limb function and possibly also change in shape	• The skin overlying the fracture may be broken (open fracture) or intact (closed fracture). • There may be substantial blood loss, or damage to nerves or internal organs.	• Fractures may be caused by a heavy blow hitting or crushing a bone (direct force) or twisting, bending, or compressing a bone (indirect force). • In adults, the break is usually through the full thickness of the bone. • In children, the bones are more springy and incomplete ("greenstick") fractures are common.	• Fracture • See your doctor at once. • First aid involves controlling bleeding, covering any open wounds, and supporting the limb in its most comfortable position.
Broken bone following repeated activity that stresses a particular site (usually a limb).	• Symptoms usually include pain that initially starts after the exercise or activity, then begins to occur during the activity, and later is present at other times as well. • There is no deformity and the limb functions normally.	• Usually occurs in otherwise healthy, active people.	• Stress fracture • See your doctor at once.
Broken bone following slight trauma	• Pain with loss of limb function and possible deformity. • Elderly women with long-term multiple fractures of the vertebrae may develop a hump at the top of the back. • Pain is aggravated by any movement of the affected part.	• Pain or deformity of the bone before the fracture occurred suggests a pre-existing abnormality of the bone. • Fractures of the hip, wrist, and vertebrae are most commonly associated with osteoporosis. • Osteoporosis is more common in women than men. • Factors that increase risk of developing osteoporosis include inherited factors, hormonal factors, calcium deficiency, alcohol abuse, smoking, lack of exercise, or drug treatment.	• Fracture due to an abnormality making the bone more fragile (pathological fracture). • Causes include osteoporosis, cancer with metastasis, blood cell abnormalities. • See your doctor at once.

SYMPTOM CHARACTERISTIC	ASSOCIATED SYMPTOMS	OTHER RELEVANT FACTORS	POSSIBLE CAUSE AND ACTION

Bowel, bleeding from

Bleeding from the bowel always warrants a visit to the doctor to determine the cause and, if blood loss has been substantial, to support the circulation. The most common cause is hemorrhoids, but more serious conditions such as bowel cancer are common and may be responsible for a bleed of any description. Serious pathologies need to be excluded by a doctor even if a minor ailment such as hemorrhoids is present. Some of the important causes in adults, such as hemorrhoids and rectal cancer, do not occur in children.

This table deals with bleeding through the rectum. Bleeding from the upper gastrointestinal tract often causes vomiting of blood, which may either be fresh blood or darkened blood resembling coffee grounds.

SYMPTOM CHARACTERISTIC	ASSOCIATED SYMPTOMS	OTHER RELEVANT FACTORS	POSSIBLE CAUSE AND ACTION
Squirt or drip of fresh blood in toilet or on toilet paper after defecation.	• Hemorrhoids are usually painless unless there are complications. • A sensation of unsatisfied defecation or urgency signals a problem inside the rectum.	• Straining to defecate predisposes to hemorrhoids.	• Hemorrhoids, polyps, tumors • See your doctor.
Small amount of fresh blood on toilet paper after defecation with sharp pain in the anus during or after defecating.		• Straining to defecate may predispose to anal injuries.	• Anal tear or break in tissue (fissure), tumor. • See your doctor.
Fresh blood mixed with stool or dark blood in stool, with or without blood clots	• No other symptoms or there may be diarrhea, constipation, abdominal pain, abdominal distension, weight loss, loss of appetite, general feeling of being unwell, symptoms of anemia (tiredness, lack of energy, breathlessness on exertion, pallor of skin, gums, or fingernails). • Mucus may be passed with stools.	• More likely in a person with personal or family history of polyps or bowel cancer. • Conditions that predispose to bowel cancer include benign polyps, ulcerative colitis.	• Bowel tumor, polyp, other problems of rectum or bowel • See your doctor.
Blood in feces or red-colored stool in an infant	• Unwell with vomiting and bouts of abdominal pain and screaming.	• Most common in children under 2 years.	• Intussusception • This is a medical emergency. • Go to the emergency room or call an ambulance at once.

SYMPTOM CHARACTERISTIC	ASSOCIATED SYMPTOMS	OTHER RELEVANT FACTORS	POSSIBLE CAUSE AND ACTION
Bowel, bleeding from (cont.)			
Episodes of bloody diarrhea	• Abdominal pain, aches, general feeling of being unwell, weight loss, fever, symptoms of anemia (tiredness, lack of energy, breathlessness on exertion, pallor of skin, gums, or fingernails). • Mucus may be passed in stool.	• Inflammatory bowel disease is most common in Jewish people (3–6 times higher than for non-Jewish racial or ethnic groups), and is higher among white racial groups than in black people or Asians. • It commences most commonly at age 15–35. • The cause of inflammatory bowel disease is unknown.	• An inflammatory bowel disease such as Crohn's disease or ulcerative colitis • See your doctor at once.
Bloody diarrhea of sudden onset	• Abdominal pain	• Most common in elderly people with a history of other cardiovascular disease, but no contact with dysentery.	• Loss of blood supply to a portion of the bowel (ischemic colitis) • This is a medical emergency. • Go to the emergency room or call an ambulance at once.
Heavy bleeding from the rectum.	• Lightheadedness or fainting, confusion, severe thirst, feeling cold and clammy, pallor of skin, gums, and nails.	• More common in men than in women. • Heavy bleeding is a rare complication of diverticulitis, which is usually associated with invisible or light bleeding.	• Diverticulitis • This is a medical emergency. • Go to the emergency room or call an ambulance at once.
Melena (black tarry stools)	• May also vomit blood, either as fresh blood or darkened blood resembling coffee grounds.	• Peptic ulcer can occur in any age group, but is relatively common in elderly people taking nonsteroidal anti-inflammatory drugs for painful inflammatory conditions such as osteoarthritis or rheumatoid arthritis.	• Bleeding from the esophagus, stomach, or duodenum, caused by a peptic ulcer, or gastritis, or dilated veins in the esophagus. • This is a medical emergency. • Go to the emergency room or call an ambulance at once.

SYMPTOM CHARACTERISTIC	ASSOCIATED SYMPTOMS	OTHER RELEVANT FACTORS	POSSIBLE CAUSE AND ACTION

Breast lump

Most women will detect a lump in their breast at some time in their life. Most of these lumps are found to be benign on further assessment. Breast cancer has been found responsible for up to 25 percent of breast lumps and remains the most common cancer in women.

SYMPTOM CHARACTERISTIC	ASSOCIATED SYMPTOMS	OTHER RELEVANT FACTORS	POSSIBLE CAUSE AND ACTION
General breast lumpiness, tender breast lumps usually affecting both breasts, but occasionally a single lump only; lumps are soft, round, and smooth; pain and tenderness may occur during menstrual cycles, usually of both breasts; lumps may come and go with menstrual cycles.	• Pain and tenderness worse premenstrually	• Usually affects premenopausal women. • Symptoms usually diminish or disappear after menopause. • Sometimes symptoms may be alleviated by reducing intake of coffee, tea, and chocolate.	• Fibrocystic disease of the breast • See your doctor. • If fibrocystic disease is diagnosed a well-fitted bra may help provide adequate support. • Vitamin E supplementation may reduce symptoms.
Firm, elastic breast lump that moves freely; usually a single, smooth lump approximately $1^1/_2$ inches (3–4 centimeters) in diameter.	• Usually painless	• Usually occurs in women 20–40 years of age.	• Fibroadenoma • See your doctor.
Hard or irregular breast lump which may be fixed in one position or attached to nearby tissues; usually only one breast is affected.	• No other symptoms or pain (5 percent of patients), bloody or nonmilky discharge from nipple, dimpling of the skin, tenderness or lumps under the armpit.	• More common in women than in men. • Most common after 40 years of age but may occur much earlier in women with a genetic predisposition.	• Breast cancer • See your doctor at once.
Firm, single, irregular lump, which may be attached to the skin.	• Lump is painful and tender.	• Gradual onset (over months) after breast trauma; the trauma involved may be subtle.	• Abnormality in fat tissue • See your doctor.
Single, tender breast lump that develops rapidly in a women who is breast feeding.	• Pain and tenderness of lump, redness of the overlying skin, fever, chills.	• Occurs during lactation.	• Breast abscess • See your doctor.

SYMPTOM CHARACTERISTIC	ASSOCIATED SYMPTOMS	OTHER RELEVANT FACTORS	POSSIBLE CAUSE AND ACTION

Breathing problems

Shortness of breath can be a gradually developing phenomenon or may be rapidly developing and life-threatening. Some of the more prominent causes of each are described in this section.

SYMPTOM CHARACTERISTIC	ASSOCIATED SYMPTOMS	OTHER RELEVANT FACTORS	POSSIBLE CAUSE AND ACTION
Long-term or repeated shortness of breath on exertion; onset gradual.	• Shortness of breath on lying flat, waking up at night with shortness of breath that is relieved by sitting up, swelling of the legs, cough, wheeze may be present, gradual decline in exercise capacity, shortness of breath at rest.	• Chronic heart failure is most common in the middle-aged and elderly, and rarely occurs in younger people. • Heart failure may follow rheumatic fever, heart attack/coronary artery disease, hypertension, arrhythmia, or heart valve abnormalities.	• Chronic heart failure • See your doctor at once.
Long-term or repeated shortness of breath on exertion; may progress gradually over many years.	• Wheezing • Cough with sputum, often worse in the morning. • Gradual decline in exercise capacity. • Sometimes alleviated by leaning forward or lying on stomach with head down.	• Chronic obstructive lung disease is very much more common in smokers and in the elderly.	• Chronic obstructive lung disease, bronchitis, emphysema • See your doctor.
Long-term or repeated shortness of breath on exertion, occurring in episodes with no symptoms at other times.	• Wheezing, persistent cough, usually without sputum. • Symptoms may be triggered by exposure to cigarette smoke, air pollution, pollens, or pets, cold air, exercise, respiratory infections.	• History of hay fever or eczema; family history of asthma, eczema, or allergies. • Usually more frequent in winter, following a common cold, or when pollen levels are high.	• Asthma • See your doctor. • If breathing is difficult, or you suspect oxygen shortage, this is a medical emergency. • Go to the emergency room or call an ambulance at once.
Acute shortness of breath at rest; onset may be sudden.	• Chest pain if the episode is precipitated by a heart attack. • May have a background history or recent worsening of shortness of breath on exertion and/or when lying flat, and/or waking up short of breath at night.	• Pulmonary edema is more likely in a person with a history of this problem. • May be caused by cardiac disease, lung infections, shock (e.g. cardiac surgery, serious infections, inhaled toxins).	• Pulmonary edema • This is a medical emergency. • Go to the emergency room or call an ambulance at once.

SYMPTOM CHARACTERISTIC	ASSOCIATED SYMPTOMS	OTHER RELEVANT FACTORS	POSSIBLE CAUSE AND ACTION
Sudden onset shortness of breath at rest	• Fever, chills, pain around the chest that is aggravated by breathing in.	• Pneumonia is a relatively common condition in all age groups, but may be more likely with an inhaled foreign body, pre-existing lung diseases, immune suppression (e.g. other disease, medications).	• Pneumonia • See your doctor at once.
Acute shortness of breath at rest; sudden onset	• Chest pain	• May occur spontaneously or as a result of chest trauma. • Most often occurs in young, healthy people without a prior lung problem. • Predisposing factors include cystic fibrosis, chronic obstructive lung disease.	• Pneumothorax • This is a medical emergency. • Go to the emergency room or call an ambulance at once.
Acute shortness of breath at rest	• Chest pain, dizziness, faintness, loss of consciousness, rapidly turning blue.	• Predisposing factors to an embolus include the use of oral contraceptives in young women, recent surgery, recent childbirth, a long period of bed rest or immobility, history of a blood clot in the leg, abnormal heart rhythms.	• Pulmonary embolism • This is a medical emergency. • Go to the emergency room or call an ambulance at once.
Sudden shortness of breath at rest with choking; occurring when inhalation of food or a small object is unlikely.	• Severe distress with inability to take in air despite desperate efforts, collapse, blue skin and gums, loss of consciousness (in complete obstruction). • Rapid labored breathing with loud harsh noise while breathing in, gagging (in partial blockage of the airway).	• Inflammation of the airways may be caused by an insect bite or sting, an allergic reaction to food or a new medication, inhalation of hot gases or other toxins. • Eating peanuts is a relatively common cause of severe reaction, which may occur in a person without a known history of peanut allergy.	• Upper airway obstruction due to swelling of structures in the throat. • This is a medical emergency. • Go to the emergency room or call an ambulance at once.

SYMPTOM CHARACTERISTIC	ASSOCIATED SYMPTOMS	OTHER RELEVANT FACTORS	POSSIBLE CAUSE AND ACTION

Chest pain

Chest pain typically arises from the heart, lungs, chest wall, or esophagus. Occasionally, chest pain is due to disease of upper abdominal structures such as the stomach or gallbladder. Any chest pain warrants full examination by a doctor.

SYMPTOM CHARACTERISTIC	ASSOCIATED SYMPTOMS	OTHER RELEVANT FACTORS	POSSIBLE CAUSE AND ACTION
Repeated episodes of central chest heaviness, pressure, tightness, or discomfort occurring on exertion and relieved by rest; often the pain radiates into the neck and/or left arm, or the arm may feel numb or tingly; attacks usually last a few minutes, sometimes up to 15 minutes.	• Shortness of breath or anxiety may be present.	• Angina pectoris occurs most frequently in the middle-aged or elderly. • Alleviated by rest or nitrate tablet taken under the tongue. • Rarely occurs in children with congenital heart problems.	• Angina pectoris • See your doctor.
Severe central chest heaviness, pressure, tightness, or discomfort that usually lasts more than 30 minutes; often the pain radiates into the neck and/or left arm; onset at rest or on exercise but without relief after resting.	• Shortness of breath, faintness, sweating, cold, nausea or vomiting, palpitations • Symptoms may be relatively mild in a person with a past history of heart disease.	• Myocardial infarction most commonly occurs in middle-aged or older people.	• Myocardial infarction (heart attack) • This is a medical emergency. • Go to the emergency room or call an ambulance at once.
Sharp, stabbing chest pain within a well-defined area, occurring on or worsened by breathing in.	• Fever, cough with sputum that may contain pus, shortness of breath. • Pain is aggravated by taking a deep breath, or by coughing.	• Pneumonia may result from a bacterial infection in previously healthy people, or due to lowering of resistance resulting from another illness.	• Pneumonia • See your doctor at once.
Sharp, stabbing chest pain within a well-defined area, worse on breathing in; may affect one or both sides of the chest.	• Shortness of breath • Coughing up blood. • Pain is aggravated by taking a deep breath, or by coughing.	• Predisposing factors include a period of immobilization or bed rest, recent surgical procedure, recent childbirth, history of a blood clot in the leg, the use of oral contraceptives.	• Loss of blood supply to an area of lung tissue (pulmonary infarct), usually due to blockage of blood vessels by a clot. • See your doctor at once.
Sudden onset of sharp, localized pain on one side of the chest, worse on breathing in.	• Shortness of breath • Pain is aggravated by taking a deep breath, or by coughing.	• Pneumothorax may occur in adults and children, either spontaneously or as a result of injury to the chest. • Most often occurs in young healthy people without a prior lung problem. • Predisposing factors include cystic fibrosis, chronic obstructive lung disease.	• Pneumothorax • See your doctor at once.

SYMPTOM CHARACTERISTIC	ASSOCIATED SYMPTOMS	OTHER RELEVANT FACTORS	POSSIBLE CAUSE AND ACTION
Sudden, severe central chest pain	• Severe shortness of breath, fainting • Skin may be cold, clammy, pale.	• Predisposing factors include a period of immobilization or bed rest, recent surgical procedure, recent childbirth, history of a blood clot in the leg, use of oral contraceptives.	• Pulmonary embolism • This is a medical emergency. • Go to the emergency room or call an ambulance at once.
Sudden onset of severe, tearing pain of the front or back of the chest that moves as time progresses; the pain often radiates through to the back or may be felt predominantly in the back of the chest; pain may also involve the arms, neck, trunk, or legs.	• Shortness of breath, fainting • Complications may cause symptoms in other bodily systems (e.g. paralysis of the legs, mental disturbances, bloody diarrhea).	• Most often occurs in middle-aged or elderly men with hypertension.	• Tearing along the inside wall of the aorta (dissecting aneurysm) • This is a medical emergency. • Go to the emergency room or call an ambulance at once.
Central chest tightness or discomfort	• Shortness of breath, wheezing • May be exacerbated by exposure to pets or cigarette smoke.	• Asthma occurs more commonly in adults or children with a history of hay fever or eczema, or a family history of asthma, eczema, or allergies. • Usually more frequent in winter, following a common cold or when pollen levels are high.	• Asthma • See your doctor at once.

Cough

Coughing is the body's way of clearing the airways of secretions or foreign material. A cough may be caused by any irritation or inflammation of the airways or lungs. Any persistent cough should be assessed by a doctor.

Acute onset of moist cough productive of phlegm.	• Fever, shortness of breath, fatigue, sharp chest pain on breathing in due to inflammation of the pleura (pleurisy). • Sputum produced is yellow or green and may contain some blood. • Cough may be aggravated by cold air, smoky environment, or exercise.	• May follow recent cold or flu. • More common in smokers.	• Bronchitis or pneumonia caused by an infection. • If the cause is bacterial, you may require treatment with antibiotics. • See your doctor.

SYMPTOM CHARACTERISTIC	ASSOCIATED SYMPTOMS	OTHER RELEVANT FACTORS	POSSIBLE CAUSE AND ACTION
Chest pain (cont.)			
Episodic dry cough with tight feeling in chest, often occurring at night or after exercise.	• Wheeze, shortness of breath • May be exacerbated by exposure to pets or cigarette smoke.	• History of hay fever or eczema • Family history of asthma, eczema, or allergies • Usually more frequent in winter, following a common cold, or when pollen levels are high.	• Asthma • Any episodic cough, especially if associated with wheeze, should be assessed by a doctor. • Management of acute episodes of asthma usually requires inhaled medications to open the airways (bronchodilator or "reliever" medication). • Continuous preventative medication may be required even when cough is not present. • See your doctor if having difficulty breathing.
Chronic cough, often productive of phlegm	• Shortness of breath on exertion, wheezing	• History of smoking or exposure to cigarette smoke.	• Chronic bronchitis or chronic lung disease • See your doctor. • Any change in the nature of a chronic "smoker's cough" should be checked by a doctor at once to rule out cancer of the airways (lung cancer).

Diarrhea

Diarrhea refers to the passing of frequent and abnormally loose or watery bowel actions. Diarrhea is often the result of gastrointestinal infection and lasts only a few days. Sometimes diarrhea can be caused by a more serious disease process. The consistency, color, and frequency of the diarrhea will give valuable clues with regard to the underlying cause. Any prolonged or unusual diarrhea should be assessed by a doctor to rule out an underlying medical condition.

Abrupt onset of watery diarrhea; mucus often present but rarely blood; lasts 1–3 days.	• Vomiting, mild abdominal pain, low-grade fever • May follow an upper respiratory tract infection. • Eating and drinking may aggravate diarrhea.	• Very common in children, especially in winter. • Highly contagious, spread by fecal–oral transmission.	• Viral gastroenteritis • Oral rehydration is usually sufficient treatment. • Medications to reduce nausea and vomiting, and acetaminophen (paracetamol) can be helpful. • Contact your doctor for advice.

SYMPTOM CHARACTERISTIC	ASSOCIATED SYMPTOMS	OTHER RELEVANT FACTORS	POSSIBLE CAUSE AND ACTION
Abrupt onset of diarrhea containing blood or mucus, with an offensive odor.	• High fever, aches and pains, nausea and vomiting, moderate to severe abdominal cramps, frequent urge to pass stool with little result. • Eating and drinking may aggravate diarrhea.	• May be a history of travel to a developing country or simultaneous illness in people who have shared contaminated food.	• Food poisoning • Antidiarrheal agents are not recommended for this type of diarrhea. • See your doctor at once.
Abrupt onset of watery diarrhea; symptoms may persist for weeks.	• Mild abdominal pain, bloating, flatulence, malabsorption (passage of larger volume stools than normal due to inability to absorb food), low-grade fever. • Eating and drinking may aggravate diarrhea.	• Often a history of travel • Common in children	• Infection with protozoal organism called giardia (giardiasis). • See your doctor.
Chronic intermittent diarrhea with mucus, sometimes alternating with constipation.	• Abdominal pain, flatulence, bloating • Emotional stress can aggravate symptoms.	• Most common in young to middle-aged women. • Increasing dietary fiber can improve symptoms.	• Irritable bowel syndrome • See your doctor.
Loose, bulky, offensively smelling stools	• Abdominal bloating and discomfort, flatulence, weight loss, signs of malnutrition or poor growth • Worse with certain types of food such as dairy or wheat products.	• May be a history of gallbladder disease, disorders of the pancreas.	• Poor absorption of food from the bowel due to lactose intolerance or celiac disease. • Malabsorption • See your doctor.

Ear pain (earache)

Ear pain is usually due to increased pressure in the middle ear or infection of the ear or surrounding structures. In addition to significant distress and discomfort, ear infections can lead to serious complications such as perforation of the eardrum or spread of an infection to the covering of the brain (meningitis). It is important to seek a doctor's advice, since antibiotic therapy may be needed.

Dull, deep ear pain, feeling of fullness or pressure	• Nasal and sinus congestion, sore throat	• Swelling and blockage of the tubes draining the middle ear results in increased pressure. • Swallowing can relieve the pressure in the middle ear.	• Fluid and pressure build-up in the middle ear. • Treatment with decongestant medication may help improve the drainage from the middle ear. • See your doctor if you are worried or if symptoms persist.

SYMPTOM CHARACTERISTIC	ASSOCIATED SYMPTOMS	OTHER RELEVANT FACTORS	POSSIBLE CAUSE AND ACTION
Ear pain (earache) (cont.)			
Rapid onset of severe, deep ear pain	• Fever, hearing loss, feeling of pressure in the ear, discharge if perforation of eardrum occurs, dizziness may occur if inner ear is involved.	• Fever may be the only symptom in young children. • Spontaneous perforation of the eardrum will relieve pain but prolong hearing loss.	• Infection of the middle ear (otitis media) caused by a virus or bacteria. • See your doctor.
Rapid onset of severe pain behind the ear	• Fever, swelling and tenderness over the mastoid (bony process behind the ear containing air cells.	• May be a history of otitis media as infection can spread from the middle ear to the mastoid. • Meningitis is a complication of mastoid infection.	• Bacterial infection of the mastoid air cells (mastoiditis) • See your doctor at once.

Eye problems

Mild redness and irritation of the eye is common and usually has a simple underlying cause. Symptoms to beware of include sudden or severe eye pain and changes in vision, especially when associated with a red eye. If in doubt, it is safer to have a doctor check the eye early to prevent any long-term complications. See also entry on *eyesight problems*.

Sore, dry, gritty eyes	• May be mild redness, normal vision. • Tiredness and the wearing of contact lenses can increase eye irritation.	• More common in the elderly. • Some medications can predispose to dry eyes (decongestants, diuretics, sleeping pills).	• Dry eyes due to reduced production of tears. • Artificial lubricants (or tears) can relieve symptoms. • See your doctor.
Sore, red eye, yellow sticky discharge, affecting one or both eyes.	• Sensitivity to light, itching, burning, gritty feeling on moving the eyes, normal vision.	• Highly contagious, often spreads from one eye to the other. • Rubbing eyes increases itching and discharge.	• Conjunctivitis caused by a bacteria or virus. • Gently cleanse eyes with cool water and avoid wearing contact lenses. • See your doctor if you are worried or if symptoms persist.
Rapid onset of red, extremely painful eye, on one side	• Nausea, vomiting, and blurred vision, cornea may be hazy due to swelling, pupil becomes fixed and dilated in severe cases. • Episode may be preceded by vision disturbances such as halos around lights. • Symptoms worsen at night when the pupil dilates.	• May be a family history of glaucoma. • More common in older people with diabetes or hypertension. • Some medications can make glaucoma worse, such as some antidepressants or steroids.	• Glaucoma • See your doctor at once.

SYMPTOM CHARACTERISTIC	ASSOCIATED SYMPTOMS	OTHER RELEVANT FACTORS	POSSIBLE CAUSE AND ACTION

Eyesight problems

Disturbances of vision are very common. Whether vision is just blurred or there is complete or partial loss of eyesight, visual disturbance is a very distressing and disabling symptom. The eyes are extremely delicate organs and can be damaged by many diseases involving other parts of the body. This section will concentrate on conditions localized to the eye. See also entry on *eye problems*.

SYMPTOM CHARACTERISTIC	ASSOCIATED SYMPTOMS	OTHER RELEVANT FACTORS	POSSIBLE CAUSE AND ACTION
Gradual onset of distorted vision, blurred vision especially when reading.	• Central vision is affected but peripheral vision remains intact.	• Most common cause of loss of vision in the elderly.	• Macular degeneration • Have your vision tested as you may need glasses. • See your doctor if you are worried or if symptoms persist.
Gradual onset of cloudy, foggy vision, distorted color, halos around lights	• Pupil of the eye may appear milky instead of black. • Frequent changes in eyeglass prescription.	• Cataracts are frequently a complication of diabetes. • Vision is worse at night with glare from bright lights.	• Cataracts • See your doctor.
Loss of parts of peripheral vision (visual field defects)	• Vision loss is the first symptom.	• Risk factors include older age, black race, diabetes, hypertension, nearsightedness, and a family history of glaucoma. • Those people at risk should be tested for glaucoma. • Early detection and treatment is the only way to prevent permanent vision loss.	• Glaucoma • See your doctor at once.
Sudden painless loss of vision in one eye, like a curtain falling down.	• Recent visual disturbances such as flashing lights or spots before the eyes. • Pressure on eye may aggravate vision loss.	• May be a history of trauma, glaucoma, cataract surgery, or nearsightedness.	• Detached retina. • This is a medical emergency. • Go to the emergency room or call an ambulance at once.

SYMPTOM CHARACTERISTIC	ASSOCIATED SYMPTOMS	OTHER RELEVANT FACTORS	POSSIBLE CAUSE AND ACTION

Fever

Fever is defined as a body temperature higher than the normal 98.6°F (37°C) taken orally. Mild or short-term temperature rises are common with minor infections, but high or sustained fever can signal a potentially dangerous and serious infection. Fever is often accompanied by other symptoms, as outlined below, which may help identify the cause.

The first table that follows applies to fever in adults and children of all ages. The second table deals with some additional symptoms specific to infants and children.

Fever in adults and children of all ages

SYMPTOM CHARACTERISTIC	ASSOCIATED SYMPTOMS	OTHER RELEVANT FACTORS	POSSIBLE CAUSE AND ACTION

Fever—body temperature above 98.6°F (37°C)	• Flushed face, hot skin, sore throat, mild headache	• May follow exposure to people with similar symptoms.	• Viral infection (e.g. cold or flu)
			• Take decongestant and fever-reducing medication.
			• For any fever it is advisable to:
			  – remove excess layers of clothing
			  – drink plenty of fluids
			  – take fever-reducing medication
			  – have a lukewarm bath
			  – check temperature every 4–6 hours.
			• See your doctor if:
			  – temperature remains at 102.2°F (39°C) or above in adults or 101.3°F (38.5°C) in children
			  – fever lasts longer than 48 hours
			  – the patient is pregnant
			  – the patient is a child.
			• In children, use a non-aspirin fever-reducing medication formulated for children.
			• See your doctor if the fever does not respond quickly to medication or if the child has a sore throat or painful ear.

SYMPTOM CHARACTERISTIC	ASSOCIATED SYMPTOMS	OTHER RELEVANT FACTORS	POSSIBLE CAUSE AND ACTION

Fever	• Aches, chills, nausea, vomiting, cramping, diarrhea	• Affects all age groups, but may cause life-threatening dehydration in the very young, the very ill, and the elderly.	• Gastroenteritis or other viral infection
			• Rest and follow standard advice for any fever:
			  – remove excess layers of clothing
			  – drink plenty of fluids
			  – take a fever-reducing medication
			  – have a lukewarm bath
			  – check temperature every 4–6 hours.
			• Use anti-diarrhea and anti-vomiting medications as advised by your pharmacist or doctor.
			• See your doctor if vomiting lasts longer than 12 hours or if there is bloody diarrhea.
			• See your doctor if you are worried or if symptoms persist.

Fever	• Severe headache, neck stiffness, drowsiness, vomiting, sensitivity to light	• May be confused and unable to respond well to questioning.	• An infection in the area around the brain (meningitis) • This is a medical emergency. • See your doctor or emergency department.

Sudden onset of fever with simultaneous sore throat	• Headache	• Tends to occur during the colder months and can be precipitated by stress, overwork, exhaustion, and when the body's immune system is fighting other infections.	• Strep throat (streptococcal infection)
			• Rest and follow standard advice for any fever:
			  – remove excess layers of clothing
			  – drink plenty of fluids
			  – take a fever-reducing medication
			  – have a lukewarm bath
			  – check temperature every 4–6 hours.
			• See your doctor.

SYMPTOM CHARACTERISTIC	ASSOCIATED SYMPTOMS	OTHER RELEVANT FACTORS	POSSIBLE CAUSE AND ACTION
Fever (cont.)			
Fever	• Ear pain, hearing loss, feeling of fullness or fluid in the ear	• Fever may be the predominant sign in a child too young to indicate other symptoms.	• Middle ear infection (otitis media) or outer ear infection (otitis externa) • Ear infections can lead to more serious problems if not treated, so see your doctor for assessment. • Antibiotic treatment may be required. • See your doctor.
Fever	• Pain with urination·or low back pain, tenderness on both sides of the lower back		• Kidney infection • You may require treatment with antibiotics. • See your doctor at once.
Fever	• Open sore or wound that is red. • Red streaking on the arms or legs originating near the wound. • Surrounding skin may be tender and hot; there may also be localized swelling.		• Blood poisoning as a result of infection of the skin or lymphatic system. • See your doctor at once.

Fever in infants and children

Fever in a child aged under 3 months	• Lethargy, pale skin, irritability		• Fever in a baby should always be investigated to rule out serious infection. • See your doctor at once.

High fever over 101.3°F (38.5°C)	• Barking cough		• Croup
			• See a doctor as soon as possible.
			• For any fever it is advisable to:
			  – remove excess layers of clothing
			  – drink plenty of fluids
			  – take a fever-reducing medication
			  – have a lukewarm bath
			  – check temperature every 4–6 hours.
			• See your doctor at once.

SYMPTOM CHARACTERISTIC	ASSOCIATED SYMPTOMS	OTHER RELEVANT FACTORS	POSSIBLE CAUSE AND ACTION
Fever	• Blisters over face, back, neck, and chest	• Occurs most commonly in children. • May occur in adult not previously infected. • Follows recent contact with person with chickenpox. • Highly infectious, so common among childcare and hospital workers. • Vaccination to prevent is now available in some countries.	• Chickenpox • Chickenpox infection may be severe in adults, so seek medical advice at once. • Take children with suspected chickenpox for assessment by a doctor, and keep child away from others who have not been infected. • See your doctor. • Chickenpox is highly contagious, so warn the doctor's office of the possibility of infection before attending.
High fever, above 102°F (39°C)	• Seizure or convulsion may be triggered by high fever.	• Three percent of children have at least one febrile convulsion. • Cooling a feverish child in a lukewarm bath can help prevent a convulsion. • A rapid rise in body temperature is more likely to cause seizures than a slow rise to the same temperature.	• Febrile seizure • This requires immediate action. • Ensure the airway is clear and turn child on to the side. • Remove clothing and bathe or sponge with lukewarm water after the seizure has finished. • See your doctor as soon as possible.

Foot and ankle problems

Most people experience occasional problems with the foot or ankle. Ankle sprains are one of the most common musculoskeletal injuries, while degenerative disorders such as gout and heel spurs may often affect people as they age.

Pain and swelling, usually on one side of the ankle, following twisting injury or fall.	• Bruising, warmth • May be aggravated by walking, but pain does not stop walking.	• Most commonly affects the outer side of the ankle.	• Injury to the ligaments in the ankle (ankle sprain) • Rest, apply ice, elevate the foot, use a compression bandage. • See your doctor.
Severe pain around the ankle following a fall, twisting injury, or direct blow to the ankle.	• Swelling of the ankle and possibly foot and toes, throbbing, warmth, bruising, inability to walk.		• Fracture or severe ligament sprain • Elevate the leg, apply ice. • See your doctor at once.

SYMPTOM CHARACTERISTIC	ASSOCIATED SYMPTOMS	OTHER RELEVANT FACTORS	POSSIBLE CAUSE AND ACTION
Foot and ankle problems (cont.)			
Pain under the heel and arch of the foot	• May be stiffness in the heel. • Aggravated by prolonged walking or running.	• Can lead to development of a heel spur (see directly below).	• Plantar fasciitis • See your doctor.
Sharp pain under the heel	• May be some swelling. • Aggravated by walking, pressing on the heel.	• May be more common in those with flat feet.	• Excess growth of bone at the heel (spur) • See your doctor.
Sudden onset of severe pain in the big toe, foot, or ankle; pain often begins at night and may last for several days.	• Swelling, skin over the affected area is usually red, shiny, and very tender to touch. • May be fever and chills.	• Most commonly affects the big toe, but can affect many other joints in the body. • Usually affects one or two joints at a time, and attacks recur.	• Gout • See your doctor.
Dull ache or pain at the back of the heel that travels up the back of the ankle and lower calf.	• Mild swelling, tenderness, warmth at the back of the ankle. • Aggravated by running, jumping, walking, bicycling especially when commencing the activity.	• More common in those with flat feet or tight muscles in the calf.	• Inflammation of the Achilles tendon at the back of the ankle (tendinitis). • Avoid activities that cause pain. • See your doctor for advice.

Genital pain

Pain in the genital area is usually the result of an infection, or may occur following injury. Sexually transmitted diseases are a common cause of infection, so it is important to practice safe sex to prevent further spread of these conditions, and consult your doctor for treatment. Any pain with sexual intercourse should be investigated by a doctor. See also entry on *vaginal problems*.

Painful blisters on the genitals, which may break, weep, and form sores; the first occurrence of blisters may last from 5 days to several weeks; further episodes are usually shorter and less severe, and usually occur less frequently.	• Itchy, tingling sensation, swollen and tender lymph nodes in the groin, pain on urination, discharge from the urethra in men and vagina in women. • Weakness or constipation may occur. • Erectile dysfunction may occur in men.	• Triggers for an episode may be stress, illness, sexual intercourse, menstruation. • If a pregnant woman has an attack of genital herpes towards the end of her pregnancy, there is a risk of passing it on to the baby and causing serious problems such as brain damage; cesarean section may overcome this problem.	• Genital herpes • Wear cotton underwear and loose clothing. • See your doctor at once. • Avoid sexual contact until medical advice is obtained.
In men, tender and swollen tip of the penis	• May be aggravated by sexual contact, pressure from tight clothing.		• Infection of the head of the penis (balanitis) • See your doctor at once.

SYMPTOM CHARACTERISTIC	ASSOCIATED SYMPTOMS	OTHER RELEVANT FACTORS	POSSIBLE CAUSE AND ACTION
In men, pain with ejaculation	• Blood in the semen, tenderness with bowel movements, or pain behind the penis or scrotum	• Sudden-onset prostatitis caused by a bacterial infection is most frequent in young men. • Prostatitis may be triggered by the use of urinary catheters.	• Prostatitis • See your doctor at once.
In men, pain during urination.	• Pus or mucus may be visible in urine.	• Urethritis is the most common form of sexually transmitted disease. • Occurs most frequently in sexually active young men.	• Urethritis, caused by sexually transmitted infection. • See your doctor at once.
In women, mild pain or discomfort while urinating.	• Vaginal discharge, abdominal pain, rectal pain, sore throat	• Chlamydia is one of the most common sexually transmitted infections, and is most common in young, sexually active people. • Most chlamydial infections do not cause any symptoms, so transmission is very common and easy.	• Chlamydia • See your doctor at once. • Avoid sexual contact until medical advice is obtained.

"Glands," swollen

The term "swollen glands" usually refers to swelling of the lymph nodes (also known as lymph glands) in the neck, armpit, or groin. There are many causes of swollen lymph nodes, ranging from mild infections to serious disorders such as cancer. Swelling often goes down as the infection resolves, but if your glands stay swollen for more than two weeks consult your doctor.

Swollen lymph nodes in the neck	• Fever, fatigue, sore throat, headache	• Infectious mononucleosis occurs most commonly in young adults. • The virus is contagious and is passed via kissing, coughing, and sneezing; sometimes known as "the kissing disease."	• Infectious mononucleosis (glandular fever) • See your doctor.
Swollen, tender lymph nodes in the neck	• Sore throat, headache, fever, bad breath, white spots on the tonsils		• Streptococcal infection ("strep throat") • See your doctor.

SYMPTOM CHARACTERISTIC	ASSOCIATED SYMPTOMS	OTHER RELEVANT FACTORS	POSSIBLE CAUSE AND ACTION
"Glands," swollen (cont.)			
Enlarged, non-tender lymph nodes throughout the body	• Fatigue, night sweats, weight loss, fever, severe itching all over the body. • Lymph nodes enlarge slowly and are usually painless.	• Non-Hodgkin's lymphoma is a relatively common cancer, and occurs most frequently in children and young adults. • Hodgkin's disease occurs most commonly in young adults or in people over 50 years, and is more prevalent in men than women.	• Cancer of the lymphatic system (lymphoma) (e.g. Hodgkin's lymphoma or non-Hodgkin's lymphoma) • See your doctor at once.
Swollen lymph nodes in the neck, groin, or armpit	• Weight loss, fatigue, tendency to bruise or bleed easily, loss of appetite • Anemia is common during the early stages of leukemia.	• Leukemia may increase susceptibility to infections (e.g. tonsillitis, pneumonia). • Leukemia is most common in children and young adults.	• Leukemia • See your doctor at once.
Swollen, inflamed lymph nodes just above the angle of the jaw, on one or both sides of the face.	• Fever, fatigue, swelling of the lymph nodes under the tongue, testicular swelling in males, or abdominal pain. • Pain with chewing or swallowing.	• Mumps is contagious, but less so than other infections such as measles or chickenpox. • It is a preventable through immunization. • In countries where mumps vaccination is widely practiced, mumps occurs most frequently in adults. • Prior to routine vaccination it occurred most commonly in children. • Mumps is spread by close contact.	• Mumps • See your doctor.

Hand and wrist problems

Problems in the wrist or hand can be caused by a variety of conditions, including overuse, injury, and falls. The hands are also a common site for developing arthritis.

Pain in the wrist or hand following repeated movements	• May be some mild swelling, tingling. • Aggravated by continuing the repeated movements; alleviated by rest.	• Inflammation of the tendons in the wrist or hand (tendinitis) may be caused by overuse. • Healing is often delayed in people with arthritis, diabetes, or gout.	• Tendinitis • Try anti-inflammatory medication as advised by your pharmacist. • See your doctor.

SYMPTOM CHARACTERISTIC	ASSOCIATED SYMPTOMS	OTHER RELEVANT FACTORS	POSSIBLE CAUSE AND ACTION
Numbness, tingling, burning in the hand, wrist pain that shoots into the palm of the hand; may be worse at night.	• Weakness, may be mild swelling • Aggravated by flexing the wrist, making a fist.	• More common in women • Associated with occupations that involve repeated forceful movements of the wrist (e.g. using a screwdriver).	• Carpal tunnel syndrome • See your doctor.
Pain, swelling, stiffness in the wrist and/or finger joints, often worse after periods of inactivity.	• Affected joints may feel hot, possible chills or fever. • May be aggravated by movement. • Can progress to cause deformities of the hand.	• Usually affects both hands/wrists at the same time. • May have other joints that are affected. • Affects more women than men.	• Rheumatoid arthritis • See your doctor at once.
Temporary, patchy, red and white discoloration of the fingers, usually following exposure to cold, may last for minutes or hours.	• May be associated numbness, tingling, burning, feeling of pins and needles.	• Can also affect the feet. • Most common in young women • More likely in smokers • Alleviated by warming the hands.	• Raynaud's disease • See your doctor.

Headache

Headache is the term used to describe any form of pain or discomfort in the head. It is an extremely common problem and one that most people have experienced. While most headaches are minor and easily treated with pain relievers, some warrant further medical investigation, and occasionally signal a more serious problem.

SYMPTOM CHARACTERISTIC	ASSOCIATED SYMPTOMS	OTHER RELEVANT FACTORS	POSSIBLE CAUSE AND ACTION
Dull, non-throbbing pain that feels like a vise around the head, squeezing both temples, extending into the neck.	• Scalp or neck tenderness, tight or tender neck and shoulder muscles • Symptoms may start after working in one position for several hours, or after driving; may be related to stress or anxiety.		• Tension headache • Try relaxation techniques. • Heat may help to relax neck and shoulder muscles, and analgesics may help the pain. • See your doctor if you are worried or if symptoms persist.
Intense, throbbing, one-sided headache, may be centered around the eye; pain may last from a few hours to several days.	• Vomiting and nausea may occur. • In some people, the headache is preceded by a warning sign (aura), which may include visual disturbances such as flashing lights or spots. • Oversensitivity to light, odors, or sound may be experienced.	• Migraine may be triggered by certain foods (e.g. cheese, strawberries, chocolate). • In women, migraine may be associated with the menstrual cycle.	• Migraine • Take analgesics such as acetaminophen (paracetamol) at once on onset of symptoms, and lie down. • See your doctor if you are worried or if symptoms persist.

SYMPTOM CHARACTERISTIC	ASSOCIATED SYMPTOMS	OTHER RELEVANT FACTORS	POSSIBLE CAUSE AND ACTION
Headache (cont.)			
Throbbing pain in the front of the head and around the eyes.	• Feeling of pressure around the eyes and nose, thick nasal discharge • May worsen with bending forward; may follow a recent cold or episode of hay fever.	• Occurs with viral infections, allergies, deep sea diving, or dental infections.	• Sinusitis • Decongestants may be helpful to relieve symptoms. • See your doctor if you are worried or if symptoms persist.
Severe headache	• Stiff neck, vomiting, fever, drowsiness, delirium, unconscious, or have convulsions. • May be worsened by exposure to bright lights.	• A child may be difficult to wake and have an unusual high-pitched moaning cry.	• Meningitis • This is a medical emergency. • Go to the emergency room or call an ambulance at once.
Severe, piercing pain in and around one eye, lasting 30 minutes to several hours; headaches may occur one or more times a day for a period of weeks or months.	• The affected eye may be bloodshot and watery. • There may be associated nasal congestion and facial flushing. • Pain often occurs at night.	• Occurs much more frequently in men than in women.	• Cluster headache • See your doctor.
Persistent, throbbing headache that begins first thing in the morning, and may lessen during the day.	• Vomiting, nausea, fatigue, blurred vision, weakness • Headache may be unlike any other headache the person has had before. • May be worsened by changing positions (e.g. moving from lying down to standing.	• Hypertension causing these symptoms is relatively rare. • Brain tumors are rare, and may be inherited or associated with exposure to ionizing radiation.	• Severe hypertension or brain tumor • This is a medical emergency. • Go to the emergency room or call an ambulance at once.
Severe headache that begins suddenly.	• Vomiting, limb weakness, double vision, slurred speech, difficulty swallowing, loss of consciousness		• Cerebral hemorrhage or aneurysm • This is a medical emergency. • Go to the emergency room or call an ambulance at once.

SYMPTOM CHARACTERISTIC	ASSOCIATED SYMPTOMS	OTHER RELEVANT FACTORS	POSSIBLE CAUSE AND ACTION

Hearing loss (deafness)

Significant loss of hearing is common, and can result from damage to the ear, disease of the ear, or changes with age that damage the delicate structures enabling us to hear. Hearing loss is due to disturbances in the external or middle ear or abnormalities in the inner ear or neuronal (nerve) pathways. Any persistent loss of hearing should be assessed by a doctor to determine the type of hearing loss, possible causes, and whether treatment is available to restore hearing.

SYMPTOM CHARACTERISTIC	ASSOCIATED SYMPTOMS	OTHER RELEVANT FACTORS	POSSIBLE CAUSE AND ACTION
Intermittent hearing loss in one or both ears	• Usually no associated symptoms or pain.	• May have history of ear wax blockage. • Wax-softening drops may relieve hearing loss.	• Ear wax blockage (ceruminosis) • See your doctor. • Do not try to remove the blockage yourself, as you may damage the eardrum or small bones in the ear. • Your doctor has special instruments to do this safely.
Sudden hearing loss in one or both ears	• Fever, symptoms of a cold, discomfort ranging from a feeling of pressure in the ear to persistent, severe ache in one or both ears. • Nausea and vomiting, dizziness, or ringing in the ears.	• Most commonly occurs following a cold. • Most common in children aged 3 months to 3 years, due to narrow eustachian tubes (tubes that allows pressure between the mouth or nose and ears to be equalized). • Eustachian tubes that are not fully developed may block easily with inflammation of the nose or throat, or with allergies. • Decongestants can aid drainage from the inner ear.	• Otitis media (viral or bacterial ear infection) commonly caused by a cold virus. • Symptoms are caused by build-up of fluid. • Perforation of the eardrum is a potential complication, which can lead to more prolonged but usually temporary hearing loss. • See your doctor. • See your doctor at once if the patient is a child.
Recurrent episodes of hearing loss, mainly low tone sounds, usually lasting 20 minutes to several hours.	• Dizziness, ringing, rushing or buzzing sound in the ears (tinnitus), nausea, feeling of movement, or dizziness (vertigo). • Occurs intermittently.	• Affects only one ear in the majority of people with the disease. • May be caused by fluid in the canals of the inner ear. • Reducing dietary salt, caffeine, alcohol may help control episodes.	• Menière's disease • See your doctor. • Antiemetics may provide symptomatic relief.

SYMPTOM CHARACTERISTIC	ASSOCIATED SYMPTOMS	OTHER RELEVANT FACTORS	POSSIBLE CAUSE AND ACTION
Hearing loss (deafness) (cont.)			
Gradual onset of hearing loss on one side	• May have facial weakness on the same side.	• Acoustic neuromas grow slowly and are more common in older people.	• Acoustic neuroma (a benign tumor of nerve cells) • Tumors may grow large enough to put pressure on other structures. • See your doctor.
Hearing loss in childhood	• Delayed language development	• Infection during pregnancy (e.g. rubella or cytomegalovirus) can cause congenital hearing loss in children. • Repeated ear infections may cause deafness in children. • Meningitis can lead to hearing loss in one or both ears. • Any child with suspected hearing loss or delayed language development should have their hearing tested.	• Congenital deafness, meningitis, or otitis media • See your doctor.

Heartburn

Heartburn does not involve the heart, but is a traditional name given to a symptom of a digestive problem that can often be relieved by indigestion medications. However, it is important to make sure the chest pain is not caused by angina or a heart attack.

Painful, burning sensation in the chest, behind the breast bone, which may rise up to the throat.	• Bitter taste in the mouth • Large meals, fatty or spicy foods may cause symptoms. • Lying down or bending may worsen symptoms. • Smoking or alcohol may aggravate symptoms. • Tight clothing or belts may make symptoms worse.	• Typically occurs after food. • Heartburn may occur during pregnancy but usually resolves after the baby is born.	• Back-washing of food and stomach acid upward into the esophagus (gastroesophageal reflux). • Take an antacid as advised by your pharmacist or doctor. • Avoid foods that seem to cause the symptoms, and do not eat within 2 hours of going to bed. • Quit smoking and reduce alcohol intake. • Raise the head end of the bed 4 inches (10 centimeters). • Lose weight if you are overweight. • See your doctor if you are worried or if symptoms persist.

SYMPTOM CHARACTERISTIC	ASSOCIATED SYMPTOMS	OTHER RELEVANT FACTORS	POSSIBLE CAUSE AND ACTION
Intermittent pain behind the breast bone, which at first may not be easily distinguished from heartburn.	• Worse with exercise, relieved by resting or rapidly-acting nitrate drugs.	• Most common in those with previous history of coronary heart disease.	• Angina pectoris, a form of coronary heart disease • See your doctor at once • Failure to treat the cause of angina may result in a heart attack.
Intense chest pain, which may initially be mistaken for severe heartburn.	• Pain spreading to left arm or both arms; pain in jaw; feeling of chest being squeezed.	• Increased likelihood in a person who has previously had a heart attack or with a history of coronary heart disease.	• Heart attack (myocardial infarction) due to sudden loss of blood supply to a section of the heart muscle due to blockage of the coronary arteries supplying the heart muscle. • This is a medical emergency. • Go to the emergency room or call an ambulance at once.

Hip problems

Hip problems often occur following a fall, especially in the elderly, or because of arthritis. Other causes of hip pain and stiffness include frequent running or problems with the cartilage in the hip joint.

Intense hip pain following a fall.	• Leg may be held in an abnormal position, may develop swelling. • Aggravated by standing, straightening the leg, lifting the leg.	• More likely to occur in the elderly.	• Hip fracture • See your doctor at once.
Stiffness and pain in one or both hips.	• May have swelling and redness around the joints, stiffness and pain in other joints. • Stiffness often aggravated by long periods in one position. • Pain aggravated by lots of walking or standing.	• More likely in older people.	• Arthritis • See your doctor for advice.
In infants, clicking of the hip	• May be some pain when the hip is stretched, movement may be limited.	• More common in girls than boys, also more common in babies born breech (buttocks first), or in those with a relative who has the same disorder.	• Congenital dislocation of the hip • See your doctor.

SYMPTOM CHARACTERISTIC	ASSOCIATED SYMPTOMS	OTHER RELEVANT FACTORS	POSSIBLE CAUSE AND ACTION
Hip problems (cont.)			
In teenagers, stiffness in the hip, pain, and limping.	• May also have pain in the knee or thigh. • Affected leg may be twisted outward. • Aggravated by walking.	• More common in overweight teens. • Affects boys more than girls.	• Dislocation of the top of the thigh bone (slipped capital femoral epiphysis). • See your doctor at once.
In children, gradual onset of hip pain and stiffness; symptoms progress slowly.	• Limping, wasting of thigh muscles, limited movements • Aggravated by walking.	• Most common in 5–10 year olds. • Affects boys more than girls.	• Degeneration of the top of the thigh bone (Perthes' disease). • See your doctor.
Shooting or burning pain in the back of one hip or buttock; pain may travel down the back of one leg.	• May also have low back pain, numbness or tingling in the foot. • Aggravated by coughing, sneezing, bending, lifting.	• More common in people with stiff backs or past back injury.	• Sciatica • See your doctor.

Knee problems

Knee pain is a common symptom in all age groups. Problems range from mild pain under the kneecap to ligament tears requiring surgery. The knee is also a common site for developing arthritis.

Intermittent pain inside the knee or along one side of the knee, often starts following a twisting injury.	• Knee may lock, may feel blocked and unable to straighten it fully, may have clicking of the knee; some swelling may be present. • Aggravated by squatting or twisting.	• More commonly occurs on the medial (inside) part of the knee.	• Tearing of the cartilage in the knee (meniscus) • Apply ice and rest. • See your doctor.
Knee pain following a fall, twisting injury, hyperextension (knee forced straight) injury, or direct blow to the knee.	• Popping sound at the time of injury, swelling that develops soon after the injury, giving way of the knee.	• Common sporting injury and one of the most serious, often requiring surgery and extensive rehabilitation.	• Tearing of one of the ligaments running through the knee joint from front to back (anterior cruciate ligament). • See your doctor.
Long-term aching and stiffness in the knee that has become worse over a period of months.	• Limited movement; the person may be unable to bend or straighten fully. • Pain is alleviated by rest, aggravated by a lot of activity. • Stiffness is often worse in the morning.	• More common in people over 50 years.	• Osteoarthritis • See your doctor.

SYMPTOM CHARACTERISTIC	ASSOCIATED SYMPTOMS	OTHER RELEVANT FACTORS	POSSIBLE CAUSE AND ACTION
Pain along the inner or outer knee, usually following a direct force to one side of the knee while the foot remains planted on the ground.	• Tenderness on one side of the knee	• Often occurs during sports such as soccer, football, skiing.	• Damage to one or more of the ligaments on the inside or outside of the knee (collateral ligaments) • See your doctor.
Red, swollen knee	• Constant ache, swelling, fever, generally feeling unwell. • May be aggravated by movement.	• Infection may spread through the blood, through a penetrating injury.	• Osteomyelitis or joint infection (septic arthritis) • See your doctor at once.

Leg problems

Leg problems can arise from a variety of conditions, from simple muscle strains to fractures and serious circulation disorders. See also entries on *knee problems*, *hip problems*, and *foot and ankle problems*.

SYMPTOM CHARACTERISTIC	ASSOCIATED SYMPTOMS	OTHER RELEVANT FACTORS	POSSIBLE CAUSE AND ACTION
Sudden pain in the leg associated with quick movement of the leg (e.g. kicking, sprinting, change of direction).	• Swelling, bruising • Pain when stretching or bending the leg, but can still move it.	• Common sporting injury, often affects hamstrings (muscles at back of thigh), quadriceps (muscles on the front of thigh), and calf muscles.	• Muscle strain or tear • Rest and apply ice to the area. • See your doctor.
Severe, constant leg pain following an injury, fall, or direct blow to the leg.	• Swelling, may be some deformity of the leg • Aggravated by attempts to walk or move the leg.		• Fracture • This is a medical emergency. • Go to the emergency room or call an ambulance at once.
Intermittent pain over the front of the shin.	• May be some pain when the shin is pressed. • Aggravated by repetitive motion (e.g. running, bicycling, jumping, walking up and down hills).	• More common in those with flat feet, bow legs, knock knees.	• Shin splints • See your doctor.
Cramping pain in the calves, feet, or hips while walking.	• May develop numbness or tingling in the feet. • Aggravated by walking, especially quickly or up hills, usually alleviated by rest.	• Risk is increased in people with abnormal cholesterol or triglyceride levels, diabetes, hypertension, men over 55 years old, women over 65 years old, those with a family history of cardiovascular disease, cigarette smokers, and those with obesity or little physical activity.	• Peripheral vascular disease • See your doctor.

SYMPTOM CHARACTERISTIC	ASSOCIATED SYMPTOMS	OTHER RELEVANT FACTORS	POSSIBLE CAUSE AND ACTION
Leg problems (cont.)			
Shooting or burning pain in the buttock and down the back of one leg.	• May also have low back pain, numbness or tingling in the foot. • Aggravated by coughing, sneezing, bending, lifting.	• More common in people with stiff backs or past back injury.	• Sciatica • See your doctor.
Pain and swelling in the back of the calf.	• Warmth, pain when touched.	• Most common following surgery or following long air flights or bus trips.	• Deep venous thrombosis • This is a medical emergency. • Go to the emergency room or call an ambulance at once.

Menstrual problems

Most menstrual problems warrant a full medical investigation, since they may indicate the presence of a disease that requires treatment. Although period pain or premenstrual syndrome may respond to simple self-treatment, it is important to ask your doctor's advice if symptoms persist or worsen.

Temporary emotional instability just prior to menstrual period.	• Bloating or discomfort in lower abdomen, irritability, depression, tearfulness, inability to concentrate, sleep disturbances, fatigue, lethargy. • Symptoms usually disappear when menstruation begins. • Caffeine may worsen irritability.	• Approximately one-third of fertile women experience some premenstrual symptoms. • The full premenstrual syndrome occurs in about 3–10 percent of fertile women.	• Premenstrual syndrome (PMS) • There is no standard treatment. • Ask your pharmacist's advice on over-the-counter medication for bloating or pain. • Vitamin B_6 supplements may help ease the symptoms. • See your doctor if you are worried or if symptoms persist.
Mild to moderate cramping pain during menstrual period.	• Pain may be aggravated by flatulence or constipation.	• Period pain sufficiently severe to cause missed school or work days is common in teenagers and young women. • Pain may be alleviated by heat applied to the lower abdomen (e.g. a hot water bottle or bath). • Severe symptoms may suggest endometriosis.	• Period pain (menstruation) • Try analgesic or non-steroidal anti-inflammatory drugs as recommended by your pharmacist. • See your doctor if you are worried or if symptoms persist.

SYMPTOM CHARACTERISTIC	ASSOCIATED SYMPTOMS	OTHER RELEVANT FACTORS	POSSIBLE CAUSE AND ACTION
Gradual onset of more pain than usual during and just before menstrual period.	• Low back pain, period pain lasting more than 2–3 days and starting before the onset of bleeding. • Spotting of small amounts of blood for 1–3 days prior to onset of period. • Menstrual bleeding may be heavier than usual. • Pain in pelvic area may worsen during sexual contact.	• Endometriosis occurs in approximately 5–10 percent of women, and is more likely in women with a mother or sister with the disease, and in women who have never become pregnant. • Pregnancy may temporarily resolve the problem, though symptoms may recur months or years later. • Endometriosis may result in infertility.	• Endometriosis • See your doctor.
More pain than usual during and just before menstrual period.	• Fever, vaginal discharge with offensive odor, abnormal vaginal bleeding, abdominal pain, pain during urination. • Onset of symptoms is usually gradual when caused by an intrauterine device (IUD).	• Pelvic inflammatory disease occurs almost exclusively in sexually active women. • Pelvic inflammatory disease may result from infections (usually sexually transmitted infections), uterine surgery (e.g. dilation and curettage, insertion of IUD, cesarean section), or childbirth.	• Pelvic inflammatory disease • See your doctor at once.
Increased volume and length of menstrual bleeding in women with an intrauterine device (IUD).	• Spotting of blood between menstrual periods, increased pain during periods.	• IUD may change the pattern of menstrual bleeding.	• IUD related adverse effect • See your doctor.
Excessive menstrual bleeding	• Pain during menstrual bleeding, longer than usual menstrual periods.	• Fibroids are most common in women over 35 years old or who have had several pregnancies.	• Uterine fibroids • See your doctor.
Irregularity or cessation of menstrual periods in a woman who is not pregnant.	• Fatigue or lethargy may occur with thyroid disease.	• A menstrual period may occasionally be missed in some women during the use of oral contraceptives. • Excessive exercise or weight loss (e.g. during athletic training or anorexia nervosa) may cause cessation of menstrual periods.	• Hormonal abnormality due to an ovarian problem, oral contraceptive use or a thyroid problem • See your doctor.

SYMPTOM CHARACTERISTIC	ASSOCIATED SYMPTOMS	OTHER RELEVANT FACTORS	POSSIBLE CAUSE AND ACTION
Menstrual problems (cont.)			
Cessation of menstrual periods in sexually active women	• Breast tenderness, abdominal bloating or feeling of fullness, nausea	• All methods of contraception carry a slight chance of failure leading to pregnancy.	• Pregnancy • Use pregnancy test kit— if positive, see your doctor. • See your doctor if you are worried or if symptoms persist.
Cessation of menstrual periods in women aged over 35 years	• Irritability, hot flashes (flushes)	• The onset of menopause most commonly occurs between the ages of 40–55. • Early menopause may occur from 35 years, or younger in rare cases.	• Menopause • See your doctor.
Recommencement of menstrual bleeding in a woman who has already gone through menopause.	• Abdominal swelling or discomfort, vaginal discharge	• Some hormonal medications may cause uterine bleeding.	• Uterine tumor or vaginal infection • See your doctor.

Nausea and vomiting

Nausea and vomiting occur with many medical conditions, and the cause is not always obvious. Since some conditions that may cause these symptoms are potentially serious, it is advisable to consult a doctor if the problem does not resolve quickly. If symptoms recur, and/or are accompanied by any other unusual symptoms, you may need medical tests to find the problem. When a person vomits blood or has severe pain, the situation should be treated as an emergency and a doctor consulted at once.

Nausea and vomiting after eating.	• Diarrhea may follow. • Unable to tolerate food or liquids.	• Symptoms occur after eating food that may have been kept too long or at incorrect temperature such as hot food kept warm several hours, or cold food that has been kept at room temperature or uncovered for several hours.	• Bacterial contamination of food (food poisoning) • Take frequent small amounts of fluid, if tolerated. • Typical cases of food poisoning will usually pass in under 12 hours. • See your doctor if the person is severely ill and unable to drink fluids, if you are worried, or if symptoms persist.
Intermittent nausea and vomiting	• Burning pain high in the abdomen. • Worse after eating, especially spicy foods. • Bland foods may relieve pain.	• Use of anti-inflammatory medications for pain (prescription or non-prescription) may damage stomach lining. • Ulcers are commonly caused by a bacterial infection and require antibiotics.	• Gastritis, or ulcer of stomach or esophagus • If symptoms are mild and not persistent, use an antacid (your pharmacist may advise you on a suitable choice). • See your doctor.

SYMPTOM CHARACTERISTIC	ASSOCIATED SYMPTOMS	OTHER RELEVANT FACTORS	POSSIBLE CAUSE AND ACTION
Recent onset nausea and vomiting	• Fever and cold or flu symptoms, diarrhea • Inability to tolerate food of liquids.		• Viral gastroenteritis • Rest and take frequent small amounts of fluids if tolerated (e.g. diluted soft drink or an electrolyte sachet from your pharmacist). • Your doctor or pharmacist may advise you further about treating specific symptoms. • See your doctor if the person is unable to tolerate fluids, if you are worried, or if symptoms persist.
Nausea and vomiting with intermittent severe pain	• Pain in the upper right abdomen, fever • Pain may worsen after eating greasy foods.		• Gallbladder inflammation or gallstones • See your doctor if you are worried or if symptoms persist.
Nausea and vomiting with steady worsening pain	• Recent onset abdominal pain in middle or lower right, fever	• Pain may begin as dull discomfort centrally and become more severe and localized to the right side.	• Appendicitis or a bowel obstruction • This is a medical emergency. • Go to the emergency room or call an ambulance at once.
Persistent nausea and vomiting over more than a week in women of child-bearing age.	• Missed menstrual period • Certain foods or smells may worsen symptoms. • Symptoms may be consistently worse at certain times of day.	• Non-predictable; a woman may experience morning sickness with one pregnancy but not a subsequent pregnancy.	• "Morning sickness" of pregnancy • See your doctor. • Avoid an empty stomach by eating frequent small meals. • Nibbling dry crackers between meals and before getting out of bed may help.
Vomiting in a baby or young child	• Crying, irritability, or quietness, inability to become interested in toys.	• Child under 2 vomiting for more than 6 hours, or child over 2 vomiting for more than 12 hours	• Viral infections are a common cause. • Children may rapidly become dehydrated. • See your doctor. • If you suspect severe dehydration, go to the emergency room or call an ambulance at once.

SYMPTOM CHARACTERISTIC	ASSOCIATED SYMPTOMS	OTHER RELEVANT FACTORS	POSSIBLE CAUSE AND ACTION
Nausea and vomiting (cont.)			
Vomiting in a baby or young child	• Uncontrollable crying, dark red diarrhea, unable to keep down any fluids.	• Obstruction is relatively rare.	• Intestinal obstruction • This is an emergency. • Visit your doctor or the emergency room at once.
Vomiting in a baby	• Forceful expulsion of stomach contents, persistent vomiting	• 20 percent of healthy babies vomit or regurgitate frequently enough to worry parents and cause them to seek medical advice. • Approximately 7 percent of babies show more severe symptoms suggesting gastroesophageal reflux disease.	• Stomach obstruction or reflux • Ask your doctor's advice to confirm the cause.

Neck problems

Symptoms involving the neck may result from a wide variety of conditions. Infections in the body will often lead to swelling of the neck glands, while poor posture and arthritis can cause neck pain and stiffness.

Neck stiffness that is present after sleep or periods of inactivity, gradually worsens over time.	• Pain and limitation of movement, spine may be tender to touch. • Aggravated by periods of inactivity or following exercise.	• Most common in those aged over 40. • May be alleviated by moving the neck gently.	• Osteoarthritis • See your doctor.
Intense neck pain that radiates into the shoulders and possibly down the arms.	• Tingling or numbness in the hands, arm weakness • May be aggravated by neck movements, sneezing, or coughing.	• May follow an injury or begin after regular daily activities.	• Vertebral disk injury causing pressure on a spinal nerve. • See your doctor.
Neck stiffness with a severe headache	• Vomiting, fever, drowsiness, may become delirious, unconscious, or have convulsions. • Exposure to bright lights increases pain.	• A child may be difficult to wake or have a high-pitched cry.	• Meningitis • This is a medical emergency. • Go to the emergency room or call an ambulance at once.

SYMPTOM CHARACTERISTIC	ASSOCIATED SYMPTOMS	OTHER RELEVANT FACTORS	POSSIBLE CAUSE AND ACTION

Numbness and tingling

A feeling of numbness or tingling usually results from a malfunction in part of the body's nervous system. The symptoms may be caused by an isolated problem in one nerve, or may be part of a more serious degenerative disease. See also entry on *weakness and paralysis.*

SYMPTOM CHARACTERISTIC	ASSOCIATED SYMPTOMS	OTHER RELEVANT FACTORS	POSSIBLE CAUSE AND ACTION
Numbness or tingling in one arm or one leg	• Neck or back pain, weakness of the affected limb • In serious cases, may have difficulty urinating. • Aggravated by sitting, bending forward, sneezing, coughing.	• The precise location of the symptoms defines which part of the back or neck is affected.	• Pressure on a nerve caused by swelling of a ruptured or bulging vertebral disk in the spine. • See your doctor at once.
Tingling or numbness in the arms, legs, trunk, or face	• Loss of strength or dexterity, vision disturbances, dizziness, unusual tiredness, difficulty walking, trembling, loss of bladder control. • Aggravated by very warm weather, hot bath, fever.	• More common among people who have lived in a temperate climate up to age 10. • Occurs much less commonly in those whose childhood was spent in a tropical climate, and extremely rare at the equator.	• Multiple sclerosis • See your doctor at once.
Numbness or tingling on one side of the body; symptoms usually start suddenly.	• Weakness in hands or feet, confusion, dizziness, partial loss of vision or hearing, slurred speech, inability to recognize parts of the body, unusual movements, fainting.	• More common with advancing age • Risk increased in people with cardiovascular disease (e.g. hypertension, coronary heart disease), lipid abnormalities (e.g. high cholesterol), diabetes.	• Transient ischemic attack or stroke • This is a medical emergency. • Go to the emergency room or call an ambulance at once.

Palpitations

Palpitations is the term used to describe an uncomfortable awareness of your heartbeat. The palpitations may take the form of fluttering, throbbing, pounding, or racing in the chest. The heart may feel as though it is beating irregularly. Palpitations may be harmless, but in certain cases they signal underlying heart disease.

SYMPTOM CHARACTERISTIC	ASSOCIATED SYMPTOMS	OTHER RELEVANT FACTORS	POSSIBLE CAUSE AND ACTION
Recurrent fluttering, racing, pounding, thumping in the chest; may have feeling of a strong pulse in the neck.	• Chest discomfort, weakness, dizziness, shortness of breath	• There are many types of variation from normal heartbeat rhythm, some of which are serious. • Arrhythmia is most commonly caused by heart disease, but may also occur with caffeine use, excessive alcohol, vigorous exercise.	• Arrhythmia • See your doctor at once.

SYMPTOM CHARACTERISTIC	ASSOCIATED SYMPTOMS	OTHER RELEVANT FACTORS	POSSIBLE CAUSE AND ACTION
Palpitations (cont.)			
Temporary racing, pounding, thumping in the chest; usually lasts for 10–20 minutes.	• Trembling, dizziness, shortness of breath, feeling of choking, nausea, diarrhea, an out-of-body sensation, tingling in the hands, chills, fear of dying. • May be aggravated by stress, and the fear of further attacks.	• Women are 2–3 times more likely than men to have these attacks.	• Panic attack • See your doctor if you are worried or if symptoms persist.
Racing heartbeat	• Shortness of breath on exertion, tiring easily, swelling in the legs and abdomen. • May be sudden fever and flu-like symptoms.	• Can occur as the result of infection, or in association with many diseases, or may have no identifiable cause.	• Cardiomyopathy (disease of the heart muscle) • See your doctor at once.
Awareness of forceful heartbeats, especially when lying on the left side.	• Shortness of breath on exertion, swelling of the legs, chest pain, dizziness	• More common in those who have had rheumatic fever.	• Heart valve disorder • See your doctor for advice.
Sudden, heavy pounding or thumping in the chest	• Pain in the middle of the chest that may spread down the left arm, sweating, shortness of breath, faintness, anxiousness, sense of impending doom. • Symptoms are not alleviated by rest.	• Increased likelihood in a person who has previously had a heart attack or with a history of coronary heart disease.	• Myocardial infarction (heart attack) • This is a medical emergency. • Go to the emergency room or call an ambulance at once.

Seizures (known as "fits," including convulsions)

Seizures result from an abrupt episode of abnormal electrical activity within the brain. There are many different types of seizures and many possible causes. Seizures may be generalized (generalized tonic-clonic convulsion also known as grand mal seizure) or localized to a particular part of the body (focal convulsion). Some seizures manifest as a brief aura followed by loss of awareness of surroundings. Any seizure warrants assessment by a doctor and often full medical investigation.

Repeated episodes of a generalized tonic-clonic seizure (grand mal seizure); begins with stiffness of limbs and jaw locking (tonic phase) followed by jerking of limbs (clonic phase) then a period of drowsiness and confusion (postictal phase).	• Urinary incontinence during fit • Aggravated by sleep deprivation, flickering lights, hyperventilation.	• No fever or current illness	• Epilepsy • See your doctor at once.

SYMPTOM CHARACTERISTIC	ASSOCIATED SYMPTOMS	OTHER RELEVANT FACTORS	POSSIBLE CAUSE AND ACTION
Repeated episodes of seizures that involve disturbances in the senses (sensory seizures).	• May have preceding aura involving visual and auditory hallucinations or distortions of taste and smell, followed by period of altered awareness sometimes associated with lipsmacking or repetitive movements (automatisms). • May be brought on by sleep deprivation, flickering lights, hyperventilation.		• Temporal lobe epilepsy • See your doctor at once.
Brief generalized seizure in child under 5 years	• Fever • No signs of infection of the brain (encephalitis) or covering of the brain (meningitis), no history of epilepsy • Rapid rise in temperature.	• 3 percent of children have at least one febrile convulsion.	• Simple febrile seizure of childhood • See your doctor at once.
Generalized or focal convulsion	• Headache, drowsiness, neck stiffness, oversensitivity to light, fever. • May have preceding febrile illness.	• May occur in previously healthy person.	• Meningitis or encephalitis • This is a medical emergency. • Go to the emergency room or call an ambulance at once.
Isolated generalized or focal convulsion	• Headache, nervous system abnormalities, decreased consciousness, newly developed squint • May follow head injury.	• May indicate raised pressure within the confined space of the skull.	• Brain tumor, abscess, or cerebral hemorrhage • This is a medical emergency. • Go to the emergency room or call an ambulance at once.
Repeated episodes of focal seizures (localized to a particular part of the body).	• Involuntary movements may occur in a single limb, one side of the body, or involve eyes deviating to one side.	• Indicates a localized lesion within the brain triggering the seizures.	• Head injury is the most common cause in young adults, while cerebrovascular accidents (strokes) are the most common cause in the elderly. • Congenital malformations of the brain, early meningitis, or perinatal brain damage are common causes in children. • This is a medical emergency. • Go to the emergency room or call an ambulance at once.

SYMPTOM CHARACTERISTIC	ASSOCIATED SYMPTOMS	OTHER RELEVANT FACTORS	POSSIBLE CAUSE AND ACTION

Sexual function problems

Several problems limit sexual pleasure or the proper function of sex organs. See your doctor, since many problems affecting sexual function may be treated. Untreated sex problems can lead to relationship problems, depression, and anxiety. See also entries on *genital pain* and *vaginal problems*.

SYMPTOM CHARACTERISTIC	ASSOCIATED SYMPTOMS	OTHER RELEVANT FACTORS	POSSIBLE CAUSE AND ACTION
In males, ejaculation before or immediately after intercourse begins.	• Anxiety, frustration, depression • May be aggravated by further worry about it happening.	• More common in young men. • Physical causes are rare, and most cases have a psychological cause.	• Premature ejaculation • See your doctor if you are worried or if symptoms persist.
In males, inability to have or keep an erection sufficient for sexual intercourse.	• Anxiety, frustration, depression	• More likely as men get older. • Physical disorders are the main cause, especially in men aged over 50.	• Erectile dysfunction (impotence) • See your doctor.
In males, pain during sexual contact	• May be redness or rash, or other symptoms of infection such as fever. • Aggravated by continued sexual contact.		• Infection (e.g. prostate, testes, or urethra), allergic reaction to spermicide • See your doctor.
Lack of sexual desire or inability to experience sexual pleasure.	• Anxiety, frustration • May be aggravated by stress, fatigue, anxiety, relationship problems.	• More common in women than men. • Physical and psychological causes can lead to this problem.	• Arousal dysfunction • See your doctor.
In females, pain during intercourse; pain may be in the vaginal area or deeper in the pelvis.	• Vaginal discharge, itching, dryness • May be aggravated by continued sexual contact.	• Pelvic inflammatory disease and infections are most common among young, sexually active women. • Endometriosis is most common among women with a family history of the disease, and in women who have never been pregnant. • Hormonal imbalances causing vaginal dryness are more common following menopause.	• Pelvic inflammatory disease, infections, hormonal imbalance, endometriosis • See your doctor.
In females, inability to have intercourse due to contraction of the vaginal muscles.	• Fear, anxiety, pain	• This is an involuntary response, outside the woman's control.	• Vaginismus • See your doctor.

SYMPTOM CHARACTERISTIC	ASSOCIATED SYMPTOMS	OTHER RELEVANT FACTORS	POSSIBLE CAUSE AND ACTION

Skin problems

The skin can show a very wide range of noticeable changes. It is important to check your skin regularly and report any changes to your doctor, since it is often difficult to tell the difference between significant changes (e.g. early skin cancers or eruptions due to other diseases) and unimportant ones, by appearance alone.

Changes in skin color

SYMPTOM CHARACTERISTIC	ASSOCIATED SYMPTOMS	OTHER RELEVANT FACTORS	POSSIBLE CAUSE AND ACTION
A new, growing or changing, brown or blue-black pigmented lesion, usually irregular or asymmetric in shape and color; often over ¼ inch (0.5 centimeter) in diameter; may be bleeding or ulcerated.	• Usually painless • May be itchy. • Surrounding skin may be inflamed.	• Melanoma occurs in all adult age groups but is rare in pre-pubescent children. • Risk is increased in people with fair skin or hair, many freckles or moles, or moles of unusual appearance. • Up to 50 percent of melanomas develop from moles. • Sun exposure, especially before age 10, may predispose to melanoma.	• Melanoma • See your doctor at once.
Sharply defined white (depigmented) patches of skin		• Vitiligo may occur in people with a family history of the disease or in those with immune disorders.	• Vitiligo • See your doctor.

Changes in shape of skin surface

SYMPTOM CHARACTERISTIC	ASSOCIATED SYMPTOMS	OTHER RELEVANT FACTORS	POSSIBLE CAUSE AND ACTION
Rash of pimples and pustules on face, chest, and back	• Inflamed, raised, red spots, excessive oiliness, blocked pores (whiteheads, blackheads), scarring. • May be exacerbated by some foods or medications.	• Acne is most common in teenagers but also occurs in 10–20 percent of adults.	• Acne (acne vulgaris) • See your doctor.
Unusual growth or ulcerated, raised lump on the face	• May be itchy or painful.	• Incidence increases with age. • Skin cancers are most common on the face, but can develop on other sun-exposed areas of the body. • Growth rate depends on the type of cancer; the growth may develop over a month or so, or slowly over many months.	• Skin cancer • See your doctor.

SYMPTOM CHARACTERISTIC	ASSOCIATED SYMPTOMS	OTHER RELEVANT FACTORS	POSSIBLE CAUSE AND ACTION
Skin problems–changes in shape of skin surface (cont.)			
Small (pinpoint or pin head size), raised, round, pink or pearly, shiny bumps with pits in the center		• Molluscum contagiosum is most common in children. • Contacts (e.g. family members or friends) may also be affected. • Commonly occurs on the face, eyelids, or genitals, but may develop on any area.	• Molluscum contagiosum • See your doctor.
Small, red, warm, tender bump around a hair follicle, that develops suddenly.	• Painful	• Boils may occur singly, or several may appear at the same time. • Occasionally multiple boils in the same area result in inflammation of the whole area. • Conditions that may predispose to boils include scratching of the skin, which allows bacteria to enter, illnesses that lower the body's resistance (e.g. diabetes).	• Boil • See your doctor.
Red bumps or elevated red patches that appear suddenly, each of which lasts from a few hours to 2 days; may be white in the center.	• Itching and tingling or a pricking sensation. • Swelling around the mouth or throat, difficulty breathing. • Scratching may worsen inflammation and may cause open, weeping sores.	• Hives most often appear on the arms, legs, or waist, but any part of the body may be affected. • Common causes include food allergies, exposure to dusts, medicines, infections, heat or cold.	• Hives (urticaria) • If needed, calamine or other soothing lotions as recommended by your pharmacist may relieve itching. • See your doctor at once.
Sudden appearance of bright red or dark red-blue, tender, deep-seated bumps or raised areas about 1–2 inches (2–5 centimeters) in diameter; usually on the front of both legs, occasionally on the outer forearms.	• Lumps are painful. • Fever, feeling of being generally unwell, joint pains, sore throat.	• Most often affects 20–30 year olds, more commonly females. • Erythema nodosum may occur as a symptom of infection, drug reaction, or an underlying illness.	• Skin eruption caused by inflammation within the skin (erythema nodosum). • See your doctor.

SYMPTOM CHARACTERISTIC	ASSOCIATED SYMPTOMS	OTHER RELEVANT FACTORS	POSSIBLE CAUSE AND ACTION
Rashes			
A tender, red, warm, swollen area of skin with an undefined border.	• Fever	• May occur where skin is broken (e.g. a cut or scratch).	• Cellulitis • See your doctor.
Small, purplish-red, bruise-like spots, may be flat or slightly raised.	• Associated symptoms, if present, will depend on the underlying cause.	• Purpura may be due to bruising, inflammation of capillaries, the use of cortisone-type medications (e.g. ointments or oral medications), diseases affecting the platelets, or serious infections.	• Bleeding into the skin (purpura) due to medical condition affecting the blood or blood vessels. • See your doctor at once.
Red rash with tiny, fluid-filled blisters; lesions tend to be dry and fragmented; may be swollen, scaly, or develop painful cracks; the margins of the rash are often ill defined.	• Severe itching • Skin may be dry in general. • Other symptoms of allergies • Itching is exacerbated by changes in temperature, mood, and contact with irritating materials.	• Symptoms may commence at any age, may occur intermittently or long-term, and may fluctuate in severity. • Allergic dermatitis is most common in people with a family history of allergic diseases. • May occur with other allergic conditions (e.g. asthma or hay fever).	• Eczema or dermatitis • See your doctor.
Red rash with tiny, fluid-filled blisters (vesicles) in an area exposed to an irritating substance; may be swollen or scaly.	• May be itchy or sore. • Scratching may worsen inflammation and may cause open, weeping sores.	• Contact dermatitis may occur following exposure to clothing, cosmetics, household detergents, occupational exposure to petroleum-based products, oils, solvents, paint, cement, rubber, resins, plants, or medicines which are applied directly to the skin. • Contact dermatitis may occur on the shoulders, neck, and scalp if the irritant is in the form of dust.	• Contact dermatitis • Try to identify the cause by eliminating suspected substances. • Avoid contact with the substance by wearing protective clothing. • See your doctor if you are worried or if symptoms persist.
Red rash or ring-like area; may be scaly.	• Itchy • If the area affected is the scalp, hairs within the affected area tend to be broken	• May affect the nails, feet, hands, groin, trunk, or scalp. • Ringworm may follow contact with pets (e.g. dogs, cats, or horses). • Usually chronic	• Fungal infection (e.g. ringworm) • See your doctor.

SYMPTOM CHARACTERISTIC	ASSOCIATED SYMPTOMS	OTHER RELEVANT FACTORS	POSSIBLE CAUSE AND ACTION
Skin problems–rashes (cont.)			
Reddish plaques covered with silvery scales; the margins are well defined.	• Usually not itchy • Arthritis may occur • Trauma, infections, or emotional upsets may predispose to symptoms.	• May affect any age group, but uncommon before 10 years of age and most common at 15–30 years of age. • Usually develops gradually.	• Psoriasis • See your doctor.
Rapid onset of well defined, reddish, slightly scaly patches on the trunk; usually a single patch precedes the development of others by a week or so.	• Patches may be slightly itchy. • May also affect the arms and legs.	• The cause of pityriasis rosea is unknown. • Symptoms occur most commonly in spring and autumn. • Any age group may be affected, most commonly young adults.	• Pityriasis rosea • See your doctor.
Sudden development of small, round, red, target-like patches and bumps that are darker in the center than the outside of the lesion; may have blistering.	• Feeling of being generally unwell, fever, sore throat, diarrhea.	• Attacks may be triggered by medications, infections, cancer, pregnancy. • Most commonly affects the back of the hands and forearms in a symmetrical fashion; may affect other areas.	• Erythema multiforme, an inflammatory disease of the skin, which is usually a reaction to infection or medication. • See your doctor.
Flat, blotchy, red rashes, which begin 4 days after symptoms of a cold in a child; rashes may join up to form one larger red area.	• Before the rash develops there may be a general feeling of being unwell, loss of appetite, fever, cough, runny nose, red watering eyes.	• Measles is preventable by vaccination. • Measles tends to be more severe in adults than in children.	• Measles • Rest in bed and avoid contact with others, especially pregnant women. • See your doctor. • Measles is highly contagious, so warn the doctor's office of symptoms before you attend.
Scaly rash across the nose and cheeks or forehead; spots are butterfly-shaped or round, with well defined margins.	• Arthritis, joint pain, fever, hair loss, kidney problems	• Usually long-term • Onset is gradual. • Women are twice as likely as males to develop lupus erythematosus. • More common in African races (e.g. Afro-Americans) than white races.	• Lupus erythematosus • See your doctor.

SYMPTOM CHARACTERISTIC	ASSOCIATED SYMPTOMS	OTHER RELEVANT FACTORS	POSSIBLE CAUSE AND ACTION
Bright red rash on the cheeks ("slapped cheek"); after a day or two, rash typically also appears on the forearms and thighs.	• Fever, general feeling of being unwell. • Arthritis may occur in adults.	• Most common in children aged 3–12. • Infection during pregnancy may cause fetal damage.	• Fifth disease (erythema infectiosum), an infectious viral disease • Avoid contact with other people, especially pregnant women. • See your doctor. • Fifth disease is infectious, so warn the doctor's office of symptoms before you attend.

Blistering conditions

SYMPTOM CHARACTERISTIC	ASSOCIATED SYMPTOMS	OTHER RELEVANT FACTORS	POSSIBLE CAUSE AND ACTION
Groups of bright red, tiny, fluid-filled blisters, which rupture and form crusts.	• Rash is itchy. • Fever (before blisters appear), feeling of being generally unwell • Adults may also experience aches and pains, headaches, serious nerve damage (rare).	• Rash mostly affects the trunk and face. • Children are most often affected, especially between 2–8 years of age; condition is usually mild, but occasionally can be fatal. • A person may be affected at any age if they have not previously had the condition. • Severe illness occurs more often in adults.	• Chickenpox • Rest, use calamine lotion or other medications recommended by your doctor or pharmacist to relieve itching, acetaminophen (paracetamol) for fever, daily bathing. • Avoid scratching spots. • See your doctor. • Chickenpox is highly contagious, so warn the doctor's office of symptoms before you attend.
Groups of tiny, fluid-filled blisters in a band-like distribution on one side of the body; the vesicles usually rupture and crust over.	• Severe pain in the area of the rash usually begins 1–2 days before the skin lesions.	• Mostly occurs in adults. • May recur.	• Shingles • See your doctor at once.
Small and larger blisters around the face and ears in a child	• Often itchy • Scratching can lead to further spread of the lesions.	• May affect adults but more common in children.	• Impetigo • Highly contagious • Other members of the household should avoid unnecessary contact with toweling or napkins that come into contact with the lesions. • See your doctor.

SYMPTOM CHARACTERISTIC	ASSOCIATED SYMPTOMS	OTHER RELEVANT FACTORS	POSSIBLE CAUSE AND ACTION

Urination, frequent

Increase in the production or frequency of urination should be assessed by a doctor to ensure the cause is not potentially serious. Most common causes for these symptoms can be overcome or controlled, so there is no need to tolerate these symptoms without asking medical advice. See also entry on *urination, painful*.

SYMPTOM CHARACTERISTIC	ASSOCIATED SYMPTOMS	OTHER RELEVANT FACTORS	POSSIBLE CAUSE AND ACTION
Involuntary leaking of urine	• Occurs with coughing, sneezing, or exercise. • May be worse during a bladder infection (cystitis), after drinking more fluids than usual, or in cold weather.	• Common in women after childbirth or with ageing.	• Stress incontinence (weakness of bladder muscles causing leakage of urine) • See your doctor. • Exercises may help strengthen the surrounding muscles. • Severe cases may require surgery.
Producing more urine than usual.	• Discolored urine • Waking at night to urinate. • Puffy swelling of extremities. • Generally feeling unwell.	• May occur with high blood pressure.	• Kidney disease • See your doctor at once.
Producing more urine than usual.	• Excessive thirst, frequent urination, increased appetite, weight loss, nausea, blurred vision • In women, frequent vaginal infections; in men, impotence; recurring yeast infections in both sexes.	• Diabetes is more common in people with obesity, hypertension, or a family history of diabetes.	• Diabetes • See your doctor at once.
In men, waking several times during the night to urinate.	• Difficulty starting a urine stream. • Urine dribbling after urinating.	• More common with ageing.	• Prostate problems (e.g. enlargement, prostatitis, or prostate cancer) • See your doctor at once.

Urination, painful

Any new occurrence of pain when urinating should be fully investigated by a doctor. For intermittent problems with which you are already familiar, like cystitis or genital herpes, your doctor or pharmacist can give advice on how to manage the problem when you recognize a new episode. See also entry on *urination, frequent*.

SYMPTOM CHARACTERISTIC	ASSOCIATED SYMPTOMS	OTHER RELEVANT FACTORS	POSSIBLE CAUSE AND ACTION
Discomfort or burning pain on urination.	• Pain under scrotum; difficulty urinating.		• Kidney infection or kidney stones • See your doctor at once.

SYMPTOM CHARACTERISTIC	ASSOCIATED SYMPTOMS	OTHER RELEVANT FACTORS	POSSIBLE CAUSE AND ACTION
Burning pain on urination	• Frequent urge to urinate, even when bladder is empty. • Urinating only small amounts. • Burning sensation in lower abdomen, urine with a strong odor. • Lack of adequate fluid intake may prolong symptoms. • Sexual contact may aggravate pain.	• Cystitis is more common in women than in men; infection may be triggered by sexual contact, use of diaphragm for birth control, or use of urinary catheters.	• Cystitis • Drink plenty of water. • See your doctor.
Painful urination	• Cloudy urine, ache or stabbing pain in lower back, fever	• More common with ageing.	• Prostate problems (e.g. prostatitis or prostate cancer) • See your doctor at once.
Painful urination in men or women.	• Inflamed genitals after sexual intercourse • In males, discharge of pus from the penis. • In females, vaginal discharge, urge to urinate frequently, abnormal menstrual bleeding. • The infection may spread from the genitals to the urethra, rectum, conjunctiva, pharynx, or cervix.	• The incidence of gonorrhea is much higher in the USA than in other industrialized countries. • Potential complications include inflammation of reproductive organs, peritonitis, inflammation around the liver, inflammation of the Bartholin's gland in women and epididymitis or abscess around the urethra in men, arthritis, dermatitis, endocarditis, meningitis, myocarditis, or hepatitis.	• Gonorrhea • See your doctor at once.
Burning pain on urination	• Blisters or sores on external genital areas (may later become scabby) • Sore, raw-feeling area inside vagina or on labia (women) • Vaginal discharge (women) • Discharge from infected sores • Burning pain in lower abdomen	• New sores after others heal • Sores may become infected. • First infection may cause flu-like symptoms. • Urinating in a warm bath may alleviate scalding sensation. • Outbreaks may be triggered by other viruses or stress.	• Genital herpes • Wear cotton underwear and loose clothing. • See your doctor at once. • Avoid sexual contact until you obtain medical advice.

SYMPTOM CHARACTERISTIC	ASSOCIATED SYMPTOMS	OTHER RELEVANT FACTORS	POSSIBLE CAUSE AND ACTION

Vaginal problems

Some vaginal problems may be successfully treated with medications available from pharmacists. If you are unsure of the cause or if symptoms persist, see your doctor or sexual health clinic for a full sexual health check-up and to ensure possible infections do not result in fertility problems or other complications. See also entry on *genital pain*.

SYMPTOM CHARACTERISTIC	ASSOCIATED SYMPTOMS	OTHER RELEVANT FACTORS	POSSIBLE CAUSE AND ACTION
Thick, white discharge forming clumps.	• In women, the vulva may be swollen and red, and there may be a thick, white discharge. • Sexual intercourse may increase discomfort. • Wearing tight clothing or synthetic underwear may worsen symptoms.	• Yeast infections can be a side effect of taking antibiotics, or may be triggered by stress, pregnancy, or use of the contraceptive pill. • Candidiasis is most likely in women with diabetes or as a side effect of some antibiotics.	• Yeast infection (candidiasis) • Wear cotton underwear and loose clothing to allow air to the area. • See your doctor.
Greenish-yellow discharge with unpleasant smell		• Recent sexual contact or new sexual partner in the last month.	• Infection such as bacterial vaginosis or trichomoniasis (a parasitic infection) • See your doctor.
Greenish-yellow discharge with unpleasant smell	• Pain in the lower abdomen, fever		• Pelvic inflammatory disease • See your doctor at once
Yellow discharge, may be thick like mucus	• Cervix bleeds when touched or scraped, abnormal menstrual bleeding, abdominal pain, fever, pain when urinating, and pain and swelling of one or both labia.	• May follow recent sexual contact with a new partner. • The incidence of gonorrhea is much higher in the USA than in other industrialized countries. • Potential complications include inflammation of reproductive organs, peritonitis, inflammation around the liver, inflammation of the Bartholin's gland in women and epididymitis or abscess around the urethra in men, arthritis, dermatitis, endocarditis, meningitis, myocarditis, or hepatitis.	• Gonorrhea • See your doctor or a sexual health clinic as soon as possible. • Antibiotic treatment is important to prevent serious complications.

SYMPTOM CHARACTERISTIC	ASSOCIATED SYMPTOMS	OTHER RELEVANT FACTORS	POSSIBLE CAUSE AND ACTION

Weakness and paralysis

Weakness and paralysis are usually caused by disorders in the nervous system. Symptoms may involve the entire body, or be limited to one part such as an arm or leg. Paralysis (loss of muscle function) is a serious symptom and should be investigated by a doctor. See also entry on *numbness and tingling*.

SYMPTOM CHARACTERISTIC	ASSOCIATED SYMPTOMS	OTHER RELEVANT FACTORS	POSSIBLE CAUSE AND ACTION
Weakness or paralysis of the arms or legs	• Progressive numbness in the arms or legs, back or neck pain • Bladder, bowel, and sexual functions may be affected. • May be aggravated by moving the neck or back.	• Can occur following injury (e.g. broken neck or back), or due to a tumor, disease, or infection.	• Spinal cord damage • See your doctor at once.
Weakness in one arm or one leg	• Neck or back pain, numbness and tingling of the affected limb • Aggravated by bending, sitting, coughing, sneezing.	• The precise location of the symptoms defines which part of the neck or back is affected.	• Ruptured vertebral disk causing nerve compression. • See your doctor at once.
Weakness or paralysis on one side of the body, usually starts suddenly.	• Tingling, confusion, dizziness, partial loss of hearing, slurred speech, inability to recognize parts of the body, unusual movements, fainting	• More common with advancing age. • Risk increased in people with cardiovascular disease (e.g. hypertension, coronary heart disease), lipid abnormalities (e.g. high cholesterol), diabetes.	• Transient ischemic attack or stroke • See your doctor at once.
Profound weakness in both legs, then progresses upwards to both arms.	• Tingling, numbness	• In the majority of cases, symptoms begin 3–21 days after a mild infection or surgery.	• Guillain-Barré syndrome • This is a medical emergency. • Go to the emergency room or call an ambulance at once.
In males, progressive muscle weakness throughout the body, usually beginning in the muscles of the pelvis.	• Muscles often enlarge. • May also have trouble climbing stairs or getting out of a chair, frequent falls.	• Usually first occurs in boys aged 3–7 years.	• Muscular dystrophy • See your doctor at once.
Weakness or paralysis on one side of the body.	• Constant headache, poor balance and coordination, dizziness, double vision, loss of sensation, loss of hearing	• Most common in people with cancer in another part of the body.	• Brain tumor • See your doctor at once.

SYMPTOM CHARACTERISTIC	ASSOCIATED SYMPTOMS	OTHER RELEVANT FACTORS	POSSIBLE CAUSE AND ACTION

Wounds or sores, non-healing

Any wound that takes longer than normal to heal warrants investigation by your doctor, as slow-healing wounds may indicate an infection or significant problem affecting the whole body.

SYMPTOM CHARACTERISTIC	ASSOCIATED SYMPTOMS	OTHER RELEVANT FACTORS	POSSIBLE CAUSE AND ACTION
Area where normal skin surface has been lost and fails to heal (ulcer), especially if overgrown or sealed-looking at the edges.	• May be itchy and/or painful.	• Typically occurs in sun-exposed areas of the skin.	• Skin cancer • See your doctor at once.
Area where normal skin surface has been lost and fails to heal (ulcer) on the lower leg.	• Long history of discomfort, swelling, and skin changes of the lower leg.	• History of disease of the leg veins (e.g. abnormal clotting or deep venous thrombosis). • Affects women more often than men.	• Persistent ulcer that does not heal due to long-term, abnormally high pressure in the leg veins (venous ulcer). • See your doctor at once.
Area where normal skin surface has been lost and fails to heal (ulcer) on the tips of the toes or fingers or in areas subject to pressure.	• Painful • Discomfort is aggravated by pressure to the area.	• History of disease in the arteries supplying blood to the affected area, or an injury affecting blood supply to the area.	• Persistent ulcer that does not heal due to severe reduction of blood supply to the area (ischemic ulcer). • See your doctor at once.
Area where normal skin surface has been lost and fails to heal (ulcer) in an area that suffers repeated trauma or pressure.	• Painless • Numbness of the surrounding skin	• History of injury or disorder of the nerves supplying the affected area. • The most common underlying cause is diabetes.	• Persistent ulcer caused by repeated trauma to an area where the individual is unable to sense pain (neuropathic ulcer). • See your doctor at once.
Area where normal skin surface has been lost and fails to heal (ulcer) on foot or lower leg in a person with diabetes.		• Embedded foreign bodies of which the patient was unaware are commonly found in people with diabetic ulcers. • Close control of blood sugar may help minimize the risk of foot ulcers.	• Diabetic ulcer • See your doctor at once. • Keep feet clean and dry at all times and wear properly fitted shoe. • Inspect feet daily for callus, infection, abrasions, or blisters.

Index

Page numbers in **bold** print refer to main entries. Page numbers in *bold italics* refer to main entry illustrations. Page numbers in *italics* refer to illustrations.

W

walking, 121, 455, 456, 457
wart virus, 36, 37
warts, **73**, **73**
waste products
  blood &, 252, 277
  breathing &, 321, 328
  CSF &, 197
  digestive system &, 337, 360,
   361, 364, 365, 368
  fetus &, 449
  lymph node &, 289
  urinary system &, 377, **378**, 383
water
  absorption of, 423
  CSF &, 197
  digestion &, 338, 357, 360, 361,
   368
  intake of, 338, 383
  lymph &, 288
  reabsorption of, **383**
  saliva &, 348, 349

water balance, 349, 377, 383
water levels, 50, 51, 55
Watson, James, 26
Wernicke's area, 191, 236
white blood cells, 14, 24, 25, **256**–7,
  256, **257**, 282, 283, 288, 292,
  294, 296, **300**
whooping cough, **471**
windpipe see trachea
wisdom tooth, 88, 93, 342–3
women see female
wound healing, **24**, 24
wrinkles, 32, 33, 464, 467
wrist, 278
  bones of, 79, 80, 84, 85, 109,
   110, **110**, 111
  movement of, 155, 160
  muscles of, 154, 155, 155, 156,
   157, **160**, **160**
  nerves of, **218**, **219**

X

xiphoid process, 98, 99, 325, 326

Y

yellow bone marrow, 80

Z

zinc deficiency, 67
zona fasciculata, 430–1, 431
zona glomerulosa, 430, 431
zona pellucida, 13, 29, 407, 444, 445
zona reticularis, **431**, **431**
zygomatic arch, 137, 138
zygomatic bone, 78, 89, 91, 91, 92,
  93, 94, 96, 140
zygomatic branch of facial nerve,
  208, 209
zygomatic muscles, 28, 131, 136,
  137, 209
zygomaticus major, 128, 131, 136,
  137, 236
zygomaticus minor, 128, 136, 137,
  236
zygote, 13, **13**, 445, 446, 448, 448
zymogen cells, 354

Produced by Global Book Publishing Pty Ltd
Level 8, 15 Orion Road, Lane Cove, NSW 2066, Australia
Ph: (612) 9425 5800 Fax: (612) 9425 5804
Email: rightsmanager@globalpub.com.au
Illustrations from the Global Illustration Archives © Global Book Publishing Pty Ltd 2002
Text © Global Book Publishing Pty Ltd 2002